TRIBOLOGY ON THE SMALL SCALE

Tribology on the Small Scale

A Modern Textbook on Friction, Lubrication, and Wear

Second Edition

C. Mathew Mate *and*
Robert W. Carpick

OXFORD
UNIVERSITY PRESS

Great Clarendon Street, Oxford, OX2 6DP,
United Kingdom

Oxford University Press is a department of the University of Oxford.
It furthers the University's objective of excellence in research, scholarship,
and education by publishing worldwide. Oxford is a registered trade mark of
Oxford University Press in the UK and in certain other countries

First Edition published in 2008
Second Edition published in 2019

Impression: 1

Published in the United States of America by Oxford University Press
198 Madison Avenue, New York, NY 10016, United States of America

British Library Cataloguing in Publication Data
Data available

Library of Congress Control Number: 2019941494

ISBN 978–0–19–960980–2
DOI: 10.1093/oso/9780199609802.001.0001

Printed in Great Britain by
Bell & Bain Ltd., Glasgow

Front cover image: Molecular dynamics simulation of an asperity covered with an oxide layer making contact with a flat metal surface where the contact area consists of only a few atom-to-atom contacts. In this image, oxygen atoms are represented by blue spheres, oxidized platinum atoms by white spheres, and metallic platinum atoms by grey spheres. Image courtesy of Prof. Ashlie Martini and Rimei Chen at the University of California, Merced.

Preface

The importance of friction, lubrication, adhesion, and wear in technology and everyday life is well known; they are encountered whenever two surfaces come into contact, such as when you walk across a room, push a pencil across a piece of paper, or stroke your favorite pet. While many excellent books have been written on tribology, most have focused on analyzing the macroscopic aspects, with only slight attention paid to the rich interplay between the atoms and molecules at the contacting surfaces, as these have historically been poorly understood. In recent decades, however, many talented physicists, chemists, engineers, and materials scientists have begun to decipher the nanoscale origins of tribological phenomena. Given the tremendous progress and excitement generated by this endeavor, now seems the opportune time for a more modern approach on tribology emphasizing how macroscopic tribological phenomena originate at the atomic and molecular level.

The goal of this book is to incorporate a bottom up approach to friction, lubrication, and wear into a modern textbook on tribology. This is done by focusing on how these tribological phenomena occur on the *small scale*—the atomic to the micrometer scale. We hope to demonstrate that focusing on the microscopic origins leads to a more scientifically rigorous understanding of tribology than typically achieved by tribology books that take a macroscopic empirical approach. It is also hoped that the reader becomes enthused with the same excitement as those working in the field have for unraveling the mysteries of friction, lubrication, and wear, as well as an appreciation for the many challenges that remain.

This book covers the fundamentals of tribology from the atomic scale to the macroscale. The basic structure—with chapters on topography, friction, lubrication, and wear—is similar to that found in conventional tribology texts. These chapters cover the microscopic origins of the macroscopic concepts commonly used to describe tribological phenomena: roughness, elasticity, plasticity, friction coefficients, and wear coefficients. Some macroscale concepts (like elasticity) scale down well to the micro- and atomic-scale, while other macroscale concepts (like hydrodynamic lubrication and wear) do not. This book also has chapters on surface energy and surface forces, and covers other topics not typically found in tribology texts, but which become increasingly important at the small scale: capillary condensation, disjoining pressure, contact electrification, molecular slippage at interfaces, and atomic scale stick-slip.

Tribology is a continually evolving field, and nanoscale studies of tribology have an especially rapid pace of progress. These factors, combined with feedback from many readers of the first edition, including students and university teachers, have motivated the writing of a substantially revised second edition of *Tribology on the Small Scale*. For the second edition, all the chapters have had numerous new sections added and the rest of the chapter updated and revised. Some of the new sections add examples from recent experiments that illustrate modern nanoscale tribological concepts. Other new sections incorporate the most significant advancements that have occurred in nanoscale tribology since the publication of the first edition, such as Persson's contact theory; the power spectrum treatment of surface roughness; and the application of transition state theory to wear, viscosity, and friction. Another important enhancement of the second edition over the first edition is the addition of problems at the end of each chapter. These problems are drawn from classes taught at University of Pennsylvania by Prof. Carpick, which were reviewed and improved by both authors; also many new problems were specifically created for the second edition.

This book is intended to be suitable as a textbook for tribology courses taught at the advanced undergraduate and graduate level in many engineering programs. In terms of the scientific and mathematical background expected of the reader, no special knowledge is assumed beyond that typically encountered by science and engineering students in their first few years at a university.

In addition to college students learning about tribology for the first time, this book is intended for several other audiences:

- Academics and scientists who wish to learn how friction, lubrication, and wear occur at the microscopic and atomic scales.
- Engineers and technicians who do not consider themselves tribologists, but who work with technologies (such as MEMS, disk drives, and nanoimprinting) where a good grasp of how tribological phenomena occur on the small scale is essential.

We would like to thank all those who provided help and encouragement during the writing of this book:

- Oxford University Press for providing us the opportunity to publish this book with them and for their encouragement and patience during the writing of the first and second editions of this book.
- Professor Steve Granick at the University of Illinois, Urban-Champaign and Professor Curt Frank at Stanford University who hosted one of us (C. M. Mate) as a visiting scholar at their universities during the writing of the first edition (S. Granick) and the second edition (C. Frank).
- Our colleagues who were kind enough to comment on various draft chapters and provide advice on particular aspects of tribology:
  - o First edition – Peter Baumgart, Tsai-Wei Wu, Run-Han Wang, Robert Waltman, Bruno Marchon, Ferdi Hendriks, Bernhard Knigge, Qing Dai, Xiao Wu, Barry Stipe, Bing Yen, Zvonimir Bandic, Kyosuke Ono, and Yasunaga Mitsuya.
  - o Second edition – Andrew Jackson, Robert Smith, Greg Rudd, Tevis Jacobs, Joel Lefever, Harman Khare, Ashlie Martini, Jackie Krim, Nicholas Spencer, Mark Robbins, Lars Pastewka, Arup Gangopadhyay, and all of the students of Professor Carpick's nanotribology class.
- Our families and especially our spouses, who have always been constant sources of support and encouragement.

Contents

1

Introduction

Starting in childhood, we all acquired a sufficient working knowledge of tribology to lead happy, productive lives. Crawling as infants, we mastered the frictional forces needed to get us where we wanted to go. Eventually, we graduated from crawling to walking to school, where an occasional icy sidewalk, if one lived in a cold climate, provided a challenging lesson on slippage and traction. If our teacher at school asked us to move our chair backwards, we knew intuitively that the chair would be easier to slide if no one was sitting on it. Mastering friction was also a critical component in many of the games that we played, whether it was gripping a bat, maintaining traction when running and stopping, or putting a devilish spin on a ping-pong ball.

In addition to friction, we also encountered wear and adhesion at a young age. The detrimental aspects of wear may have been first learnt when our favorite toys wore out quicker than we felt they should. The more positive aspects of wear may have been first appreciated as we sanded our first woodworking projects and polished our first art sculptures into their final artistic shapes. We also quickly learnt that crayons and pencils become dull when rubbed against paper, but can be sharpened back up by grinding in a pencil sharpener. Our first awareness of adhesion may have been on a humid day when someone commented on how "sticky" it feels. Or perhaps it was when we first wondered why spiders and flies can walk on the ceiling, but dogs and cats cannot.

Once in college, science and engineering classes usually only cover the topics of friction, lubrication, adhesion, and wear at a rudimentary level. For friction, all that is usually taught is that static friction is greater than kinetic friction, friction is proportional to the normal contact force with different coefficients for different materials, and viscous friction in a fluid is proportional to velocity. The principles of thin film lubrication are covered only in an advanced class on fluid mechanics as an example of solving Reynolds' equation. Wear is only briefly covered in specialized engineering classes on tribology, despite its pervasiveness as a failure mechanism, and adhesion is covered incidentally in courses on chemical bonding or polymer physics.

Considering how much we encounter the tribological phenomena of friction, lubrication, adhesion, and wear in our daily lives and the wide extent of these phenomena in industry and technology, one might be puzzled why these topics are only marginally covered in our current education system. In the authors' opinion, the fault lies not with colleges and universities, but rather with the inherently complicated and interconnected

Tribology on the Small Scale: A Modern Textbook on Friction, Lubrication and Wear. Second edition. C. Mathew Mate and Robert W. Carpick. © Oxford University Press 2019. Published in 2019 by Oxford University Press.
DOI: 10.1093/oso/ 9780199609802.001.0001

physical origins of most tribological phenomena. This multifaceted nature has made it difficult for scientists and engineers to develop predictive theories for most tribological phenomena. Instead, empirically derived trends (for instance, that friction is proportional to the loading force) are often the only predictive tools available. These empirical approaches have the drawback of being predictive only over a limited range of parameters. Since the underlying physical mechanisms are not well understood, often one does not even know what the important parameters are or over what range the observed trends are valid. Similarly, if a purely analytical approach is attempted, the lack of knowledge of the relevant parameters often leads to inaccurate predictions of tribological behavior. This poor predictive power has led to the field of tribology being perceived in many scientific quarters as more of a "black art" than as a scientific discipline. This lack of predictive power may also be the reason why educators are reluctant to spend much time on tribology concepts whose application may be dubious in many situations.

For example, if one wanted to analyze the friction force acting on a chair sliding across a hardwood floor, the most expedient approach would be to take advantage of past empirical studies that have shown that friction is generally proportional to the loading force (the weight of the chair and the person sitting on it) with a proportionality constant called the coefficient of friction or μ. Using this approach, the next step is to determine from experiment how μ depends on the parameters suspected of influencing friction: sliding velocity, hardness of the wood, type of floor wax, etc. After a few hours of experiment, one would begin to have a good idea how friction depends on these parameters, but would have trouble predicting without further experimentation how the friction might change if new parameters were introduced, for example, by adding felt pads to the bottom of the chair legs to prevent them from scratching the hardwood floor.

While many tribology problems are still best approached through empirical investigations, these types of investigations are not the focus of this book. Instead, the focus is on the physical origins of tribology phenomena and how understanding these can be used to develop analytical approaches to tribological problems. In essence, the goal is to make tribology less of a black art and more of a scientific endeavor. This will be done by emphasizing how the tribological phenomena of friction, lubrication, adhesion, and wear originate at the small scale. Or, equivalently, how physical phenomena occurring at the atomic to micron scale eventually lead to macroscale tribological phenomena. The hope is that, once readers have gained a solid understanding of the nanoscale origins of tribological phenomena, they will be well equipped to tackle new tribology problems, either by applying analytical methods or developing better empirical approaches.

1.1   Why is it called tribology?

The pursuit of the microscopic origins of friction, lubrication, adhesion, and wear is not a recent scientific activity. Over the centuries, many have pondered on these origins, and in recent decades it has become quite fashionable for leading scientists to take up the challenge. One of the early pioneers and champions of the microscopic approach was Prof. David Tabor (1913–2005) of Cambridge University. One of the frustrations faced by Tabor and others working in this field at that time was the lack of a scientific name for the area of study encompassing all the phenomena occurring between contacting objects. It was felt that this lack of terminology was depriving the field of a certain level of status and respect within the scientific community. (For example, some of Tabor's colleagues at the Cavendish Laboratory in Cambridge would disparagingly refer to his research group as the "Rubbing and Scrubbing Department" (Hahner and Spencer 1998).) To counter this, Tabor coined the name *tribophysics* for the research group that he headed while investigating practical lubricants, bearings, and explosives at Melbourne University during the Second World War, which he derived from the Greek word *tribos*, meaning rubbing.

In 1966, H. Peter Jost led a Committee of the British Department of Education and Science to produce the "Jost Report," which officially launched the word *tribology* to describe the entire field and which was derived from Tabor's earlier word, tribophysics (Jost 1966). While the literal translation of tribology is "the science of rubbing," in the Jost Report, this definition was adopted: "The science and technology of interacting surfaces in relative motion and of associated subjects and practices." After the Jost Report, the term tribology quickly became established as the field's official name, and the word now commonly appears in the titles of papers, books, journals, professorships, and institutions concerned with this topic.

While the name tribology has certainly increased the credibility of the field as a valid discipline of scientific and engineering, the term remains somewhat unknown outside the field. So, tribologists need to be prepared to explain the word to those who have not heard of it before, or who mistake it for the study of tribes or of the number three.

1.2　Economic and technological importance of tribology

One of the goals of the 1966 Jost Report was to document the potential economic savings that could be achieved through the development and adoption of better engineering practices for minimizing the unnecessary wear, friction, and breakdowns associated with tribological failures. The possible savings within the United Kingdom were estimated to be roughly equivalent to 1% of its GNP. Since the Jost Report, other agencies have also evaluated possible savings, and the consensus view now is that between 1% and 2% of most industrialized nations' GNP could be gained through proper attention to tribology (Dake et al. 1986, Jost 1990, Chattopadhyay 2014); for example, this would correspond to $186–$371 billion for the United States in 2017.

While these potential economic benefits have long been recognized, this has not always been followed up with the level of investment in tribology research and development felt to be warranted by many tribologists. Possibly a major factor in this reluctance to back tribology projects comes from its historical "black art" character, which tends to cast doubt on how successful a proposed tribology project will be. Hopefully, this book will help diminish these doubts by demonstrating a rational and scientific basis for approaching tribological problems. Another way of diminishing doubts about the science of tribology is through a few success stories.

1.2.1 Tribology success story #1: reducing automotive friction

Improving the fuel economy of cars and trucks has long been a major technology goal of the automotive industry (Tung and McMillan 2004, Holmberg et al. 2012, Lee and Carpick 2017). An obvious way for improving fuel efficiency is reducing the energy loss through friction. For the average passenger car, the analysis by Holmberg et al. (2012) indicates that only about 21.5% of the fuel energy goes to actually moving the car along the road, while 33% is lost overcoming friction within the automobile (Figure 1.1).

Due to the fairly mature nature of automotive technology, generally only incremental improvements are achieved each design cycle; however, the cumulative effect of these incremental improvements has been quite substantial. For example, from 1980 to 2016, the average fuel economy of cars and trucks sold in the United States increased from 19.2 to 24.7 miles per gallon, despite the average horsepower more than doubling from 104 to 230 (U.S. Environmental Protection Agency 2017). In addition, the reduction in automotive friction and improvements in the wear resistance of automotive components has led to the median age of automobiles on the road increasing from 5.1 years in 1969 to 11.6 years in 2016 (Bureau of Transportation Statistics 2017).

As shown in Figure 1.1, automotive frictional losses come from

- the friction within the engine,
- the friction within the transmission system,
- the rolling resistance and traction of the tires against the road, and
- the friction during braking.

Even though automobiles powered by internal combustion have been around for over a century, the automotive industry still continues to find ways to lower friction losses

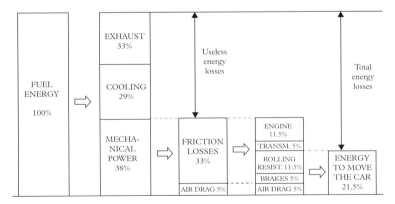

Figure 1.1 *Breakdown of how the fuel energy in the average passenger car is used and dissipated as it is converted into useful work to move the car. Reproduced from Holmberg et al. (2012) with permission from Elsevier, copyright 2012.*

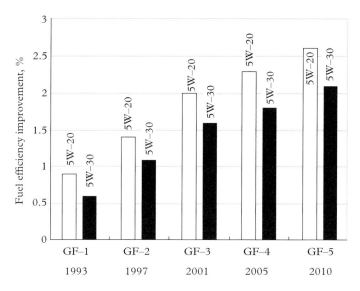

Figure 1.2 *Improvement in fuel efficiency due to the reduction in engine friction achieved through changes in engine oils for gasoline fueled (GF) vehicles. GF-1 to GF-5 engine oil specifications correspond to changes in base oil and additive chemistry. Reducing the oil viscosity from 5W-30 to 5W-20 provides an additional 0.5% improvement. The baseline is Energy Conserving II engine oils available prior to 1993. Courtesy of Arup Gangopadhyay at Ford Powertrain Research and Advanced Engineering.*

in automotive powertrains and tires. An example of this is shown in Figure 1.2, which illustrates the impact that improved engine oils have had on fuel efficiency. From 1993 to 2010, changes in engine oil, including reducing their viscosity, resulted in a 1.6% improvement in fuel efficiency through the reduction of non-viscous engine friction. Along with the improvement in fuel economy, these newer engine oils also achieve higher wear protection, higher resistance to oxidation, and less formation of sludge and varnish within the engines. Over the 1993–2006 period, an additional 1% improvement in automotive fuel efficiency was achieved by using lower viscosity transmission fluids and another 1% by using lower viscosity gear lubricant (Gangopadhyay 2006).

Due to the desire to reduce the transportation sector's contribution to global CO_2 emissions, the automotive industry continues to strive to improve fuel economy, with further reduction in automotive friction still expected to be an important contributor to this goal. The amount of possible gains in fuel economy that could potentially be achieved by reducing friction has been analyzed by Holmberg et al. (2012), who estimate that friction could be reduced by upwards of 18% if one could build a car out of components with friction coefficients as low as the lowest demonstrated in research labs.

A 2017 report to the U.S. Department of Energy (DOE; Lee and Carpick 2017, Chapter 2) discusses a wide variety of technological opportunities for further improving fuel economy related to tribology. These include further improving engine and drivetrain lubricants, optimizing component design (e.g., through thin film coatings and surface finishes of parts, or improving the design of pistons), designing lubricants in tandem with the engine and drivetrain, incorporating advanced engine sensing and actuation, improving computationally-aided design and modeling, and developing advanced coatings, finishes and lubricants. This DOE report "estimates that continued efforts in this area . . . could easily achieve 2–5% additional fuel economy gains."

1.2.2 Tribology success story #2: solving adhesion in MEMS devices

It has long been realized that miniaturization of machines can result in major new technologies. In recent years, the most promising way for fabricating microscale mechanical devices has been to use processes originally developed for fabricating semiconductor electronic devices. By using these fabrication processes, mechanical functions (such as actuation, fluid flow, thermal response, etc.) can be integrated on a small area of a chip along with electronic signal processing. A major advantage of fabricating these *microelectromechanical systems* (MEMS) with semiconductor processing techniques is that they achieve excellent economies of scale, since many devices are fabricated simultaneously onto a single chip. This low unit cost, along with the integration of mechanical and electrical functions into a small space, enables whole new types of technologies to become commercially viable.

MEMS devices can be categorized as follows based on how their mechanical constituents move and contact (Romig et al. 2003):

- Class I—no moving parts (e.g., pressure sensors, inkjet printer heads, and microphones);
- Class II—moving parts, but no rubbing or impacting surfaces (e.g., accelerometers, gyros, and radiofrequency (RF) oscillators);
- Class III—moving parts with impacting surfaces, (e.g., digital micromirror devices (DMDs), RF contact switch);
- Class IV—moving parts with impacting or rubbing surfaces (e.g., micromotors).

These classes are listed in order of increasing tribology complexity, which typically corresponds to an increasing tendency for failure from tribological phenomena such as adhesion, friction, mechanical stress, wear, and fracture. Most of the MEMS devices that have been successfully commercialized belong to Class I and II, with only a few in Class III, and none in Class IV. As the lack of tribological interactions contributes to higher reliability, the best way to avoid a tribology reliability issue in a MEMS device is to design it so that it moves as little as possible and without impacting contacts!

The most widely used Class III MEMS device is the DMD, developed by Texas Instruments (TI) and used in digital light processing (DLP) video projection devices such as large screen televisions (Hornbeck 2011). The development of the DMD provides a good success story of how solving microscale tribology issues can enable

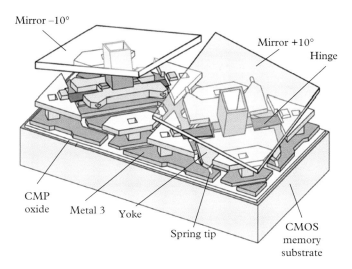

Figure 1.3 *Two of the mirrors in a digital micromirror device (DMD). Electrostatic attraction is used to rotate the mirrors $\pm 10°$ to the mechanical stops where the spring tips make contact. Reprinted from Hornbeck (2011) with permission from Cambridge University Press, copyright 2011.*

a new technology to gain sufficient reliability for commercialization. In a DMD, an array of mirrors, each about 16 μm across, is used to project an image onto a video screen. As illustrated in Figure 1.3, the intensity of each pixel is controlled by rotating the individual micromirrors through $\pm 10°$ by using electrostatic attraction. Before shipping its first DMD product in 1996, TI carried out extensive reliability engineering and testing (Douglass 1998, Van Kessel et al. 1998), and a number of reliability issues were addressed: stuck mirrors, fatigue of the mirror hinge, excessive hysteresis in hinge deflection, mirrors breaking as result of vibration and shock, and particles preventing mirrors from rotating. Here we focus on the sticking of the mirrors against their mechanical stop, which was a persistent tribology problem for which the TI engineers implemented a combination of clever solutions based on a thorough micro-understanding of the adhesive mechanism.

In a DMD, the individual micromirrors are rotated from the on to off positions. To ensure that each mirror has the correct angular position at the end of the rotation, the mirror yoke is designed to come to rest against a mechanical stop, as illustrated in Figure 1.3. During the development of the DMD, it was found that adhesive forces acting at this contact would sometimes be large enough to result in the mirror sticking against the stop, making it non-functional. These adhesive forces originate from the meniscus force due to water vapor condensing around the contact (discussed in Chapter 6) and from van der Waals forces (discussed in Chapter 7). A number of design changes were implemented in the DMD to reduce the magnitude of these adhesive forces and to improve the release function:

- The DMD was hermetically sealed in a dry environment to minimize the capillary condensation of water.
- The contacting parts were covered with a low surface energy "anti-stick" material to minimize the van der Waals force.
- Miniature springs were added to the parts of the mirror yoke that makes contact— the "spring tips" shown in Figure 1.3. These spring tips store elastic energy when the parts come into contact, which helps push the mirror away from the surface when the electrostatic attractive force is released.

These design modifications dramatically reduced the tendency of the micromirrors to stick, greatly improving the DMD operating margins (Douglass 1998, Van Kessel et al. 1998). From 1996 to 2010, TI sold over 20 million DLP systems with DMDs, demonstrating that, when proper attention is paid to the microscale tribology issues, a reliable product can include a MEMS device with contacting components.

The RF MEMS switch is another Class III MEMS device that was first commercialized by Analog Devices. These MEMS devices use a conductive gold cantilever that is pulled by electrostatic forces into contact with another electrode, thereby closing an electronic path for RF conductivity (Goggin et al. 2015). The small size of the MEMS switch is not only beneficial in terms of space and weight, but also consumes far less power than larger conventional switches, as well as providing a number of other improvements in electrical performance (Rebeiz 2004).

Initial attempts to commercialize MEMS switches were unsuccessful largely due to tribological reliability issues of the contact, which must survive billions of switching cycles to be commercially viable. By engineering the packaging environment to include hermetic sealing to prevent contamination and electrostatic charging from the environment and by developing materials at the contact junction that minimized wear and adhesion, Analog Devices was able to produce a reliable commercial device.

As discussed in Section 1.4.2, miniaturization of MEMS contact switches to have sub-100 nm feature sizes (a nanoelectromechanical contact switch) is being developed as a new technology for potentially competing with complementary metal–oxide semiconductor (CMOS) transistors, once the reliability issues associated with the repeated contact and the manufacturing issues can be solved.

For Class IV MEMS devices where surfaces rub against each other, solving the tribology issues is much more challenging (Williams and Le 2006, Achanta and Celis 2015). For example, much fanfare was made about the first working MEMS micromotor in 1988 (Fan et al. 1989). While these micromotors rotated as desired, the rotors need to supported by a bearing that is typically made by silicon microfabrication technologies; this typically results in the contacting surfaces having high friction and wear, severely limiting the reliability lifetimes of the micromotor. While micromotors have been demonstrated with a low friction liquid bearing, a gas-lubricated bearing, a contactless magnetic bearing, and an electrostatic bearing (Shearwood et al. 2000, Wong et al. 2004, Chan et al. 2012, Sun et al. 2016), micromotors with sufficient reliability for commercialization still have not been demonstrated.

1.2.3   Tribology success story #3: slider–disk interfaces in disk drives

The hard disk drive (HDD) industry may be the most striking example of where an exponential improvement of a technology over many decades has been sustained through the continual solving of tribological problems.

The first disk drive was introduced in 1956 as part of the IBM RAMAC computer. The RAMAC disk drive stored an impressive 4.4 megabytes, for its day, in a space the size of a small refrigerator. Disk drive technology has advanced to the point that a terabyte or more of data can now be stored on an HDD in a laptop computer. This tremendous increase in storage capacity has come about largely from the exponential growth in the areal storage density (the number of bits that can be stored per square inch on a disk surface): from 1956 to 2018, the areal storage density of disk drives increased by a factor of 10^7, with the density annual growth rate ranging from 30 to 100%. To sustain this tremendous growth in areal density over a period of time spanning many decades, all aspects of disk drive technology had to be continually improved. Here we focus on the tribological challenges faced by the disk drive industry in the recent past to sustain this rapid rise in areal storage density.

Inside a disk drive, a slider a with read/write recording head flies over a rotating disk, as illustrated in Figure 1.4. The information is stored as magnetic bits in a thin layer of magnetic material on the disk surface. Since the magnetic field from these bits decays rapidly away from the disk magnetic medium with a decay distance that scales with bit size, the magnetic spacing between the head sensor and the magnetic medium on the disk needs to scale with the lateral size of the magnetic bit on the disk (Marchon and Olson 2009): so, as the areal density goes up, the spacing must go down.

Since the recording head slider flies at high speeds (typically around 10 m/s) with a clearance of just a few nanometers over the disk surface, careful attention needs to be paid during the product development and manufacture towards minimizing the risks from high speed contacts between the slider and disk. To minimize the number of these contacts, the slider is designed to fly on an air bearing generated by the disk pulling air underneath the slider and over a series of steps and pockets precisely fabricated onto the bottom surface of the slider. These surface features form an air bearing surface (ABS) that generates a lifting force, which balances the loading force from the suspension enabling the slider to fly over the disk with its trailing edge a few nanometers above disk surface. To protect against occasional impacts, the disk and slider surfaces are coated (about 3–4 nm for 2005 drives) with a hard material, usually diamond-like carbon. To further ensure that the slider–disk interface is not damaged by the high speed impacts, a molecularly thin film of lubricant (typically about 1 nm thick) is applied over the carbon overcoated disk.

Even though the space available between the head sensor and the disk magnetic layer for these protective layers has been rapidly diminishing, the dramatic reduction in magnetic spacing from 96 nm in 1995 to ~10 nm in 2013 (Marchon et al. 2013) was not achieved through any major technological innovations, but rather through careful attention to tribological detail, in particular:

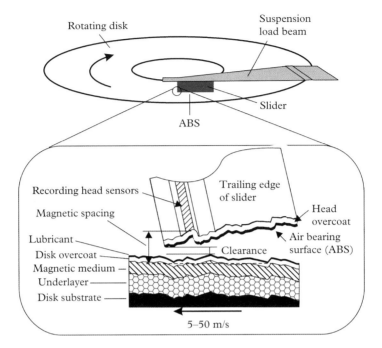

Figure 1.4 *A recording head slider flying over a rotating disk surface in a disk drive (top). An enlarged view of the trailing edge of the slider where the recording head is located (bottom). The cross section illustrates the various components of the recording head and disk that contribute to the magnetic spacing, which is defined as the distance from the top of the disk magnetic medium to the bottom of the head sensor. The higher the areal storage density, the smaller the magnetic spacing needs to be.*

- introduction of denser carbon overcoats through changes in composition and deposition processes;
- design of lubricant systems that preserve at least a monolayer of lubricant on disk surfaces over a disk drive's lifetime (Mate 2013);
- careful control of disk topography to achieve sub-nanometer roughness (see Section 2.4.3.1);
- improved fabrication processes and better tolerance control for the slider ABS, resulting in better control of the slider–disk spacing;
- development of "flying height control" where a small heater is integrated into the recording head to actively control the final of the slider–disk clearance (Suk et al. 2005, Mate et al. 2015).

With these improvements, disk drives achieve excellent long term reliability while their sliders fly over the disk surfaces at high speeds with incredibly small clearances, ~1.5 nm for 2013 drives (Marchon et al. 2013). The disk drive industry plans to continue increasing areal storage densities, which will mean reducing the magnetic spacing to less than 7 nm (Wood 2000, Mate et al. 2005, Marchon et al. 2013). However, the various constituents of the magnetic—overcoats, lubricant, roughness, and clearance—are approaching their physical limits as to how thin they can be made and still provide the desired level of protection (Mate et al. 2005, Marchon et al. 2013). This will prove to be particularly challenging for the two new technologies that industry has been intensely developing for achieving future increases in areal density:

- *Heat assisted magnetic recording (HAMR)*—In HAMR, a small laser and waveguide are incorporated into the recording head and are used to briefly spot-heat a tiny area of the disk surface during the recording write process. The magnetic media in this area is temporarily heated above the Curie temperature, making it possible to write an individual magnetic bit that is only a few nanometers across (Shiroishi et al. 2009). Since the magnetic media is heated locally to >400°C during the HAMR write process, this high temperature can negatively impact the disk drive tribology by thermally degrading the carbon overcoat and lubricant, and by creating thermal protrusions on the head and disk surfaces that greatly increase the frequency and severity of intermittent contacts (Marchon et al. 2014, Kiely et al. 2018).
- *Bit patterned media (BPM)*—In BPM, rather than storing the bits as magnetic domains in a continuous magnetic media as is conventionally done, each bit is stored on a small island of magnetic material. The weak magnetic coupling between neighboring islands makes it is possible to pack more bits into a unit area in BPM than with continuous media. The patterning process to fabricate a BPM disk, however, creates in a much rougher disk surface than for conventional continuous media, making it difficult for the recording head slider to fly over the disk without excessive wear at a small enough magnetic spacing to take advantage of the increased areal density (Albrecht et al. 2015).

1.3   A brief history of modern tribology

The practice of tribology goes as far back as the prehistoric humans, who first used wear to fashion tools and friction to start fires. Ancient Egyptian artwork from 4000 years ago shows slaves dragging of sledges bearing heavy statues and includes the depiction of an earlier tribologist pouring a liquid at the sliding interface to reduce friction (Dowson 1998, Ayrinhac 2016). In China's Forbidden City, the huge stones there are conjectured to have been transported in the fifteenth and sixteenth centuries by sliding them over an ice path (Li et al. 2013).

Scientific investigation of friction began with Leonardo da Vinci who recorded in his notebooks the observation that friction was proportional to load and independent of the apparent area of contact. This law of friction was rediscovered by the French physicist Guillaume Amontons, who published his findings, now referred to Amontons' Laws of Friction, in 1699. Amontons' results immediately provoked controversy about the microscopic origins of friction, which even today has not been fully resolved. The possible origins of friction are discussed in Chapter 4 and 11.

With the industrial revolution, machinery of all sorts came into widespread use and with it a growing need for a better control of friction, lubrication, and wear. During this period, the principles of hydrodynamic lubrication were first discovered through the experimental work of Beauchamp Tower (1884) and the theoretical work of Osborne Reynolds (1886). The subsequent development of this theory of hydrodynamic lubrication enabled reliable bearings to be designed for lubricating the machinery of the modern age. Lubrication is discussed in Chapters 9 and 10.

During the twentieth century, enormous industrial growth and development of new technologies further fueled demand for better tribology. To meet this demand, numerous tribological engineering solutions have been developed over the last century, notably:

- hydrodynamic bearing design (Chapter 9);
- theory of contact mechanics (Chapter 3);
- synthetic lubricants;
- solid lubricants;
- wear-resistant materials.

It has long been thought that a lack of understanding of the microscopic origins of tribological phenomena has impeded the development of the best tribology technology. In the middle of the twentieth century, several scientists conducted pioneering studies on the microscopic origins of friction, lubrication, and wear producing:

- 1925—Hardy's studies of boundary lubrication (Chapter 10)
- 1940s—Bowden and Tabor's theory of molecular adhesion for friction (Chapter 4 and 11)
- 1953—Archard's law for adhesive wear (Chapter 12)
- 1966—Greenwood and Williamson's analysis of multi-asperity contact area (Chapter 3)

Since the latter part of the twentieth century there has been an accelerated effort to determine the micro- to atomic scale origins of tribology phenomena. Besides the desire to reduce detrimental friction and wear in technological applications, this effort has also been driven by new experimental and theoretical techniques for characterizing materials at the nanometer scale, facilitating the discovery of the atomic and molecular origins of friction, lubrication, adhesion, and wear. This emerging subfield of tribology has been christened *nanotribology* as it seeks to understand tribology phenomena at the atomic and nanometer scale.

1.3.1 Scientific advances enabling nanotribology

The nature of contact makes its microscopic and nanoscopic origins difficult to study. Generally contact occurs at a multitude of contact zones at the apexes of small protrusions or asperities on opposing surfaces, as illustrated in Figure 1.5(a). Since these contacts are sandwiched between two solids, they are inaccessible to most scientific characterization techniques. Adding to this difficulty, the contact occurs principally at the summits of the surface roughness, meaning that the material volume affected by contact tends to be nanoscopically small and difficult to detect. The push and pull of the asperities rubbing against each other further complicates matters as the contacting microstructures are not static but evolve as sliding progresses.

Fortunately, numerous techniques have been developed over the past half century for characterizing surfaces and nanoscale amounts of materials. These newer techniques have led to a wealth of information on how the structure and chemistry of surfaces evolve during contact and how they influence tribological phenomena.

For one category of surface analytical techniques (illustrated in Figure 1.5(b)), the individual surfaces are placed in high vacuum and irradiated with electrons, ions, or X-rays, and the kinetic energy of the ejected electrons or ions are measured with an energy analyzer. Since only electrons or ions near the surface can be ejected toward

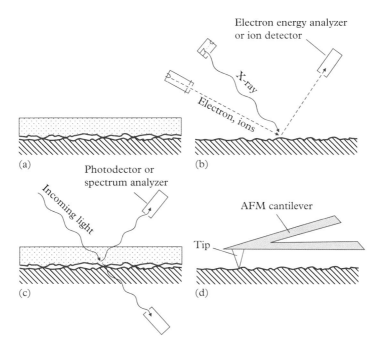

Figure 1.5 *(a) Due to surface roughness, contact between two solid surfaces occurs primarily at the summits of the surface asperities. (b) Vacuum techniques for analyzing surfaces before or after contact. AES: impinging electrons, energy of ejected electrons analyzed; ESCA/XPS or XAS: impinging X-rays, energy of ejected electrons analyzed; SIMS: impinging ions, energy and charge-to-mass ratio of scattered and ejected ions measured. (c) Optical techniques like Raman spectroscopy, fluorescence spectroscopy, surface plasmon resonance, FTIR, and sum frequency generation can now be used to analyze a buried contact interface. (d) An AFM tip rubbing against a surface can be used to simulate a single asperity contact.*

the detector, these are very sensitive techniques for determining surface structure and chemistry, but the necessity of vacuum operation means these techniques are only used for analyzing surfaces before and after contact. Some of the vacuum techniques valuable for characterizing the chemical composition and molecular structure of tribological surfaces include:

- Auger electron spectroscopy (AES);
- electron spectroscopy for chemical analysis (ESCA), also known as X-ray photo-electron spectroscopy (XPS);
- X-ray absorption spectroscopy (XAS);
- secondary ion mass spectrometry (SIMS).

Fuller descriptions of these techniques can be found in Briggs and Seah (1990) and Somorjai (1994, 1998).

As illustrated in Figure 1.5(c), several optical and X-ray probes have been developed that allow for in situ characterization of the buried contacting surfaces:

- Fourier transform infrared (FTIR) spectroscopy (Mangolini et al. 2012);
- fluorescence microscopy (McGhee et al. 2018);
- surface plasmon resonance spectroscopy (Krick et al. 2013);
- Raman spectroscopy (Campion and Kambhampati 1998, Wahl et al. 2007);
- sum frequency generation (Shen 1994);
- X-ray diffraction and X-ray reflectivity (Mate et al. 2000).

For a review of these and other techniques used to study buried tribological interfaces, see Sawyer and Wahl (2008).

Since contact typically occurs at the summits of asperities, characterizing the topography of contacting surfaces has always been an important starting point for characterizing a tribological surface. Initially this was done with stylus profilometers that only measured profiles with micron resolution along single lines. Next came the scanning electron microscope (SEM), which is capable of imaging surfaces with nanometer lateral resolution, but is unable to quantify the heights of the surface features. Optical interferometry provides very high vertical resolution, but is limited by diffraction to ~1 μm in the lateral directions. More recently, surface topography measurements have been dominated by scanning probe techniques—the scanning tunneling microscope (STM) and, more importantly, the atomic force microscope (AFM)—which can generate three-dimensional topography images with true atomic resolution. The use of optical interferometry and the AFM for topography measurements is discussed in Chapter 2.

In addition to topography measurements, the AFM is also able to measure the contact forces acting on a single asperity tip, as illustrated in Figure 1.5(d). Consequently, since its invention in 1986, the AFM has become one of the principal tools for investigating nanoscale contact and friction forces. Use of the AFM for measuring contact forces is discussed in Chapters 6, 7, 8, 10, and 11.

The surface force apparatus (SFA) is another important tool for measuring the forces between contacting surfaces; in an SFA, the forces are measured between two atomically smooth mica sheets in a variety of chemical environments. SFA force measurements are discussed in Chapters 7, 8, and 10.

The dramatic increase in computer performance over the past few decades has led to the prevalence of many computer-based simulation techniques for predicting physical phenomena. Several of these simulation techniques have been adapted to studying tribological phenomena:

- Molecular dynamics simulations have been used to directly address the atomic origins of friction, lubrication, and adhesion, but are limited to small volumes and short time scales (Thompson and Robbins 1990, He et al. 1999, Gao et al. 2004, Dong et al. 2013).
- Finite element analysis simulations are routinely used to analyze the contact mechanics of macro-, micro-, and nanoscale multi-asperity contacts.
- Most modern fluid film bearings are designed using computer aided design (CAD) software. In many instances, analysis programs have been developed to handle bearings with sub-micron dimensions.

1.4   Breakthrough technologies relying on tribology at the small scale

Perhaps the most exciting aspect of modern tribology's push towards the nanoscale is the potential payoff that this research can have for enabling breakthrough micro- and nanoscale technologies.

As device components miniaturize, they become more susceptible to the forces and to atomic scale phenomena occurring at their contacting surfaces. As a consequence, established engineering solutions that might work well for macro-tribology situations tend to be inadequate for micro- and nanoscale devices.

For example, for macroscopic machines, fluid film bearings are used to provide lubrication for the moving parts. The thickness of the lubricant film in these bearings is typically microns to millimeters; but, when devices are miniaturized to the sub-millimeter scale, the gap between the moving parts of the bearing's lubricant film becomes sub-micron. This small gap leads to more frequent contact between the asperities on the moving surfaces, as well as to confinement effects within the lubricant film that can dramatically increase the lubricant's viscosity. Both the friction and wear from contact and the enhanced viscosity can reduce the bearing's effectiveness if the bearing is not properly redesigned for the small scale.

Further, as machines become micro-sized, the capillary and molecular adhesion forces begin to dominate over gravity and inertia forces. So, while loading forces and bulk hardness may be the major factors determining friction and wear of macro-sized objects running in contact, they become relatively less important for minute objects where molecular adhesion forces are comparable to loading forces.

Previously, we discussed the two examples of the DMDs and disk drives where successful products were shipped once solutions were found to the problems associated with micro- and nanoscale tribology. In the next few sections, we discuss a few exciting new technologies where implementation is being held up by problems of tribology on the small scale.

1.4.1   Nanoimprinting

Driven by the difficulties and high tooling costs of extending photolithography techniques to the production of sub-100 nm features, a tremendous research effort is being expended on developing alternatives to photolithography for fabricating nanoscale structures. *Nanoimprinting* is a promising new technology for replicating features as small as <10 nm in size and, due to its low cost and simplicity, should be an attractive alternative to not only photolithography, but also to e-beam lithography and extreme UV lithography (EUVL) (Chou et al. 1996, McClelland et al. 2005, Schift 2008, Traub et al. 2016).

Figure 1.6 illustrates the typical UV nanoimprint lithography (UV-NIL) process. In this technique, a transparent template stamp (usually quartz) is first fabricated with the sub-100 nm features that are the inverse of the structures to be replicated. Next, the template stamp is aligned over the substrate, and a low viscosity, UV-curable resist material is injected between them (Figure 1.6(a)). The template and substrate are pressed together, and the resist exposed to UV light through the template (Figure 1.6(b)). The template stamp is withdrawn, leaving the pattern replicated in the resist material (Figure 1.6(c)). Finally, this pattern is etched into the substrate to create a permanent inverse replica of the original template pattern (Figure 1.6(d)).

Thermal nanoimprint lithography (thermal-NIL) is another version of nanoimprinting. This method differs from UV-NIL in that a thermoplastic polymer is used as the resist rather than a UV-curable polymer. After the thermoplastic polymer is spin-coated onto the substrate, the template stamp is pushed against the polymer film as they are heated above the polymer's glass transition temperature so that the polymer's viscosity is low enough for it to flow around the template features. After the polymer has cooled down below the glass transition temperature, the template is withdrawn to leave a patterned resist ready for the etch step.

As the old adage says "the devil is in the details" and, for nanoimprinting, a very critical detail is the level of defects generated during the imprint process. For example, if features are spaced 40 nm apart, this corresponds to a density of 6.25×10^{10} per square centimeter. Since the features are only a few tens of nanometers across, even few cubic nanometers of material going astray now and then can easily add up to an unacceptable number of defects. A major problem faced in implementing nanoimprinting as a manufacturing process is minimizing these defects to an acceptable level.

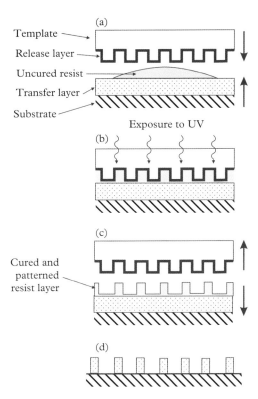

Figure 1.6 *Schematic of the UV nanoimprint lithography process. (a) Alignment of the template stamp and deposition of a UV curable resist material that will serve as the etch barrier. (b) Imprint and expose to UV. (c) Withdrawal of the template stamp. (d) After etching and removal of the resist mask, the substrate has the inverse pattern of the template.*

For a nanoimprinting process such as the one illustrated in Figure 1.6, there are two general categories of mechanisms that generate tribology related defects:

1. When the template stamp and substrate are pushed together, the uncured resist material needs to flow around and wet all of the nanoscale features on the template surface, as any uncovered portion becomes a defect void in the replicated pattern. The spreading of the uncured material into the nanoscale features on the template surfaces is governed by the flow of liquids in tight spaces, which is discussed in Chapters 9 and 10. Wetting is discussed in Chapter 5.

2. When the template stamp is withdrawn, it has to separate without damaging the resist material. As described by Schift (2008), these withdrawal related defects include:

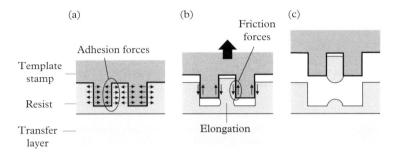

Figure 1.7 *Schematics of how, when the nanoimprint template stamp is withdrawn from the cured resist, friction and adhesion on the sidewalls can lead to a defect being generated in the patterned resist structure by the elongation of a resist feature, followed by it becoming detached from the resist and adhering to the template stamp.*

- elongation of resist features with the potential to detach these features (Figure 1.7),
- delamination of resist from the substrate,
- shrinkage of the resist,
- generation of rims on the resist features due to side motion of template during withdrawal, and
- relaxation of strain that was frozen into the resist during curing.

Figure 1.7 illustrates the first of these withdrawal related mechanisms where a defect is created when a piece of cured resist adheres so strongly to the template during withdrawal that it elongates and detaches from the rest of the resist. This not only creates a void defect for that particular replica, but also this defect can propagate to future replicas if the adhered material is not cleaned off the template stamp. A common way to prevent the cured resist from adhering to the template stamp is to coat the stamp surface with a thin film, called a *release layer*, which has a low surface energy. This release layer is typically one molecule thick, as a thicker film would adversely impact the dimensions of the replicated nanoscale features and a thinner film would provide inadequate coverage. The release layer works by lowering the adhesion energy acting over the resist–template surface area below the resist–resist and resist–substrate cohesion energy. The physical origins of and the interrelationships between adhesion force and surface energy are discussed in Chapters 5, 6, and 7.

Even with a low surface energy release layer, proper release of the smallest features can be problematic as high friction forces on the sides of high aspect ratio features can dominate over cohesive forces within the cured resist, which is also illustrated in Figure 1.7. This occurs as the surface area of the features does not scale down as fast as their volume. This combination of the surface forces of friction and adhesion eventually becomes larger than the internal cohesive forces as structural features are made microscopic.

As of the writing of this book, nanoimprinting is still not used as a mass fabrication technique, but is being actively developed for those manufacturing processes where relatively large areas need to be patterned with high resolution patterns that have simple geometries. These include fabrication BPM (Section 1.2.3; Albrecht et al. 2015) and optical components such as high brightness LEDs, displays, and subwavelength polarizers (Schift 2008, Traub et al. 2016).

1.4.2 Nanoelectromechanical contact switch

As mentioned in Section 1.2.2, MEMS devices are used in inkjet printer heads, RF oscillators, accelerometers, sensors, and optical displays. In this section we highlight how the miniaturization of another MEMS device may one day provide a breakthrough technology for making smaller logic circuits, with better performance characteristics, than can be made by scaling the CMOS circuits. The MEMS device being developed for this purpose is a miniaturized version of the MEM contact switch, typically referred to as a nanoelectromechanical (NEM) contact switch or relay. (A *nanoelectromechanical system* (NEMS) device differs from a MEMS device in that the characteristic length of a MEMS device is between 1 mm to 100 nm, while a NEMS device has a characteristic length <100 nm.)

Figures 1.8(a) and 1.8(b) schematically show the simplest design of a MEM or NEM switch, which has a three-terminal architecture that enables it to operate as a transistor. In this type of switch, an electrostatic force is used to deflect a conducting mechanical element into contact with an electrode. Once contact is made, the switch is in the "on" position, and current can flow between the source and the drain, with the current–voltage response illustrated in Figure 1.8(c). For a review of the variety of MEM/NEM switches that are under development, see Loh and Espinosa (2012).

The most attractive aspect of a NEM switch transistor over a CMOS transistor is the near zero current between the source and the drain electrodes when the NEM switch is the "off" state (in contrast, CMOS transistors have substantial leakage currents in this state). Also, due to its small size, the NEM switch consumes low power in the "on" state. Thus, NEM transistors are being viewed as an attractive alternative to CMOS transistors for applications requiring low power consumption, which becomes ever more important as the number of transistors per unit volume increases with future miniaturization. Another advantage of NEM devices is that they can be better suited for harsher environments than CMOS devices, as they are relatively insensitive to radiation, temperature, and external electric fields. This aspect of NEMS devices should make them attractive for those technologies such as automotive, aerospace, and oil and gas drilling that operate in harsh environments.

To enable the widespread use of NEM switches, two main technical challenges need to be overcome:

1. Developing MEMS manufacturing processes that scale down well to the sub-100 nm dimensions required for NEMS devices. As the fabrication techniques for making extremely small NEM devices are, for most part, still under development,

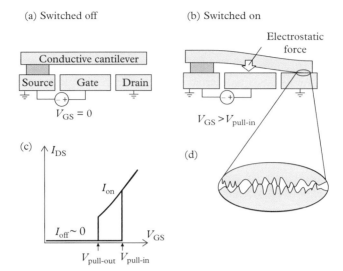

Figure 1.8 *(a) Schematic of an NEM switch with a three-terminal architecture when switched off. (b) The switch is on when a voltage V_{GS} is applied between the gate and the source so that the electrostatic force is great enough to pull the cantilever into contact with the drain, thereby providing a conductive path between the source and the drain. (c) Dependence of the drain–source current I_{DS} on the gate–source voltage V_{GS}. (d) Expanded view of the contact region illustrating how, due to the surface roughness, the true area of contact is only a small fraction of the apparent area of contact.*

    most contact switch transistors demonstrated to date have been MEMS devices rather NEMS devices. But, if NEM transistors are going to eventually replace CMOS transistors, it is necessary to make NEM switch transistors as small as these future CMOS transistors.

2. Developing NEM switch designs that can reliably function for up to 10^{15} on–off contact cycles (Peschot et al. 2015). As many of the failure mechanisms for NEM switches are associated with these repeated contacts, the rest of this section focuses on the tribology issues of these contacts that need to be resolved for there to be widespread implementation of NEM contact switches.

    As illustrated in Figure 1.8, in a NEM switch contact occurs when the voltage V_{GS} between the source and the drain is greater than a critical "pull-in" voltage $V_{\text{pull-in}}$. When contact is made, the current from the source to the drain rises sharply (much sharper than for a typical CMOS device, another advantage of the NEM transistor). Since the electrostatic force scales inversely with square of the gap distance, it is desirable to have the gap as small as possible to minimize the magnitude of V_{GS} that needs to be applied (preferably to just a few hundred millivolts, so as to minimize the consumed power, which scales with the square of the voltage). The magnitude of $V_{\text{pull-in}}$ can be further reduced by designing the cantilever to be as flexible as possible, but to still be stiff enough that, when V_{GS} is reduced back to zero, the elastic restoring force can overcome adhesion forces that developed during contact. This adhesion results in a hysteretic response of the current to the gate voltage, as shown in Figure 1.8(c).

The surface forces that underlie this adhesion, such as the van der Waals and capillary forces, are discussed in detail in Chapters 6 and 7. As mentioned previously, as mechanical devices are made smaller, adhesive forces become more problematic, since surface forces scale with the surface area and do not scale down as fast as those characteristics such as stiffness that scale with the volume. So, if the NEM switch is not properly designed to account for this as it is scaled down in size, at some size the surface forces dominate over the elastic restoring force and the beam is stuck in the contact position when $V_{GS} = 0$. Even if the structure can overcome adhesion, the hysteresis induced by adhesive forces increases the power dissipated in each switching cycle; minimizing adhesion is key to increasing the efficiency of a NEM switch (Pott et al. 2010).

The roughness of the two surfaces that make contact also influences the reliability and performance of NEM switches. As illustrated in Figure 1.8(d), contact only occurs at the summits of the surface roughness, so the real area of contact is only a small fraction of the apparent area. The true area of contact has a major influence the electrical performance of a NEM switch as the contact resistance between the source and the drain scales inversely with true area of contact. When the switch is used as a transistor, the contact resistance, along with the capacitance, determines the value of the RC delay. As NEM switches are scaled down in size, this contact resistance is predicted to rise *exponentially* with the inverse of the diameter of the contact zone (Dadgour et al. 2011), due to the combined effects of both the smaller apparent area and the smaller loading forces from the adhesion and the applied electrostatic force. (Chapter 3 discusses the true area of contact and how it depends on the loading force.) High contact resistance can also lead to excessive conductive heating at the contact junction, causing rapid diffusion of metal atoms around the contacting asperities (which increases the adhesion force) or, worse, sudden evaporation of material at the contacting interface.

Another major reliability problem is contact wear due to all the repeated contacts that occur from the high number of on–off contact cycles. This contact wear is aggravated by the high impact velocity reached when the flexible beam snaps towards the drain electrode once the pull-in voltage is exceeded. This wear tends to increase the roughness of the contact surfaces, adversely impacting the contact resistance.

Another tribology related problem with MEM/NEM switches is the formation of insulating layers on the contacting electrode surfaces (Hermance and Egan 1958, Czaplewski et al. 2012). These layers form after vapor-phase hydrocarbon species in the surrounding atmosphere adsorb on the electrode surfaces. When subjected to the combination of contact stresses, electric fields, and current flows that occur during the repeated contacts, these adsorbates then catalytically react with the metal surfaces and with each other to form insulating polymer films, sometimes called "tribopolymer" or "frictional polymer" films. Finding materials and conditions that prevent the formation of such layers is a key challenge for NEM switches.

1.4.3   Nanotechnology

Much of the current excitement about nanotechnology and nanoscience is driven by the development of new techniques that allow, not only the manipulation of matter at the atomic scale, but also the integration of these nanoscale structures into complicated systems or devices. If this bottom up approach is to become a viable technology, new manipulation methods are needed to move nanoscale objects reliably to places where they can be assembled into nanoscale devices. This will require a good understanding of how friction and adhesion occur on the atomic and molecular scale, in essence: *nanotribology*.

Many people, when promoting nanotechnology, like to quote the following passage from Richard Feynman's famous 1959 article *There's Plenty Room at the Bottom* (Feynman 1960): "The principles of physics, as far as I can see, do not speak against the possibility of maneuvering things atom by atom. It is not an attempt to violate any laws; it is something, in principle, that can be done; but, in practice, it has not been done because we are too big."

What is usually neglected are Feynman's caveats in that paper on how friction and adhesion at the atomic and molecular level present major challenges for developing atomic scale machines:

- "All things to do not simply scale down in proportion. There is the problem that materials stick together by molecular (Van der Waals) attractions."
- "Lubrication involves some interesting points... Let the bearings run dry!"

Now, over a half-century later, in the emerging field of nanotechnology, people are on the verge of fulfilling Feynman's vision of building machines atom by atom. With our accumulated knowledge over the last half-century of the atomic origins of friction and adhesion, we are also beginning to appreciate the difficulties and subtleties of manipulating nanoscopic pieces of matter: Even with his caveats, things are not as simple as Feynman suggested!

While Feynman was right that molecular adhesive forces dominate over external forces when things are scaled down to the molecular level, he neglected to mention the many other molecular forces that occur besides van der Waals attraction. As we will see in Chapters 6 and 7, in addition to van der Waals forces, other forces—electrostatic forces, hydrophilic, hydrophobic, structural, double-layer, and capillary adhesive forces—contribute to the adhesion between small objects. We also now appreciate better the major role that molecular adhesion plays in generating the friction as two materials slide over each other, as discussed in Chapter 11.

Feynman's phrase "Let the bearings run dry!" was based on the assumption that friction originates from viscous forces between atoms acting incoherently across a sliding interface and, therefore, should be negligible at small sliding velocities. As discussed in Chapter 11, we now know that molecules trapped in between two solids can give rise to a static friction force by interlocking contacting surfaces. The negligible friction envisioned by Feynman has only been observed in a few carefully constructed experiments designed to bring into contact two solid surfaces with incommensurate atomic structures, with weak interactions across their sliding interface (Chapter 11). This phenomenon of *superlubricity* has given hope to the concept that nano-machines can be built with little or no friction between their moving parts.

1.5 PROBLEMS

1. Consider a small fairyfly that sits towards the outer edge of a disk-shaped MEMS gear that has a diameter of 800 μm and rotates at 100,000 rpm. The fairyfly remains stuck to the gear despite the centripetal acceleration. This fairyfly, one of the smallest known insects at ~140 μm long, has a mass of 100 nanograms and eight legs.

   (a) Modeling the fairyfly as a point mass at the gear edge, find the value of the friction force per leg that is holding the fairyfly to the gear.
   (b) Would this amount of force be enough to keep the fairyfly from slipping down a vertical wall?
   (c) If the area of contact each leg forms with the gear is a circle 0.5 μm in diameter, find the shear stress (friction force per unit area of contact) in units of Pa.

2. Let us assume that the speed with which a recording head slider in a disk drive flies over a particular radius of a disk is 10 m/s and that the clearance is 1.5 nm. Now consider scaling this up to an airplane flying over the ground with the same relative tolerance. A typical recording head slider is ~0.85 mm long, while the Boeing 787 Dreamliner is ~57 m long.

   (a) Find the equivalent ground clearance and flying speed of the 787 assuming all dimensions scale by the same ratio as the hard disk head length and the airplane length.
   (b) Using online research tools, find the typical peak air pressure in the confined space between the recording head slider and the disk during normal operation.

3. Find an example of a Class IV MEMS device other than those described in the article by Romig et al. (2003). The device could be described either in a published patent or in a published journal article. Describe the device and discuss the tribological issues that hamper reliability or performance of the device. Include figures if this helps to illustrate your points.

4. Consider Feynman's argument that at small scales, we can "let the bearings run dry!" We can do this virtually by imagining two sliding parts in a nanoscale machine and comparing them to sliding parts in a macroscale machine. For simplicity, model the two parts as two cubes made out of silicon in contact with one another for both the macro and the nano cases. The macroscale parts are 1 cm × 1 cm × 1 cm in dimension, and the nanoscale parts are 10 nm × 10 nm × 10 nm in dimension. They

are brought into contact and slid back and forth at an amplitude equal to 1% of their lateral size, at a frequency of 100 Hz, for 5 minutes. Assume the friction coefficient between the two parts is $\mu = 0.1$ and that a loading force is applied to the interface such that the normal contact pressure is 10 MPa. Find the average temperature rise in the macroscale and the nanoscale situations. For this calculation, you can use the average speed to calculate the power dissipated by friction as (power) = (friction force) × (sliding speed) and assume that no heat is dissipated into the environment through a lubricant or by any other means such as thermal radiation. Note: you may need to look up other material properties of silicon.

1.6   REFERENCES

Achanta, S. and J.-P. Celis (2015). "Nanotribology of MEMS/NEMS." In: *Fundamentals of friction and wear on the nanoscale*, E. Gnecco and E. Meyer, Eds. Cham, Switzerland: Springer International Publishing, pp.631–56.

Albrecht, T. R., H. Arora, V. Ayanoor-Vitikkate, J.-M. Beaujour, D. Bedau, D. Berman, A. L. Bogdanov, Y.-A. Chapuis, J. Cushen, E. E. Dobisz, G. Doerk, H. Gao, M. Grobis, B. Gurney, W. Hanson, O. Hellwig, T. Hirano, P.-O. Jubert, D. Kercher, J. Lille, Z. Liu, C. M. Mate, Y. Obukhov, K. C. Patel, K. Rubin, R. Ruiz, M. Schabes, L. Wan, D. Weller, T.-W. Wu and E. Yang (2015). "Bit patterned magnetic recording: theory, media fabrication, and recording performance." *IEEE Transactions on Magnetics* **51**(5): 0800342.

Ayrinhac, S. (2016). "The transportation of the Djehutihotep statue revisited." *Tribology Online* **11**(3): 466–73.

Briggs, D. and M. P. Seah (1990). *Practical surface analysis* (2nd ed.). Chichester, UK: John Wiley & Sons.

Bureau of Transportation Statistics (2017). Table 1–26: Average age of automobiles and light trucks in operation in the United States. U.S. Department of Transportation.

Campion, A. and P. Kambhampati (1998). "Surface-enhanced Raman scattering." *Chemical Society Reviews* **27**(4): 241–50.

Chan, M. L., B. Yoxall, H. Park, Z. Kang, I. Izyumin, J. Chou, M. M. Megens, M. C. Wu, B. E. Boser and D. A. Horsley (2012). "Design and characterization of MEMS micromotor supported on low friction liquid bearing." *Sensors and Actuators A: Physical* **177**: 1–9.

Chattopadhyay, R. (2014). *Green tribology, green surface engineering, and global warming*. Materials Park, OH: ASM International.

Chou, S. Y., P. R. Krauss and P. J. Renstrom (1996). "Nanoimprint lithography." *Journal of Vacuum Science & Technology B* **14**(6): 4129–33.

Czaplewski, D. A., C. D. Nordquist, C. W. Dyck, G. A. Patrizi, G. M. Kraus and W. D. Cowan (2012). "Lifetime limitations of ohmic, contacting RF MEMS switches with Au, Pt and Ir contact materials due to accumulation of 'friction polymer' on the contacts." *Journal of Micromechanics and Microengineering* **22**(10): 105005.

Dadgour, H. F., M. M. Hussain, A. Cassell, N. Singh and K. Banerjee (2011). Impact of scaling on the performance and reliability degradation of metal-contacts in NEMS devices. In: *2011 International Reliability Physics Symposium*.

Dake, L. S., J. A. Russell and D. C. Debrodt (1986). "A review of DOE ECUT tribology surveys." *Journal of Tribology* **108**(4): 497–501.

Dong, Y., Q. Li and A. Martini (2013). "Molecular dynamics simulation of atomic friction: a review and guide." *Journal of Vacuum Science & Technology A: Vacuum, Surfaces, and Films* **31**(3): 030801.

Douglass, M. R. (1998). Lifetime estimates and unique failure mechanisms of the digital micromirror device (DMD). In: *36th Annual International Reliability Physics Symposium*, Reno, NV, pp.9–16.

Dowson, D. (1998). *History of tribology* (2nd ed.). Hoboken, NJ: John Wiley & Sons.

Fan, L. S., Y. C. Tai and R. S. Muller (1989). "IC-processed electrostatic micromotors." *Sensors and Actuators* **20**(1–2): 41–7.

Feynman, R. P. (1960). "There's plenty of room at the bottom." *Engineering and Science Magazine* **208**: 22.

Gangopadhyay, A. (2006). Personal communication.

Gao, J. P., W. D. Luedtke, D. Gourdon, M. Ruths, J. N. Israelachvili and U. Landman (2004). "Frictional forces and Amontons' law: from the molecular to the macroscopic scale." *Journal of Physical Chemistry B* **108**(11): 3410–25.

Goggin, R., P. Fitzgerald, B. Stenson, E. Carty and P. McDaid (2015). Commercialization of a reliable RF MEMS switch with integrated driver circuitry in a miniature QFN package for RF instrumentation applications. In: *2015 IEEE MTT-S International Microwave Symposium (IMS)*, IEEE.

Hahner, G. and N. Spencer (1998). "Rubbing and scrubbing." *Physics Today* **51**(9): 22–7.

He, G., M. H. Muser and M. O. Robbins (1999). "Adsorbed layers and the origin of static friction." *Science* **284**(5420): 1650–2.

Hermance, H. and T. Egan (1958). "Organic deposits on precious metal contacts." *Bell System Technical Journal* **37**(3): 739–76.

Holmberg, K., P. Andersson and A. Erdemir (2012). "Global energy consumption due to friction in passenger cars." *Tribology International* **47**: 221–34.

Hornbeck, L. J. (2011). "The DMD^{TM} projection display chip: a MEMS-based technology." *MRS Bulletin* **26**(4): 325–7.

Jost, H. P. (1966). *Lubrication (tribology) education and research. A report on the present position and industry's needs.* Department of Education and Science, Her Majesty's Stationary Office, London.

Jost, H. P. (1990). "Tribology – origin and future." *Wear* **136**(1): 1–17.

Kiely, J. D., P. M. Jones and J. Hoehn (2018). "Materials challenges for the heat-assisted magnetic recording head–disk interface." *MRS Bulletin* **43**(2): 119–24.

Krick, B. A., D. W. Hahn and W. G. Sawyer (2013). "Plasmonic diagnostics for tribology: in situ observations using surface plasmon resonance in combination with surface-enhanced Raman spectroscopy." *Tribology Letters* **49**(1): 95–102.

Lee, P. M. and R. W. Carpick, Eds. (2017). *Tribological opportunities for enhancing America's energy efficiency.* A report to the Advanced Research Projects Agency-Energy (ARPA-E) at the U.S. Department of Energy.

Li, J., H. Chen and H. A. Stone (2013). "Ice lubrication for moving heavy stones to the Forbidden City in 15th- and 16th-century China." *Proceedings of the National Academy of Sciences* **110**(50): 20023–7.

Loh, O. Y. and H. D. Espinosa (2012). "Nanoelectromechanical contact switches." *Nature Nanotechnology* **7**(5): 283–95.

Mangolini, F., A. Rossi and N. D. Spencer (2012). "In situ attenuated total reflection (ATR/FT-IR) tribometry: a powerful tool for investigating tribochemistry at the lubricant-substrate interface." *Tribology Letters* **45**(1): 207–18.

Marchon, B. and T. Olson (2009). "Magnetic spacing trends: from LMR to PMR and beyond." *IEEE Transactions on Magnetics* **45**: 3608–11.

Marchon, B., X. Guo, B. K. Pathem, F. Rose, Q. Dai, N. Feliss, E. Schreck, J. Reiner, O. Mosendz, K. Takano, H. Do, J. Burns and Y. Saito (2014). "Head–disk interface materials issues in heat-assisted magnetic recording." *IEEE Transactions on Magnetics* **50**(3): 137–43.

Marchon, B., T. Pitchford, Y.-T. Hsia and S. Gangopadhyay (2013). "The head–disk interface roadmap to an areal density of Tbit/in^2." *Advances in Tribology* **2013**: 521086.

Mate, C. M. (2013). "Spreading kinetics of lubricant droplets on magnetic recording disks." *Tribology Letters* **51**(3): 385–95.

Mate, C. M., Q. Dai, R. N. Payne, B. E. Knigge and P. Baumgart (2005). "Will the numbers add up for sub-7-nm magnetic spacings? Future metrology issues for disk drive lubricants, overcoats, and topographies." *IEEE Transactions on Magnetics* **41**(2): 626–31.

Mate, C. M., H. Deng, G.-J. Lo, I. Boszormenyi, E. Schreck and B. Marchon (2015). "Measuring and modeling flash temperatures at magnetic recording head–disk interfaces for well-defined asperity contacts." *Tribology Letters* **58**(2): 27.

Mate, C. M., B. K. Yen, D. C. Miller, M. F. Toney, M. Scarpulla and J. E. Frommer (2000). "New methodologies for measuring film thickness, coverage, and topography." *IEEE Transactions on Magnetics* **36**(1): 110–14.

McClelland, G. M., C. T. Rettner, M. W. Hart, K. R. Carter, M. I. Sanchez, M. E. Best and B. D. Terris (2005). "Contact mechanics of a flexible imprinter for photocured nanoimprint lithography." *Tribology Letters* **19**(1): 59–63.

McGhee, E. O., A. A. Pitenis, J. M. Urueña, K. D. Schulze, A. J. McGhee, C. S. O'Bryan, T. Bhattacharjee, T. E. Angelini and W. G. Sawyer (2018). "In situ measurements of contact dynamics in speed-dependent hydrogel friction." *Biotribology* **13**: 23–9.

Peschot, A., Q. Chuang and L. Tsu-Jae King (2015). "Nanoelectromechanical switches for low-power digital computing." *Micromachines* **6**(8): 1046–65.

Pott, V., H. Kam, R. Nathanael, J. Jeon, E. Alon and T. J. K. Liu (2010). "Mechanical computing redux: relays for integrated circuit applications." *Proceedings of the IEEE* **98**(12): 2076–94.

Rebeiz, G. M. (2004). *RF MEMS: theory, design, and technology*. Hoboken, NJ: John Wiley & Sons.

Romig, A. D., M. T. Dugger and P. J. McWhorter (2003). "Materials issues in microelectromechanical devices: science, engineering, manufacturability and reliability." *Acta Materialia* **51**(19): 5837–66.

Sawyer, W. G. and K. J. Wahl (2008). "Accessing inaccessible interfaces: *in situ* approaches to tribology." *MRS Bulletin* **33**(12): 1145–50.

Schift, H. (2008). "Nanoimprint lithography: an old story in modern times? A review." *Journal of Vacuum Science & Technology B* **26**(2): 458–80.

Shearwood, C., K. Ho, C. Williams and H. Gong (2000). "Development of a levitated micromotor for application as a gyroscope." *Sensors and Actuators A: Physical* **83**(1–3): 85–92.

Shen, Y. R. (1994). "Surfaces probed by nonlinear optics." *Surface Science* **300**(1–3): 551–62.

Shiroishi, Y., K. Fukuda, I. Tagawa, H. Iwasaki, S. Takenoiri, H. Tanaka, H. Mutoh and N. Yoshikawa (2009). "Future options for HDD storage." *IEEE Transactions on Magnetics* **45**(10): 3816–22.

Somorjai, G. A. (1994). *Introduction to surface chemistry and catalysis*. New York: John Wiley & Sons.

Somorjai, G. A. (1998). "From surface materials to surface technologies." *MRS Bulletin* **23**(5): 11–29.

Suk, M., K. Miyake, M. Kurita, H. Tanaka, S. Saegusa and N. Robertson (2005). "Verification of thermally induced nanometer actuation of magnetic recording transducer to overcome mechanical and magnetic spacing challenges." *IEEE Transactions on Magnetics* **41**(11): 4350–2.

Sun, B., F. Han, L. Li and Q. Wu (2016). "Rotation control and characterization of high-speed variable-capacitance micromotor supported on electrostatic bearing." *IEEE Transactions on Industrial Electronics* **63**(7): 4336–45.

Thompson, P. A. and M. O. Robbins (1990). "Origin of stick-slip motion in boundary lubrication." *Science* **250**(4982): 792–4.

Traub, M. C., W. Longsine and V. N. Truskett (2016). "Advances in nanoimprint lithography." *Annual Review of Chemical and Biomolecular Engineering* **7**: 583–604.

Tung, S. C. and M. L. McMillan (2004). "Automotive tribology overview of current advances and challenges for the future." *Tribology International* **37**(7): 517–36.

U.S. Environmental Protection Agency. (2017). "Light-duty automotive technology, carbon dioxide emissions, and fuel economy trends." Available at: https://www.Epa.Gov/fuel-economy-trends.

Van Kessel, P. F., L. J. Hornbeck, R. E. Meier and M. R. Douglass (1998). "MEMS-based projection display." *Proceedings of the IEEE* **86**(8): 1687–704.

Wahl, K. J., R. R. Chromik, C. C. Baker and A. A. Voevodin (2007). "In situ tribometry of solid lubricant nanocomposite coatings." *Wear* **262**(9–10): 1239–52.

Williams, J. A. and H. R. Le (2006). "Tribology and MEMS." *Journal of Physics D: Applied Physics* **39**(12): R201–14.

Wong, C. W., X. Zhang, S. A. Jacobson and A. H. Epstein (2004). "A self-acting gas thrust bearing for high-speed microrotors." *Journal of Microelectromechanical Systems* **13**(2): 158–64.

Wood, R. (2000). "The feasibility of magnetic recording at 1 Terabit per square inch." *IEEE Transactions on Magnetics* **36**(1): 36–42.

2

Surface Roughness

To begin to understand friction, lubrication, and wear, we need to first understand how two surfaces make contact with each other. Bowden (1950) pointed out that "Putting two solids together is rather like turning Switzerland upside down and standing it on Austria—the area of intimate contact will be small" (Bowden 1950). For most tribological surfaces, the slopes of the asperities are much smaller than the mountain slopes of Switzerland and Austria. For very smooth surfaces (e.g., the slider–disk interface of a disk drive) the situation is more like putting two American states, Kansas and Nebraska, known for their flat terrain on top of each other; in this situation, the contact is just as likely to occur on the debris (such as cows) as on topographic features (such as farmhouses or the occasional hill). The widely different topography of Switzerland–Austria versus Kansas–Nebraska leads to quite different contact: the interlocking of jagged peaks versus the mashing of cows between two flat pastures.

Likewise, as we examine what happens at the atomic scale when two solids touch, much depends on the nanoscale surface texture. For atomically smooth mating surfaces, any loose atom or molecule acts like debris between the two surfaces. For sharp asperities, strong molecular adhesive forces and high contact pressures can result in extensive plastic deformation on the atomic scale when two solids touch. Understanding tribological behavior first depends on understanding the topography of the two contacting surfaces. In this chapter, we cover the general concepts of surface topography and how to apply them to tribological surfaces, and cover the experimental techniques for measuring surface topography.

2.1 Surface finish, texture, and roughness

Most manufactured surfaces have been processed or finished in some fashion to achieve a certain degree of roughness, the nature and level of which depends on the eventual function of the part. The resulting surface topography is called the *surface finish*, the *surface texture*, or *roughness*. If the eventual function is tribological, the surface texture needs to be adequately controlled to ensure the proper tribological performance. A wide variety of manufacturing processes exist for finishing surfaces: grinding, polishing, electropolishing, lapping, abrasive blasting, honing, electrical discharge machining (EDM),

Tribology on the Small Scale: A Modern Textbook on Friction, Lubrication and Wear. Second edition. C. Mathew Mate and Robert W. Carpick. © Oxford University Press 2019. Published in 2019 by Oxford University Press.
DOI: 10.1093/oso/ 9780199609802.001.0001

milling, lithography, industrial etching/chemical milling, and laser texturing (De Garmo et al. 2011).

Generally, surfaces are finished to have the smoothest surface that is economically feasible as rougher surfaces tend to have the higher wear and friction. An example of this behavior is shown in Figure 2.1, which shows a roughness parameter map delineating the safe and unsafe regions for abrasive wear, as determined in experiments by Hirst and Hollander (1974) for lubricated stainless steel surfaces.

From Figure 2.1 we can see that at high values of roughness (as measured by the standard deviation of asperity heights σ_s) abrasive wear occurs and that the boundary between the safe and unsafe regions moves toward higher loads as this roughness decreases. We also see from Figure 2.1 that the safe region has a lower boundary, indicating that, if the surfaces become too smooth, they again become susceptible to wear. That wear occurs when both surfaces become very smooth is a common phenomenon, which results the friction and adhesion forces to rising dramatically when the true contact area becomes very large.

In addition to higher wear and friction, rougher surfaces also tend to be susceptible to corrosion and crack formation as these typically nucleate at the defects and irregularities or can be induced by the high stresses, which are more common for rougher surfaces in contact. So, during the engineering development phase of a tribological surface, appropriate machining or manufacturing processes must be determined for producing the specified range of surface textures that satisfy the friction, wear, and corrosion requirements.

Not all rough surfaces are bad though. Often it is important for a surface to be textured a certain way to generate the right look to the eye or feel to the touch. For example, a

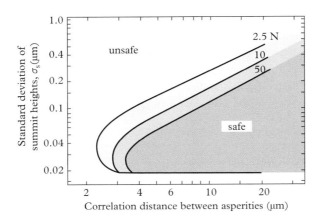

Figure 2.1 *Map delineating the safe from the unsafe regions for surface abrasion for a stainless steel ball sliding at different loads across a stainless steel plate lubricated with steric acid, as function of the standard deviation of the summit heights of asperities on the steel surfaces σ_s and the correlation distance between these asperities. Reproduced with permission from Hirst and Hollander (1974). Copyright 1974, Royal Society of London.*

very smooth surface may look too shiny to the eye or may be too difficult to slide a finger or hand over. In the case of tactile sensation, the amplitude and wavelength of the roughness tend to be the main determinants of how a surface feels to the touch (Dowson 2009, Mate and Carpick 2011, Wandersman et al. 2011, Skedung et al. 2013).

Surface texturing is also important to many other surface properties such as wettability, which will be discussed in Chapter 5. Several other examples include the traction of tires against a wet road (Persson et al. 2005), and the adhesion of thin films, which adhere best if deposited onto a rough surface.

2.2 Measuring surface roughness

Due to the importance of surface topographies to many scientific and engineering fields in addition to tribology, a wide variety of techniques have been invented over the years for measuring surface topographies. Table 2.1 compares experimental techniques commonly used to characterize the roughness of surfaces.

The choice of measuring tool often depends on the length scale over which the topographical information is desired. For example, if it is sufficient to know the surface roughness over a large area of the sample without much precision, optical methods are typically used, such as optical microscopy, optical interference, and optical reflectance. If the area of interest is less than a few millimeters in size, a simple line scan may suffice, and a mechanical stylus profilometer might be used. Another common technique used in this length scale range is scanning electron microscopy (SEM), which can examine samples at low resolution from many millimeters across and then can zoom in and image areas with lateral resolution as high as 10 nm. Generally, SEM does not provide quantitative height information of the topographic features, but some height information can be obtained by tilting the sample or doing cross-sectioning.

If it is desirable to know the surface topography with the highest possible resolution, a scanning probe such as the scanning tunneling microscope or atomic force microscope (AFM) is used, though these techniques are limited to relatively small areas ($<100 \, \mu$m across). Of these two techniques, atomic force microscopy (AFM) is now by far the most widely used. AFMs can routinely achieve lateral resolutions on the order of a few nanometers and, in the best circumstances, can achieve atomic resolution. Since a main focus of this book is what happens at the smallest possible scale, we will discuss AFM in more detail in Section 2.2.2.

Another important criterion for selecting a topographic measurement technique is how destructive it is. If the sample is to be used after the roughness measurement, then a nondestructive technique should be used. Since optical methods (except for taper sectioning) can measure topographies without contacting the sample being measured, they are frequently used during manufacturing for nondestructive characterization of roughness.

Other techniques can sometimes be destructive. For example, SEM requires the sample be subjected to a vacuum environment and to a high energy electron beam, both of which can potentially degrade the materials in the surface layers. Another example

Table 2.1 *Comparison of experimental techniques commonly used for measuring surface topographies. Adapted from Dunaevsky et al. (1997).*

Type of measurement	Resolution (nm)		Dimensions	Comment
	Lateral	Vertical		
Non-contacting methods				
Optical methods				
Optical microscopy	250–350	180–350	3D	Depends on optical system quality
Taper sectioning	250–350	180–350	2D	Destructive technique
Optical interferometry	200–500	1	2D	
Optical reflectance		2–3	1D, 3D	Very smooth surfaces, <100 nm
Electron microscopy				
Scanning electron microscopy (SEM)	10		3D	Requires vacuum, limited z-height info.
Transmission microscopy (TEM)	0.05	0.05	2D	Requires vacuum
Atomic force microscopy (AFM)	Atomic resolution		3D	No vacuum required
Contact methods				
Stylus profilometry	1300–2500	5–250	1D, 2D, 3D	Stylus size is ultimate limit on resolution
Replicas (negative impressions)	Surface details duplicated down to 2 nm			Used when surface cannot be examined directly
Capacitance	1500		1D	Between two conducting surfaces
Air gauging	100		1D	Flow between surfaces

is stylus profilometry, where a sharp tip made of a hard material is scanned in contact over the surface; even though the load on the tip is very light, it can still damage softer materials, potentially leaving behind a gouge as the tip is dragged across the surface. AFM, however, is typically considered a nondestructive technique: even though it also involves scanning a tip across the surface, the contact forces can generally be kept well below the threshold for causing damage. As discussed in Chapter 12, however, under suitable circumstances modification of surfaces by AFM is also possible and can be used to study wear processes at the atomic level.

2.2.1   Optical interferometry

One popular, versatile, and powerful surface profiling method is optical interferometry (Wyant and Creath 1992); Figure 2.2 shows a schematic of a common type of optical interferometer. A key advantage of this method is that it is non-contact, and thus does not perturb the sample in any way. Another advantage is that it is a "full field" method, which means that it measures simultaneously all the heights over the full field of the image rather going through the more time consuming procedure of scanning a probe laterally over the surface, as done in some other profilometer techniques. Optical interferometers can resolve heights as large as a few centimeters down to less than a nanometer. Being an optical method, the lateral resolution is determined by the far field diffraction limit of light of $\sim\lambda/2$, which for green light corresponds to ~200 nm.

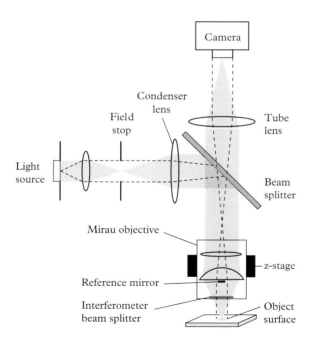

Figure 2.2 *Schematic of an optical interferometer with a Mirau objective.*

Optical interferometry imaging utilizes the interference between two waves of light, which results in an easily measured reduction in the intensity, even for small phase shifts between the two waves. This phase shift results from the path length difference between a reference light and light reflected off the sample. Figure 2.2 shows a design of an optical interferometer that has a Mirau objective, where a small mirror is placed in front of the objective lens that partially reflects some of the light back to the detector to serve as the reference light. The rest of the light reflects off the sample back towards the detector where it interferes with the reference light reflected off the mirror. A piezoelectric actuated positioning stage moves the Mirau objective vertical to surface to change the difference in path length between the reference light and the light reflected off the sample (and change the corresponding phase shift). For monochromatic light, different locations on the sample will encounter a maximum in intensity (a bright fringe) when the path length difference is an integer multiple of the light wavelength. For a large change in surface height, however, this creates an ambiguity as one cannot tell if this sudden change corresponds to a fraction of a fringe or to multiple fringes plus that same fraction. To resolve this ambiguity, most optical interferometers use a white light source that produces a wide range of wavelengths of light, enabling surfaces with a wide range in topographic heights be quantitatively measured and imaged.

2.2.2 Atomic force microscopy (AFM)

In AFM, the force acting on the last few atoms of a sharp tip is measured as that tip moves gently over the sample surface, as illustrated in Figure 2.3. This is accomplished by mounting the tip at the end of a cantilever with a known spring constant; the cantilever deflects by an amount proportional to the force. To achieve atomic resolution, it is necessary to measure deflections smaller than the size of an atom; since atoms are typically a few angstroms in diameter, the measurement sensitivity should be better than an angstrom. Several methods have been developed for measuring the small deflections of AFM cantilevers, including electron tunneling (Binnig et al. 1986), optical interference (Martin et al. 1987), optical deflection (Meyer and Amer 1988, Drake et al. 1989), capacitance (Neubauer et al. 1990), and piezoelectric resonators (Benes et al. 1995, Giessibl 2000). As of 2019, many AFM designs routinely measure deflections as small as 10^{-11} m (0.1 Å), with the most common detection method in commercial AFMs being optical deflection, as illustrated in Figure 2.4.

For determining the topography of surfaces, an AFM can use either the repulsive force when the tip is in contact with the surface or the attractive force when the tip is separated by a small distance from the sample. In the repulsive mode (Binnig et al. 1986), the tip is scanned across a hard surface while touching it with constant force, so that the tip moves up and down over the surface features in much the same way as a stylus moves in the groove of an old fashioned phonograph record. In this mode, a tip with a radius on the order of 10 nm contacts with a repulsive force of a few nanonewtons achieving a lateral resolution of a few nanometers.

To reduce tip wear and reduce the tip–sample contact area, a "tapping mode" or "intermittent contact mode" is frequently used. In intermittent contact mode, the

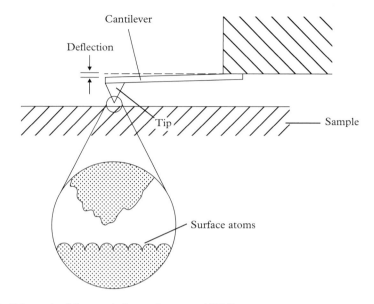

Figure 2.3 *Schematic of the atomic force microscope (AFM).*

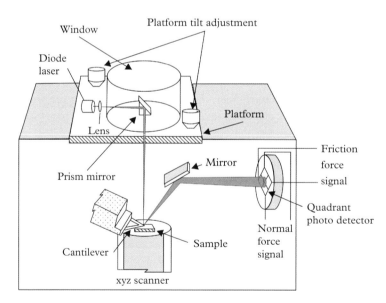

Figure 2.4 *Schematic of an optical deflection AFM. The normal and lateral displacements of the cantilever are determined by reflecting a laser beam off the cantilever onto a quadrant photo detector to measure the bending angle of the cantilever.*

cantilever is vibrated at or near its natural frequency with a relatively large oscillation amplitude, typically tens of nanometers. The tip contacts with the sample briefly during the lowest part of the oscillation cycle, but the large amplitude pulls the tip far enough away from the sample to disengage it from the adhesion force. This intermittent contact between tip and sample reduces the vibration amplitude below that of the out-of-contact amplitude by an amount directly related to the intermittent normal contact force. A feedback loop then is used to adjust the z-height of the sample so as to keep the reduction of amplitude constant as the oscillating tip scans over the surface. The intermittent contact method ensures that the repulsive contact forces are far less damaging than that for contact imaging.

To eliminate the potential for tip wear and eventual degradation of resolution, a "non-contact" force imaging mode can be used. This mode is particularly useful when imaging surfaces with high surface energies that cause excessively high adhesive forces to act on the tip in the intermittent contact mode. In non-contact mode, the tip is held typically a few nm away from the sample, where the attractive van der Waals force typically dominates (Martin et al. 1987). When an attractive force acts on the tip, the cantilever's resonance frequency shifts to lower frequency; but, if repulsive, the shift is to higher frequency. A feedback circuit adjusts the z-height so that the frequency shift or amplitude remains constant.

2.3   Characterizing surface roughness

While an infinite variety of surface topographies exist, only a small number of ways are used, in practice, to describe surface topographies. This can work well for those surfaces (which fortunately include most practical surfaces) where the topography is a mixture of forms that each dominate at different length scales. For example, Figure 2.5 illustrates a surface texture that is regular at large length scales and random at small length scales. The regular structure could have come from a surface finishing process producing a wave-like pattern over the surface with a well-defined "direction of lay." The line profile orthogonal to the direction of lay illustrates the waviness spacing (or wavelength) and waviness height. Further magnification reveals a finer irregular structure superimposed on the regular wavy structure. With further magnification the roughness from the individual atoms is revealed. The revealing of different types of roughness with increased magnification is a common occurrence in topography measurements, leading one well known tribologist (Archard 1957) to describe surfaces as "protuberances on protuberances on protuberances."

The geometric characteristics of the surface in Figure 2.5, which are common to many types of surfaces, can be categorized from the smallest scale to the largest scale as follows:

- *Atomic scale roughness*: roughness from the atoms on the surface of the solid.

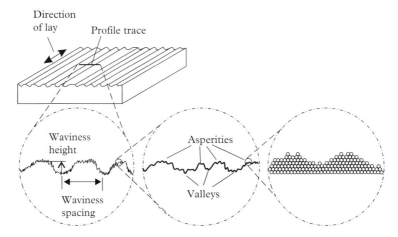

Figure 2.5 *Illustration of a regular wavy surface texture (top). Increasingly magnified view of the fine scale roughness (bottom).*

- *Roughness*: short wavelength surface irregularities characterized by hills (asperities) and valleys. These surface features typically have random spacings, heights and depths.
- *Waviness*: surface irregularities with a much larger spacing than the roughness features and with a small statistical variation of spacing and amplitude (i.e., a regular structure). Waviness can result from heat treatment, clamping deflections, vibrations, warping strains, and, for machined parts, from the grinding or polishing process used to finish the surface.
- *Errors of form*: flaws that occur during manufacture or later misuse of the part; gross deviation in shape resulting from the part not being made correctly or being damaged.

While the above characterizations of topographies are common to many surfaces, others have topographies that do not fit simply in the above categories. For example, the roughness may be regular at all length scales or random at all length scales.

2.4   Roughness parameters

As mentioned previously, profilometry and AFM are two common ways to measure surface topography. These instruments trace over a surface to provide a single line profile of the surface topography, as illustrated in Figure 2.6. By repeatedly collecting traces separated by a small distance in the orthogonal direction, a three-dimensional (3D) representation of the topography can be generated. These profile traces can then be analyzed with a computer to determine parameters that describe the surface roughness. Alternately, in optical profilometry, optical interference of a split beam is used to create a full-field height map of the surface, also producing a 3D topographic image.

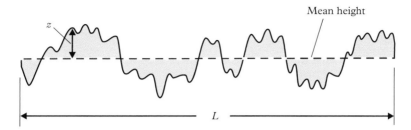

Figure 2.6 *An example of a single profile of surface topography which could be obtained by AFM or profilometry.*

2.4.1 Variation in z-height

For a single profile trace, a center line is first found by determining the line through the trace for which the mean-square-deviation of the trace about that line is a minimum. This center line also corresponds to the mean height of all data points in the trace. Then z is defined as the height of the profile relative to the mean height. For 3D collections of traces, a plane defined in the same manner is used to determine the mean height or surface mean.

The first roughness parameter, the average-roughness R_a, is defined by

$$R_a = \frac{1}{L} \int_0^L |z| \, dx, \tag{2.1}$$

where z is the height of the profile above the mean height at some position x, and L is the total length of the profile over which R_a was computed, as shown in Figure 2.6. Note that the value of R_a, along with the other roughness parameters that follow, tends to vary as the sampling length L is varied. This will be discussed further in Sections 2.4.4 and 2.5.

The root-mean-square deviation of z about the mean height (also called the R_q, the rms, or the standard deviation σ) is a statistically more meaningful parameter and is defined by

$$R_q = \text{rms} = \sigma = \left(\frac{1}{L} \int_0^L z^2 \, dx \right)^{1/2}. \tag{2.2}$$

For most surfaces, R_a and R_q are similar in magnitude; for example (Bhushan 1999), for a sinusoidal surface,

$$R_q = \left(\frac{\pi}{2\sqrt{2}}\right) R_a \approx 1.11 R_a, \qquad (2.3)$$

while for a surface with a random Gaussian distribution of surface heights,

$$R_q = \sqrt{\frac{\pi}{2}} R_a \approx 1.25 R_a. \qquad (2.4)$$

Several other parameters exist for further characterizing the variation of z-height of a line profile or 3D image:

- R_p—the distance from the mean height to the maximum peak height;
- R_v—the distance from the mean height to the maximum valley depth;
- R_t—the vertical height from the lowest valley to the highest asperity, also known as the peak-to-valley (P–V) roughness ($R_t = R_p + R_v$);
- R_z—the distance from the average of the five lowest points to the five highest points of the profile or image.

Obviously, these parameters characterize only the variation of z-height about the surface mean and do not provide information of how neighboring z's are correlated.

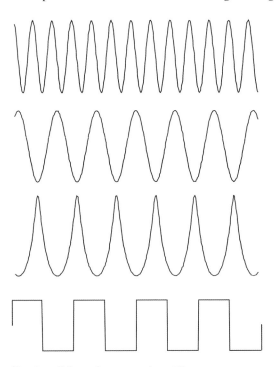

Figure 2.7 *Surface profiles that all have the same value of R_q.*

They incompletely characterize a surface's topography by not providing any information about lateral variations of topography: slopes, shapes, and sizes of surface features, or their spacing or regularity. For example, Figure 2.7 shows surface profiles that all have the same value of R_q, but different frequencies and shapes that result in widely different contact mechanics and contact areas in tribological situations. These roughness parameters for z-height variations are mainly useful for characterizing surfaces with a particular type of roughness, such as would be made by the same manufacturing method where lateral variations scale with height variations. A common mistake is to compare the z-height roughness parameters between surfaces textured by different methods and then to assume that surfaces with similar R_q values, for example, will have similar tribological behavior.

2.4.2 Surface height distributions

The z-height roughness parameters R_a, R_q, and σ only begin to tell how the surface heights vary from the surface mean. By plotting the *height distribution* (also called the *probability distribution* or *height histogram*) one can obtain a more detailed picture of how the surface heights vary. The height distribution is the probability $\phi(z)dz$ that the height of a particular point on the surface is between z and $z + dz$ and is calculated by determining the fraction of the profile line (see Figure 2.8) that is between the lines for z and $z + dz$.

Integrating the probability distribution $\phi(z)$ gives the cumulative probability function $\Phi(z)$:

$$\Phi(z) = \int_z^\infty \phi(z')\, dz', \tag{2.5}$$

generating the S-shaped curve on the right side of Figure 2.8. $\Phi(z)$ is also called the *bearing area curve* as it expresses, as a function of z, the fraction of the area underneath the surface contour and above a particular z-height (see Figure 2.8).

For many real surfaces, the height distribution is close to a Gaussian distribution, which is given by

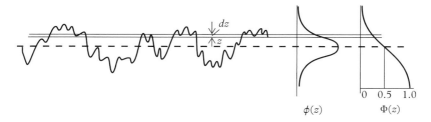

Figure 2.8 *Height or probability distribution $\phi(z)$ and cumulative probability distribution $\Phi(z)$, in this case for a Gaussian distribution of surface heights.*

$$\phi(z) = \frac{\exp\left(-z^2/2\sigma^2\right)}{\sigma\sqrt{2\pi}}. \tag{2.6}$$

A useful feature of a Gaussian distribution is that the standard deviation σ completely describes the distribution. For a Gaussian distribution, 68.3% of the surface heights lie within one standard deviation ($\pm\sigma$) of the surface mean, 95.4% within $\pm 2\sigma$, and 99.7% within $\pm 3\sigma$.

While many surfaces can be described by Gaussian distributions, it is important to remember that many surfaces are not. For example, consider a surface with a regular structure, such as the machined surface illustrated in Figure 2.5: the height distribution for this surface would be poorly described by a Gaussian distribution as the surface heights are not random, but wavy with a characteristic wavelength and amplitude generated by the surface finishing process. At length scales much shorter than the characteristic wavelength of the waviness, however, the z-height roughness may potentially be described by a Gaussian distribution.

Also, while the center portion of the height distribution may well follow a Gaussian distribution, if one goes far enough out on the tails of the distribution, eventually deviations from the Gaussian distribution become apparent. For example, surfaces finished with a grinding tool typically have Gaussian height distribution, but subsequent light polishing of that ground surface will remove the high asperities from the height distribution. If the ground surface is polished vigorously enough, however, it will remove most of the residual roughness of the grinding process and convert the originally broad Gaussian distribution into a much narrower one.

As a general rule of thumb, the high end of the z-height distribution does not extend much beyond the $+3\sigma$ point. So, if a surface with a standard deviation of roughness σ_1 is brought into contact with another surface with a standard deviation σ_2, the onset of contact will typically occur when the separation distance $\sim 3\sigma_{composite}$, where the composite standard deviation of roughness $\sigma_{composite}$ is defined by

$$\sigma_{composite} = \sqrt{\sigma_1^2 + \sigma_2^2}. \tag{2.7}$$

2.4.2.1 *Example: z-height parameters for a polished steel surface*

Steel is the most common structural material and is found in many tribological contacts. Of the many types available, 52100 steel is a high-carbon, chromium containing steel alloy used in a wide variety of mechanical applications, most commonly for bearings. A beneficial feature of this steel is that, in the annealed condition, it is relatively easy to machine, but after a suitable heat treatment it has a high hardness and abrasive resistance, making it a good choice when high wear resistance is required.

Figure 2.9 shows optical interferometry image of a small area of a flat piece of 52100 steel that has been machined flat with a rough finish, then polished to a mirror finish, while Figure 2.10 shows an intermittent contact AFM image of the same surface. After

Figure 2.9 *White light interferometric image of a 52100 steel surface that has been polished to a mirror finish. The z-height roughness parameters are $R_a = 1.2$ nm, $R_q = 1.7$ nm, $R_p = 7.0$ nm, and $R_v = -57.6$ nm. Image courtesy of H. Khare and R. W. Carpick, University of Pennsylvania.*

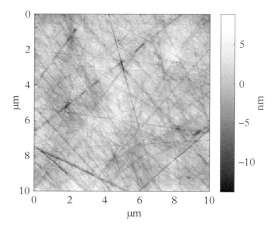

Figure 2.10 *AFM image obtained in the intermittent contact mode of a 10 μm × 10 μm area of the same 52100 stainless steel surface as in Figure 2.9. The z-height roughness parameters are $R_a = 1.7$ nm, $R_q = 2.2$ nm, $R_p = 8.8$ nm, and $R_v = -13.9$ nm. Image courtesy of H. Khare and R. W. Carpick, University of Pennsylvania.*

this vigorous polishing, most of the topographical features associated with the original machining have been removed (though some deep grooves remain), and the surface has become very smooth, with the rms roughness being ~2 nm for both type of images, which is several orders of magnitude smaller than the original machined surface before polishing.

Figure 2.11(a) and (c) show the z-height distributions for the optical and AFM images on histogram plots. As is apparent from the histograms, these distributions are close to Gaussian, though it can be difficult to tell where the deviations away from a Gaussian are occurring. A common way to make this comparison more readily is to plot

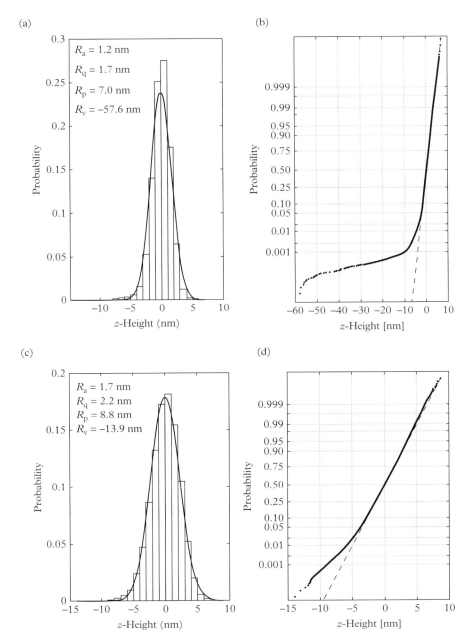

Figure 2.11 *Histogram and normal probability plots of the z-height distributions for the optical interferometry image (a, b) and for the AFM image (c, d), in Figures 2.9 and 2.10, respectively. For the histogram plots (a, c) bars show the distribution of z-heights, while the curves show a Gaussian distribution with the same standard deviation. For the normal probability plots (b, d), even though each pixel of the images is plotted as a discrete point, the high density of points generally make them look like a solid line except at the tails of the distribution. The dashed lines show where points should lie for a Gaussian distribution with the same standard deviation.*

the distributions on *normal probability plots*, as shown in Figure 2.11(b) and (d). On a normal probability plot, if a distribution is perfectly Gaussian, all the points fall on a straight line. As can be seen from Figure 2.11(b) and (d), the z-heights do indeed follow closely a Gaussian distribution except at the low values of z-heights, where the deep grooves, leftover from the original machining process, distort the tail of the distribution to lower values. These grooves also lead to an asymmetry in R_v and R_p roughness parameters with the mean-to-valley parameter being significantly larger than the mean-to-peak parameter.

2.4.3 Asperity summits roughness parameters

Since contact usually occurs at asperity summits, some of the most tribologically relevant topography parameters are those characterizing these peaks:

- n_s—the number of the asperity summits per unit area;
- σ_s—the standard deviation of summit heights;
- R_s—the mean radius of curvature of the summits.

An individual asperity summit has associated with it two radii of curvatures, R_x and R_y, measured in two orthogonal directions. The mean radius is then defined by

$$\frac{1}{R_i} = \frac{1}{2}\left(\frac{1}{R_x} + \frac{1}{R_y}\right), \tag{2.8}$$

where the i subscript refers to it being for the ith asperity; R_s is the average over all R_i values.

These summit roughness parameters are typically determined from a 3D surface topography image that is analyzed with an appropriate computer program. If such an analysis is not available, it is possible to estimate some of the summit parameters from the z-height parameters (Greenwood 1984, McCool 1986):

- For an isotropic textured surface with a Gaussian distribution of summit heights, the standard deviation of summit heights can usually be approximated by the standard deviation of z-heights,

$$\sigma_s \sim \sigma = R_q. \tag{2.9}$$

- For randomly rough surfaces, the mean height of the asperities above the surface mean is usually in the range 0.5–1.5R_q.
- The mean radius of curvature R_s of the summit asperities can be approximated with the rms curvature σ_κ of the profilometer trace,

$$R_s \sim 1/\sigma_\kappa \tag{2.10}$$

where σ_κ is determined from the profiles of z-height.

In the computation of σ_κ, the curvature κ is determined at discrete intervals along the profile. For three consecutive heights z_{i-1}, z_i, and z_{i+1} with a lateral interval spacing h, the curvature at interval i is

$$\kappa_i = (z_{i+1} - 2z_i + z_{i-1})/h^2. \tag{2.11}$$

Then the rms curvature is determined by

$$\sigma_\kappa^2 = \left(\frac{1}{n}\right) \sum_{i=1}^{n} \kappa_i^2. \tag{2.12}$$

Alternatively, the rms curvature can be expressed in terms of the rms of the second derivative the z-height topography $\nabla^2 z_{\mathrm{rms}}$:

$$\sigma_\kappa = \frac{1}{2}\nabla^2 z_{\mathrm{rms}} = \frac{1}{2}\left\langle \left|\nabla^2 z\right|^2\right\rangle^{1/2}. \tag{2.13}$$

2.4.3.1 *Example: summit parameters for a disk from a disk drive*

In modern disk drives, the disks have to be much smoother than most engineered surfaces as the recording head needs to fly as close as possible to the recording medium on the disk surface in order to store the highest possible density of bits in that medium. During reading and writing of data to the recording media, the head flies over the disk surface at speeds in excess of 10 m/s, and the clearance between the recording head and the disk is typically set (for 2013 drives) at ~1.5 nm above the tallest asperities. Since 2010, disk drives have routinely been shipped with disk media that have a mean-to-peak roughness of less than 1.5 nm (Mate 2013), and the industry continues to strive to make the disk surfaces even smoother.

To achieve this level of smoothness, disk manufacturers have developed methods for polishing and etching the disk substrates to a nearly atomically smooth finish. After polishing, a recording medium and a protective overcoat are then deposited onto these substrates (see Figure 1.4.) During deposition, the grains of the recording medium grow as vertical columns with caps that have a slight radius of curvature, which makes the final surface topography look like a random array of small bumps, as shown in Figure 2.12. These caps also result in the z-height roughness parameters of the final disk surface being somewhat higher than those of the initial substrate surface.

Since most of the contact between the head and the disk will occur at the apexes of these carbon overcoated grains, it would make sense to analyze the roughness of this surface in terms of the summit roughness parameters in addition to the z-height roughness parameters. Figure 2.13 shows the distributions of summit heights and the mean radii of curvature for the grain caps shown in Figure 2.12, along with the summit roughness parameters R_s, σ_s, and n_s. Consistent with eq. (2.9), the standard deviation

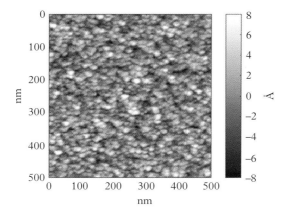

Figure 2.12 *Non-contact mode AFM image of a 500 nm × 500 nm area of a disk surface from a 2017 disk drive. The distribution of z-heights for this image is close to Gaussian, and the z-height roughness parameters are $R_a = 2.0\,\text{Å}$, $R_q = 2.5\,\text{Å}$, $R_p = 11.0\,\text{Å}$, and $R_v = -8.4\,\text{Å}$. This AFM image was collected at the Stanford Nano Shared Facilities (SNSF), supported by the National Science Foundation under award ECCS-1542152.*

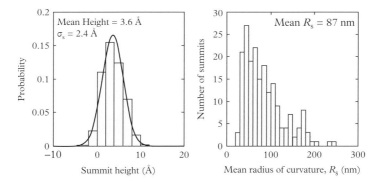

Figure 2.13 *(a) Distribution of asperity summit heights for the AFM image of the disk surface in Figure 2.12. The bars are the measured values and the solid line is a Gaussian distribution with the same standard deviation and mean height. (b) Distribution of mean radii of curvature for the summits in this image. The areal density of asperity summits is $n_s \sim 2500\,\mu m^{-2}$.*

of summit heights $\sigma_s = 2.4\,\text{Å}$ is a similar value as the standard deviation of z-heights $R_q = 2.5\,\text{Å}$, and the mean height of the summits is $1.4R_q$, which is at the upper end of the expected range of 0.5–$1.5R_q$.

It is interesting to notice that even though the disk surface is an order of magnitude smoother than the highly polished steel surface in Figure 2.10 based on the z-height roughness parameters ($R_q = 0.25$ nm for the disk versus $R_q = 2.2$ nm for the polished steel), at the nanometer lateral length-scale it appears to be much rougher. This higher nanoscale roughness shows up not only in the relatively small values of the radii of curvature of summits (~90 nm for the disk surface), but also in the typical slopes encountered in the AFM images (~4° for the disk surface versus ~1° for the polished steel surface.)

2.4.4   Surface roughness power spectrum

As mentioned previously, the roughness parameters for the z-height variations contain no information about the lateral variations of topography. However, when we utilize the theories in the next chapter for the contact mechanics of rough surfaces, we will need to have a certain amount of information about these lateral variations. The Greenwood and Williamson contact model, for example, uses the summit roughness parameters discussed in the previous section, while the Persson theory uses roughness power spectra, which we will cover in this section.

The surface roughness power spectrum $C(q)$ is related the two-dimensional (2D) Fourier transform $z(q)$ of the surface topography $z(x,y) = z(x)$ by

$$C(q) = \frac{(2\pi)^2}{A} \left\langle z(q)^2 \right\rangle, \qquad (2.14)$$

where A is the area over which $z(q)$ is calculated and $z(q)$ is given by

$$z(q) = \frac{1}{(2\pi)^2} \int_A z(x) e^{-iq \cdot x} d^2x. \qquad (2.15)$$

It is fairly straightforward to calculate $z(q)$ from a 3D topography image, such as an AFM image (Persson et al. 2005). For example, if the z-heights are measured on a square mesh with a spacing constant a and mesh locations specified by the position vector $x = (n_x, n_y)a = na$, then

$$z(q) \approx \frac{a^2}{(2\pi)^2} \sum_n z_n e^{-i(q_x n_x a + q_y n_y a)}, \qquad (2.16)$$

where $z_n = z(x_n)$. The wavevector q is given by $q = (q_x, q_y) = (2\pi m_x/L, 2\pi m_y/L)$, where L is the width of the area being imaged and m_x and m_y are the integer numbers between 0 and $N - 1$, with N being the number of data points along the x and y directions. Since there are efficient algorithms for calculating the discrete Fourier transforms, it is convenient to re-express eq. (2.16) in terms of H_m, which is the 2D discrete Fourier transform of z_n:

$$z(q) \approx \frac{a^2}{(2\pi)^2} H_m. \qquad (2.17)$$

Some useful relations for the power spectra $C(q)$ are as follows:

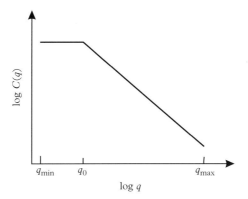

Figure 2.14 *Surface roughness power spectrum of a surface that is self-affine over the range $q_0 > q > q_{max}$. $q_{min} = 2\pi/L$, where L is the width of the area being imaged, and $q_{max} = 2\pi/a$, where a is the lateral spacing between the data points.*

- If a surface is statistically isotropic in the lateral dimensions, it follows that $C(\mathbf{q}) = C(q)$, where q is the magnitude of the wavevector $\mathbf{q}$.
- The rms of z-heights R_q is related to $C(q)$ by

$$R_q^2 = \left\langle z^2 \right\rangle = 2\pi \int_{q_{min}}^{q_{max}} qC(q)dq, \qquad (2.18)$$

where $q_{min} = 2\pi/L$ and $q_{max} = 2\pi/a$, and where L is the width of the area over which the R_q is being measured; q_{min} and q_{max} therefore correspond to the shortest and longest wavevector components of $z(q)$ over the area of the topography image (Figure 2.14).

- If ∇z is the slope at a particular location on the surface, then the rms surface slope of a topography image is given by

$$\nabla z_{rms} = \sqrt{\left\langle (\nabla z)^2 \right\rangle} = \left[2\pi \int_{q_{min}}^{q_{max}} q^3 C(q)dq \right]^{1/2}. \qquad (2.19)$$

2.4.4.1 *Example: surface power spectra of fractured rock and polished steel*

Figure 2.15 shows an AFM image of a piece of rock taken from the exposed face of the Corona Heights Fault in San Francisco, CA. The fault was produced by the motion of the tectonic plates and was exposed by quarrying in the late 1800s. As earthquakes are governed by the frictional properties of faults, the exposed surface of this fault provides useful information on the topographic properties of a fault that has been subjected to significant frictional slip. Optically, this fault surface has grooves evident along the slip

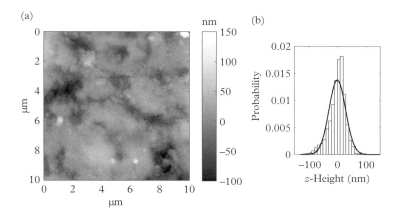

Figure 2.15 *(a) AFM image of a piece of rock from the Corona Heights Fault that was subjected to prolonged slip before being exposed. The z-height roughness parameters for this image: $R_a = 21.6$ nm, $R_q = 29.0$ nm, $R_p = 213$ nm, and $R_v = 123$ nm. (b) Bars show the distribution of z-heights for the AFM image, while the curve shows a Gaussian distribution with the same standard deviation. While the distribution is moderately Gaussian, it is also skewed somewhat to higher z-heights. AFM image and data courtesy of C. Thom and D. Goldsby, University of Pennsylvania; sample courtesy of E. Brodsky, University of California-Santa Cruz.*

direction of the fault (Kirkpatrick and Brodsky 2014, Thom et al. 2017). At the smaller scales of the AFM image in Figure 2.15, however, no such anisotropy is apparent and the surface roughness is isotropic, indicating that different modes of deformations occur at the macro- and micro-scales.

Figure 2.16 shows the surface roughness power spectra for the AFM image of this rock surface in Figure 2.15, and compares it to that for the polished steel surface in Figure 2.10. Interestingly, over this range of wavevectors, they both plot as straight lines on a log–log plot indicating that the power spectra can be expressed in the form of a power law. Similar power law behavior has also been observed for many other tribologically relevant surfaces: cleaved rocks, sand paper, asphalt, Plexiglas, etc. (Persson et al. 2005), and, as discussed the next section, usually indicates a self-affine fractal geometry.

2.5   Self-affine fractal surfaces

Surfaces with fractal topographies are particularly interesting cases, as further magnification brings out finer scale features that have the same topographical characteristics as the larger scale features.

A surface topography is considered to be *self-similar* if a magnified version of itself looks the same as the unmagnified version in that its statistical properties remain unchanged after magnification. For example, if one magnifies the view of a self-similar surface by a factor of 2, the rms roughness will appear to be the same if multiplied by 2. The surface topography is *self-affine* if the magnification factor in the z-direction needs to

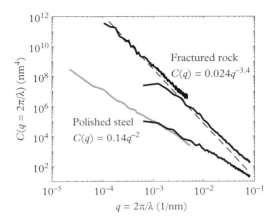

Figure 2.16 *The black lines show the surface roughness power spectra from AFM images of the fractured rock and the polished steel surfaces (there are two black lines for the rock sample since two different image sizes were used to capture the data). The gray line shows the surface roughness power spectrum from an optical interferometry image of polished steel surface. Fractured rock data courtesy of C. Thom and D. Goldsby, University of Pennsylvania for a sample courtesy of E. Brodsky, University of California-Santa Cruz. Polished steel data courtesy of H. Khare and R. W. Carpick, University of Pennsylvania.*

be different than that in the x–y lateral directions in order for the magnified and original surfaces to have the same statistical properties. More formally, if we magnify the view by a factor λ along the x- and y-directions, then the surfaces are self-affine if an H exists such that $\lambda^H z(x/\lambda, y/\lambda)$ looks statistically the same as $z(x,y)$ by a factor λ^H (as shown below in eq. (2.21), the rms roughness R_q scales as λ^H). The exponent H is called the Hurst exponent and lies between 0 and 1 for self-affine surfaces, while $H = 1$ corresponds to self-similar surfaces. These surfaces are also referred to being *fractal* in nature, with the H being related to the fractal dimension D_f by $H = 3 - D_f$.

Persson (2014) has pointed out that many engineered surfaces (e.g., produced by polishing) or natural surfaces (especially if they have been weathered) are often are self-affine and have fractal dimensions in the range 2.0–2.3 ($H = 0.7$–1.0). While no fundamental theory explains these pervasive trends, surfaces with fractal dimensions closer to 3 have higher rms slopes, indicating that these surfaces have much sharper asperities and protrusions than surfaces with fractal dimensions closer to 2. These sharper protrusions, however, tend to wear away more quickly during contact, lowering the fractal dimension to the more stable range of $D_f = 2.0$–2.3; consequently those surfaces that have undergone some polishing or weathering process tend to have lower fractal dimensions.

An important aspect of self-affine surfaces is that the power spectra of these surfaces have a power law behavior that follows

$$C(q) \propto q^{-2(H+1)}. \tag{2.20}$$

Figure 2.14 shows schematically a power spectrum that is self-affine over the range $q_0 > q > q_{max}$ and flat for wavevectors $q < q_0$ (i.e., q_0 corresponds to the longest wavelength component of roughness, while q_{min} corresponds to the wavevector associated with the image size.)

For self-affine surfaces the rms surface roughness R_q can be estimated using eq. (2.18) to obtain

$$R_q^2 = 2\pi \int_{q_{min}}^{q_{max}} q C(q) \, dq$$

$$\approx 2\pi \int_{q_0}^{q_{max}} q^{-2H-1} \, dq$$

$$\propto q_0^{-2H} - q_{max}^{-2H} \approx q_0^{-2H}$$

$$R_q \propto q_0^{-H} \tag{2.21}$$

if $q_{max} \gg q_0$. Similarly, using eq. (2.19) for ∇z_{rms}, we obtain the following estimation of square of the average slope

$$\left\langle (\nabla z)^2 \right\rangle = 2\pi \int_{q_0}^{q_{max}} q^3 C(q) \, dq$$

$$= 2\pi \int_{q_0}^{q_{max}} q^{1-2H} \, dq$$

$$\propto q_{max}^{2(1-H)} - q_0^{2(1-H)}$$

$$\propto q_{max}^{2(1-H)} \tag{2.22}$$

and

$$\nabla z_{rms} \propto q_{max}^{1-H} \tag{2.23}$$

if $q_{max} \gg q_0$.

Thus, for self-affine surfaces the rms surface slope ∇z_{rms} is determined by the shortest wavelength roughness components (i.e., the largest q), while the rms roughness or R_q is determined by the longest wavelength components (i.e., the smallest q). Since the average curvature of surface protrusions is proportional to ∇z_{rms}, eq. (2.23) implies that self-affine surfaces with lower values of H (higher values of the fractal dimension D_f) have sharper protrusions, as was mentioned earlier.

As shown in Figure 2.16, the power spectrum for the fracture rock from the Corona Heights Fault follows a power law that corresponds to a Hurst exponent $H = 0.6$. Since this exponent is in the range 0–1, the surface topography can be considered self-affine over this range of wavevectors. Indeed, extensive imaging over a wide range of length scales indicate that surfaces fractured by fault motion are self-affine over scales ranging from 5 nm to 2 m except for a change in the Hurst exponent at 10 μm (Candela et al. 2009, Kirkpatrick and Brodsky 2014). Specifically, at scales above 10 μm, a lower Hurst exponent is seen when the roughness measured along the fault sliding direction versus perpendicular to it. The anisotropy is expected given the highly directional grooves that are present. The fact that the anisotropy disappears at smaller scales indicates that a different mode of deformation, namely plastic flow, occurs for small asperities, while larger asperities undergo brittle, abrasive wear along the sliding direction (Candela and Brodsky 2016). This example illustrates the value of inspecting roughness for the sake of determining information about the processes that produced the surface roughness.

For the polished steel surface, however, the Hurst exponent is close to zero. From eq. (2.21) this implies that the rms roughness R_q does not change as the magnification is changed; this is consistent with the results shown in Figure 2.9 and Figure 2.10, where R_q ~ 2 nm for both images even though the image sizes are an order of magnitude different. A value of $H \sim 0$ is also typical of surfaces formed when a liquid is cooled below its glass transition temperature (Persson 2014). For the polished steel examined here, the low H value, however, comes about from the extensive polishing that preferentially removes the longer wavelength roughness.

....

2.6 PROBLEMS

1. In Section 2.4.1, the center line or mean height for a single profile trace is specified as the line through the trace for which the mean-square deviation of the trace about that line is a minimum. Prove mathematically that this statement is true. In other words, prove that the constant h that has the minimal mean square deviation from a profile $z(x)$ over some interval of length b will be equal to the mean value of $z(x)$ over the interval b.

2. The surface roughness power spectrum $C(\boldsymbol{q})$ defined in eq. (2.14) can be written as:

$$C(\boldsymbol{q}) = (2\pi)^{-2} \int d^2 r \exp\left[-i\boldsymbol{q} \cdot \boldsymbol{r}\right] C(\boldsymbol{r}) \qquad (2.24)$$

where $C(\boldsymbol{r})$ is the height–height correlation function, which is defined as:

$$C(\boldsymbol{r}) = \left\langle z\left(\boldsymbol{r} + \boldsymbol{r}'\right) z\left(\boldsymbol{r}'\right)\right\rangle. \qquad (2.25)$$

As stated in eq. (2.20), for a self-affine surface, the function $C(q)$ should decay as $q^{-2(H+1)}$. Prove this statement is true for an isotropic surface. To do this, you should show that the above equation for $C(\boldsymbol{q})$ can be written in the form

$$C(\boldsymbol{q}) = q^{-2(H+1)} \times \text{(a constant integral)}. \tag{2.26}$$

Hint: consider the substitutions $\boldsymbol{r} = \boldsymbol{r}^*/q$ and $\boldsymbol{r}' = \boldsymbol{r}'^*/q$, and remember that the definition of a self-affine surface means that a surface with surface heights $z(x,y)$, when magnified by a factor λ in the in-plane directions will have heights described by $\lambda^H z(x/\lambda, y/\lambda)$.

. . . .

2.7 REFERENCES

Archard, J. F. (1957). "Elastic deformation and the laws of friction." *Proceedings of the Royal Society of London, Series A: Mathematical and Physical Sciences* **243**(1233): 190–205.

Benes, E., M. Groschl, W. Burger and M. Schmid (1995). "Sensors based on piezoelectric resonators." *Sensors and Actuators A: Physical* **48**(1): 1–21.

Bhushan, B. (1999). *Principles and applications of tribology.* New York: John Wiley & Sons.

Binnig, G., C. F. Quate and C. Gerber (1986). "Atomic force microscope." *Physical Review Letters* **56**(9): 930–3.

Bowden, F. P. (1950). BBC Broadcast.

Candela, T. and E. E. Brodsky (2016). "The minimum scale of grooving on faults." *Geology* **44**(8): 603–6.

Candela, T., F. Renard, M. Bouchon, A. Brouste, D. Marsan, J. Schmittbuhl and C. Voisin (2009). "Characterization of fault roughness at various scales: Implications of three-dimensional high resolution topography measurements." *Pure and Applied Geophysics* **166**(10–11): 1817–51.

De Garmo, E. P., J. T. Black and R. A. Kohser (2011). *DeGarmo's materials and processes in manufacturing* (11th ed.). Hoboken, NJ: John Wiley & Sons.

Dowson, D. (2009). "A tribological day." *Proceedings of the Institution of Mechanical Engineers Part J: Journal of Engineering Tribology* **223**(3): 261–73.

Drake, B., C. B. Prater, A. L. Weisenhorn, S. A. C. Gould, T. R. Albrecht, C. F. Quate, D. S. Cannell, H. G. Hansma and P. K. Hansma (1989). "Imaging crystals, polymers, and processes in water with the atomic force microscope." *Science* **243**(4898): 1586–9.

Dunaevsky, V. V., Y.-R. Jeng and J. A. Rdzitis (1997). "Surface texture." In: *Tribology data handbook: an excellent friction, lubrication, and wear resource*, E. R. Booser, Ed. Boca Raton, FL: CRC Press, pp.415–34.

Giessibl, F. J. (2000). "Atomic resolution on Si(111)-(7x7) by noncontact atomic force microscopy with a force sensor based on a quartz tuning fork." *Applied Physics Letters* **76**(11): 1470–2.

Greenwood, J. A. (1984). "A unified theory of surface-roughness." *Proceedings of the Royal Society of London, Series A: Mathematical Physical and Engineering Sciences* **393**(1804): 133–57.

Hirst, W. and A. E. Hollander (1974). "Surface finish and damage in sliding." *Proceedings of the Royal Society of London, Series A: Mathematical, Physical and Engineering Sciences* **337**(1610): 379–94.

Kirkpatrick, J. D. and E. E. Brodsky (2014). "Slickenline orientations as a record of fault rock rheology." *Earth and Planetary Science Letters* **408**: 24–34.

Martin, Y., C. C. Williams and H. K. Wickramasinghe (1987). "Atomic force microscope force mapping and profiling on a sub 100-Å scale." *Journal of Applied Physics* **61**(10): 4723–9.

Mate, C. M. (2013). "Spreading kinetics of lubricant droplets on magnetic recording disks." *Tribology Letters* **51**(3): 385–95.

Mate, C. M. and R. W. Carpick (2011). "Materials science: a sense for touch." *Nature* **480**(7376): 189–90.

McCool, J. I. (1986). "Comparison of models for the contact of rough surfaces." *Wear* **107**(1): 37–60.

Meyer, G. and N. M. Amer (1988). "Novel optical approach to atomic force microscopy." *Applied Physics Letters* **53**(12): 1045–7.

Neubauer, G., S. R. Cohen, G. M. McClelland, D. Horne and C. M. Mate (1990). "Force microscopy with a bidirectional capacitance sensor." *Review of Scientific Instruments* **61**(9): 2296–308.

Persson, B. N. J. (2014). "On the fractal dimension of rough surfaces." *Tribology Letters* **54**(1): 99–106.

Persson, B. N. J., O. Albohr, U. Tartaglino, A. Volokitin and E. Tosatti (2005). "On the nature of surface roughness with application to contact mechanics, sealing, rubber friction and adhesion." *Journal of Physics: Condensed Matter* **17**(1): R1.

Skedung, L., M. Arvidsson, J. Y. Chung, C. M. Stafford, B. Berglund and M. W. Rutland (2013). "Feeling small: exploring the tactile perception limits." *Scientific Reports* **3**: 2617.

Thom, C. A., E. E. Brodsky, R. W. Carpick, G. M. Pharr, W. C. Oliver and D. L. Goldsby (2017). "Nanoscale roughness of natural fault surfaces controlled by scale-dependent yield strength." *Geophysical Research Letters* **44**(18): 9299–307.

Wandersman, E., R. Candelier, G. Debregeas and A. Prevost (2011). "Texture-induced modulations of friction force: the fingerprint effect." *Physical Review Letters* **107**(16): 164301.

Wyant, J. C. and K. Creath (1992). "Advances in interferometric optical profiling." *International Journal of Machine Tools and Manufacture* **32**(1–2): 5–10.

3

Mechanical Properties of Solids and Real Area of Contact

When two materials are pressed against each other, initial solid–solid contact occurs at the surfaces' high points. Unless the surfaces are exceptionally smooth, this area where these asperities touch is generally an extremely small fraction of the total area covered by the surfaces; yet the forces generated between the contacting atoms in this small area are responsible for most tribological phenomena—friction, wear, adhesion, etc. Consequently, understanding how forces acting on asperities distort the material around the points of contact provides an important basis for understanding tribology.

In this chapter, we study how the stresses generated when two objects are brought into contact change the surface topography. The previous chapter dealt with characterizing the topography of a non-contacting surface, but characterizing its topography when in contact is extremely difficult as these surfaces are inaccessible to most experimental techniques. Usually, we have to analyze the *contact mechanics* of the hidden contacting interface if we want to know the contact stresses, the extent of elastic and plastic deformation, and the real area of contact. With contact mechanics, we combine our knowledge of surface geometry (gained in the previous chapter) with our understanding of the mechanical properties of materials (to be gained in this chapter) to determine the stresses and deformations within the contacting materials.

3.1 Atomic origins of deformation

The mechanical properties of a material originate from how the individual atoms are bound to their neighboring atoms in the material. When two similar or compatible atoms—say of aluminum—are brought together, a bond will form between them.

Figure 3.1 plots the forces bonding these two atoms together: an attractive force, which gradually decreases in magnitude over a range of 4–5 atomic diameters as the atoms move apart; a repulsive force, which increases rapidly as the atoms are squeezed together due to the overlap of the electron clouds that surround each atom; and the resultant force, which is the sum of the attractive and repulsive forces. The attractive force can have several different origins: metallic bonding that occurs in metals, ionic bonding that occurs

Tribology on the Small Scale: A Modern Textbook on Friction, Lubrication and Wear. Second edition. C. Mathew Mate and Robert W. Carpick. © Oxford University Press 2019. Published in 2019 by Oxford University Press.
DOI: 10.1093/oso/ 9780199609802.001.0001

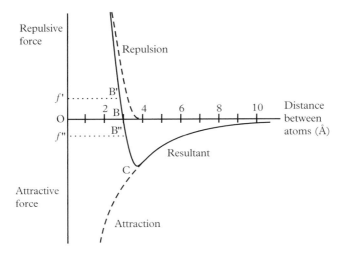

Figure 3.1 *How the force between two atoms varies with separation distance. The dashed lines show the attractive force that dominates at large distances and the repulsive force that dominates at short distances. The solid line shows the resultant sum of the attractive and repulsive forces. OB is the separation distance at equilibrium when the resultant force equals zero. Pulling on the atoms with a force f″ increases the separation to OB″, while pushing the atoms together with a force f′ decreases the separation to OB′.*

between the positive and negative charged ions like in crystals of table salt, and van der Waals bonding such as occurs between the organic compounds that might make up a piece of plastic or a lubricant film. No matter what the nature of the attractive force, the general shape of the force as a function of separation distance in Figure 3.1 holds for all types of atomic bonding, though the scales on the force and separation axes vary considerably for the different types of bonding.

When the attractive and repulsive forces are balanced, the resultant force acting between the two atoms is zero, meaning the atoms are at their equilibrium separation distance (point B in Figure 3.1). Within a solid, the atoms have found positions where the sum of all the forces acting between them and the neighboring atoms is zero. To achieve this balance, the atoms frequently order to form a lattice that minimizes the interaction energy between them, as illustrated in Figure 3.2(a), with the atoms spaced at their equilibrium separation distance. If a tensile force that is not too large is applied to the crystal, as shown in Figure 3.2(b), the bonds between the atoms stretch like small springs, increasing the spacing between the atoms so that the resultant force between each atomic pair becomes slightly negative (point B″ in Figure 3.1). If a compressive force is applied, the spacing between the atoms decreases and the resultant force becomes slightly positive (point B′ in Figure 3.1). The solid deforms until the added attractive (repulsive) forces between the atoms compensate the externally applied tensile (compressive) force. Similarly, if a shear force is applied to the solid, as shown in Figure 3.2(c), the atoms move slightly sideways until atomic level forces balance the applied shear force.

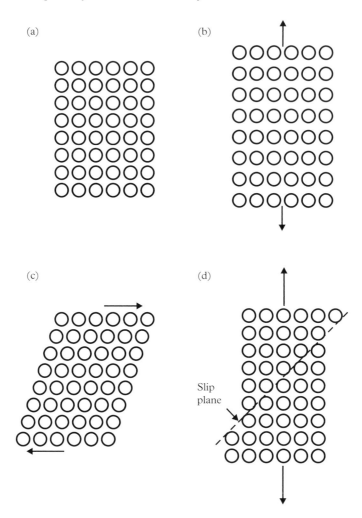

Figure 3.2 *Atoms in a crystal structure subjected to (a) no applied force, (b) to a small tensile force, (c) to a small shear force, and (d) to a tensile force large enough to cause slippage along a crystal plane.*

For small applied forces, the atoms only move a small fraction of the bond length away from the equilibrium separation distance, and the restoring force is proportional to the distance moved resulting in a linear deformation (Hooke's law). When the force is released, the atoms return to their original positions. This is an *elastic deformation*. If a large enough tensile force is applied to the solid that the force on individual atoms exceeds the value of point C in Figure 3.1, the attractive restoring force will be insufficient to hold the material together. Under these conditions, *inelastic deformation* takes place, and many atoms are no longer at their original positions when the applied force is removed.

(a)

(b)

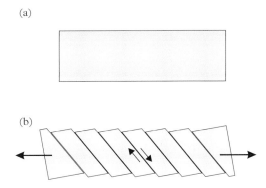

Figure 3.3 *Schematic of a single (hcp) crystal undergoing extensive slip parallel to the (0001) plane due to an applied tensile force.*

Inelastic deformation can take place several ways: For brittle materials, the atoms do not slide easily past each other. So, when an externally applied tensile force exceeds the attractive force holding the atoms together in the solid, the material fractures apart. For ductile materials, they exhibit *plastic deformation*, where atoms slide one atomic position along a slip plane, as illustrated in Figure 3.2(d). This slip motion repeats many times along a large number of slip planes and grain boundaries in the solid until all the forces on individual atoms are below the threshold value C needed to initiate plastic deformation. Figure 3.3 illustrates how slippage at a series of crystal planes in a single crystal can lead to plastic elongation of the entire crystal.

The sliding of one plane of atoms all in unison over another plane of atoms is extremely difficult as the force required is the sum of the force that needs to be exerted on each atom in the plane to slide it over the activation barrier between adjacent lattice sites. The force for slippage is much less if the motion occurs only for a small number of atoms around a *dislocation* rather than all the atoms in a crystal plane. Consequently, most ductile plastic deformation processes occur via the motion of dislocations (Haasen 1996). Figure 3.4 illustrates how one type of dislocation, an *edge dislocation*, causes motion along a slip plane: An applied shear stress pushes the upper half of the crystal to the left and lower half to right. The compression on the upper half lattice causes several planes of atoms to move one lattice spacing to the left forming an apparent extra plane (marked as full circles) within the crystal. The shear stress then causes this extra plane to move through the crystal one lattice spacing at a time until it emerges at the left. The force to move this dislocation through the crystal is much smaller than trying to slide all the atoms at once through one lattice spacing, as only a small number of atoms associated with the dislocation are moved from their low energy equilibrium positions in the crystal lattice.

From these atomic origins of elastic and plastic deformations, we are now ready to discuss some of the basic relations of contact mechanics.

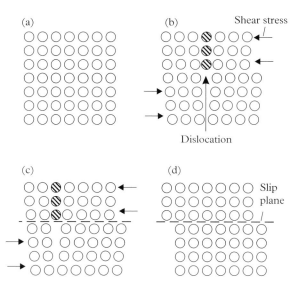

Figure 3.4 *An edge dislocation (hatched circles) moving through a crystal as it undergoes slip deformation. (a) The undeformed crystal. (b, c) Two stages as the dislocation moves one lattice spacing to the left due to an applied shear stress. (d) The undeformed crystal after the dislocation has moved all the way through it.*

3.2   Elastic deformation

3.2.1   Basic relations

When a solid is subjected to a load L, like the one applied along the axis of the cylinder in Figure 3.5(a), a *stress* is created in the solid that produces an elastic deformation of the object that is characterized by a *strain*:

$$\sigma_1 = \text{stress in axial direction} = \frac{L}{\text{cross-sectional area}}, \tag{3.1}$$

$$\varepsilon_1 = \text{strain in axial direction} = \frac{\text{change in length}}{\text{orginal length}}. \tag{3.2}$$

If the strain is small, so that the atoms move only a small fraction of their equilibrium separation distance, the strain is proportional to the applied stress:

$$\varepsilon_1 = \frac{\sigma_1}{E}$$
$$\sigma_1 = E\varepsilon_1 \tag{3.3}$$

where E is the elastic constant or *Young's modulus* of the material. A strain is considered elastic if it is reversible, that is, the solid returns to its original shape when the stress is

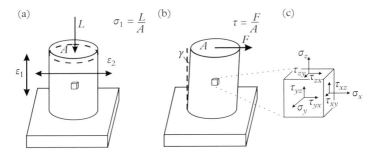

Figure 3.5 *(a) A cylinder sitting on a flat surface while being subjected to a load L resulting in an axial strain ε_1 and lateral strain ε_2. (b) The cylinder subjected to a shear stress τ resulting in a shear strain γ. (c) An expanded view of an infinitesimally small volume element within the cylinder that illustrates all the possible components of normal stress σ and shear stress τ that act on such an element. For the cylinder in (a) subjected to only an uniaxial stress, only the σ_z component is non zero with $\sigma_z = \sigma_1 = L/A$; for the cylinder in (b) subjected to only a shear stress, only the τ_{zx} component is non zero with $\tau_{zx} = \tau = F/A$.*

removed. Being reversible also means that all the work expended in elastically deforming the material is recovered when the stress is removed.

When you squeeze an object (like the cylinder in Figure 3.5(a)) in one direction, some of the material moves in the directions orthogonal to the applied stress. The elastic strain ε_2 in the transverse directions is proportional to the elastic strain ε_1 in the direction of applied stress:

$$\varepsilon_2 = -v\varepsilon_1 \qquad (3.4)$$

where v is the Poisson's ratio, which normally ranges from 0 to 0.5, and for most structural materials is between 0.2 and 0.35. (Cork is a somewhat exceptional material in that $v \sim 0$, so, when you apply an axial load to push a cork into a wine bottle, it does not expand in the transverse direction to prevent sliding into the neck of the bottle.)

When a solid is subjected to a shear stress, as shown in Figure 3.5(b), the shear strain γ is proportional to the shear stress τ:

$$\tau = G\gamma, \qquad (3.5)$$

where G is the shear modulus. The shear modulus is related to the Young's modulus for homogeneous, isotropic, linear, elastic materials by

$$E = 2G(1 + v). \qquad (3.6)$$

3.2.2   Elastic deformation of a single asperity contact

Next we will consider the contact mechanics of a single asperity being pushed against a flat surface with some load L. For this analysis, we will only consider two simple contact geometries: (1) an asperity with a spherically shaped summit contacting a flat surface, and (2) a cylindrically shaped asperity with a flat summit (i.e., a flat punch) contacting a flat surface. Later we will use some of these results for single asperity contact to derive the contact phenomena for multiple asperity contacts.

3.2.2.1 *Approximating an asperity contact as sphere on flat*

For many of our analyses of tribological contacts, we will use the crude approximation that the contact geometry for a single asperity can be modeled simply as a sphere on a flat, as this spherical approximation allows us to gain key insights into the nature of a single asperity contact. For contacts with a non-spherical geometry one can either refer to an analytical treatment in a suitable text (Johnson 1985) or use an appropriate numerical method, such as finite element analysis (FEA), to analyze the contact mechanics. Often, however, the sphere-on-flat approximation can provide a quick and reasonably accurate prediction for many of the quantities of physical interest, such as the contact areas and contact pressures.

3.2.2.2 *Elastic contact area for a sphere on a flat*

In order to better understand the Newton optical interference fringes that occur when two glass lenses are pressed together, a 23-year-old Heinrich Hertz worked out, during his 1880 Christmas vacation, an analysis of elastic deformation for an elliptical contact area (Hertz 1882, Johnson 1985). Frequently, we simplify the analysis of this type of contact mechanics problem by transforming it into an equivalent sphere-on-flat geometry.

When a spherically shaped summit is brought into contact with a flat surface with a load L, as shown in Figure 3.6, the surfaces deform to create a contact zone of radius a. According to Hertz's equations for elastic deformation for a sphere-on-flat geometry, the radius of the contact zone is given by

$$a = \left(\frac{3RL}{4E_c} \right)^{1/3},$$

$$(3.7)$$

where E_c is the composite elastic modulus of two contacting materials, with moduli E_1 and E_2 and Poisson's ratios v_1 and v_2, and is given by

$$\frac{1}{E_c} = \frac{1 - v_1^2}{E_1} + \frac{1 - v_2^2}{E_2}.$$

$$(3.8)$$

For this geometry, the real area of contact A is given by

$$A = \pi a^2$$
$$= \pi \left(\frac{3RL}{4E_c} \right)^{2/3}$$

$$(3.9)$$

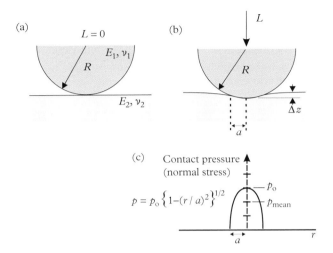

Figure 3.6 *Sphere on flat contact geometry with (a) no load and (b) load = L. (c) The distribution of contact pressure across the contact zone of (b).*

and the mean contact pressure (normal stress) at the interface p_{mean} by

$$p_{mean} = \frac{L}{A}$$

$$= \frac{1}{\pi}\left(\frac{4E_c}{3R}\right)^{2/3}L^{1/3} \qquad (3.10)$$

The contact pressure is not uniform over the circular contact area, but has a maximum at the center and falls to zero at the edge of the contact zone, as illustrated in Figure 3.6(c). For this situation, the Hertz pressure distribution

$$p = p_0\left\{1 - (r/a)^2\right\}^{1/2}, \qquad (3.11)$$

where the maximum p_0 is 3/2 times the mean pressure p_{mean}. This pressure distribution elastically deforms the two bodies in the direction perpendicular to the interface by the amount

$$\Delta z = \frac{a^2}{R} = \left(\frac{9}{16RE_c^2}\right)^{1/3}L^{2/3}. \qquad (3.12)$$

Several other types of stresses exist at the sphere-on-flat contact in addition to the normal stress plotted in Figure 3.6(c): a stress radial to the axis of symmetry and a shear stress = 0.5 × (normal stress − radial stress). Detailed equations and descriptions for how the normal and shear stresses, that make up the components of the stress tensor, decay away from the contacting interface for this and other Hertzian contact geometries can be found in the following references (Hamilton 1983, Sackfield and Hills 1983, Johnson 1985, Williams and Dwyer-Joyce 2001). We will defer further discussion of these stresses to later in this chapter, when we discuss how these stresses initiate plastic deformation underneath the contact.

3.2.2.2.1 Equivalent situations to the sphere on flat geometry

As mentioned earlier, the Hertzian contact theory has also been applied to surfaces that produce elliptical contact areas—such as two crossed cylinders of different radii, or two ellipsoidal asperities—and solutions to these situations are discussed in contact mechanics textbooks (Johnson 1985, Popov 2010). Here we briefly discuss the situation for an asperity with an ellipsoidal shape summit pressing against a flat surface, and the shape of this summit is described by two principal radii of curvature R' and R''. For situations where the ratio R'/R'' is not too large, the Hertzian contact eqs. (3.7)–(3.12) are still reasonably accurate, if we assume that R in these equations is the equivalent radius defined by

$$R = (R'R'')^{1/2}. \tag{3.13}$$

For $R'/R'' = 1$, the Herztian sphere-on-flat contact equations are exact of course, but even for $R'/R'' = 10$, these equations are still accurate to within 2.5% using eq. (3.13) (Popov 2010).

Another common situation is two spherical surfaces making contact, as illustrated in Figure 3.7 where two surfaces with slight protrusions are pushed into contact with the centers of the protrusions aligned with each other. Generally, the summits of protrusions or asperities on opposing surfaces have different radii of curvatures, which we characterize as R_1 and R_2. According to the Hertzian contact theory, however, this situation is equivalent to a sphere-on-flat contact geometry if use an effective radius for the sphere given by

$$R_{\text{eff}} = \left(\frac{1}{R_1} + \frac{1}{R_2} \right)^{-1}. \tag{3.14}$$

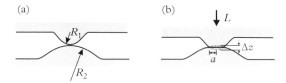

Figure 3.7 *Side view of two surfaces protrusions with spherical shaped summits with radii of curvature of R_1 and R_2 being pushed into contact. (a) Just before the two protrusions make contact and (b) during contact with a loading force L, which results in a contact radius a and an elastic deformation Δz.*

3.2.2.2.2 Limitations of Hertz's contact theory It is important to be aware of the assumptions and limits of the Hertz contact theory:

- This analysis only applies to homogeneous and isotropic materials when loading is in the linear elastic regime (in practice, this means all stress components should be below yield or fracture limits).
- Also, the theory also assumes that adhesion between the surfaces is negligible and that the surfaces are frictionless.
- Eqs. (3.7)–(3.12) are only really valid for situations when the indentation depth Δz and the contact radius a are much less than R_{eff}. This can have consequences for very compliant materials like polymers, hydrogels, and biological matter, where low moduli lead to large indentation depths and contact areas.

Solutions exist for many cases where these assumptions are violated (Johnson 1985, Popov 2010, Jacobs et al. 2014).

3.2.2.2.3 Example: Spherical steel particle sandwiched between two flat surfaces As an example, let's consider the situation illustrated in Figure 3.8 where a small stainless steel particle (radius $R = 1$ μm) is sandwiched between two ideally smooth surfaces, also made of stainless steel. This situation is a simple model for a small wear or debris particle stuck between two contacting surfaces. For conditions where elastic contact occurs, we would like to determine the contact radius a, the mean pressure p_{mean}, and the magnitude of the compression Δz where the particle touches the two flats. For 304 stainless steel, elastic modulus $E = 200$ GPa, Poisson's ratio $v = 0.3$, and hardness $H = 1.5$ GPa; so the composite elastic modulus is

$$\frac{1}{E_c} = \frac{1-v_1^2}{E_1} + \frac{1-v_2^2}{E_2} = 2\left(\frac{1-0.3^2}{200 \text{ GPa}}\right).$$

$$E_c = 110 \text{ GPa} \tag{3.15}$$

Where the particle touches the flat surface, the contact radius a, the mean pressure P_m, and the magnitude of the compression Δz are given by

$$a = \left(\frac{3RL}{4E_c}\right)^{1/3} = \left(\frac{3 \times 1\mu\text{m}}{4 \times 110 \text{ GPa}}\right)^{1/3} L^{1/3} \tag{3.16}$$

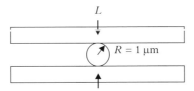

Figure 3.8 *Spherical particle between two flats.*

$$p_{mean} = \frac{1}{\pi}\left(\frac{4E_c}{3R}\right)^{2/3} L^{1/3} = \frac{1}{\pi}\left(\frac{4 \times 110 \text{ GPa}}{3 \times 1 \text{ } \mu m}\right)^{2/3} L^{1/3} \tag{3.17}$$

$$\Delta z = \left(\frac{9L^2}{16RE_c^2}\right)^{1/3} = \left(\frac{9}{16 \times 1 \text{ } \mu m \times (110 \text{ GPa})^2}\right)^{1/3} L^{2/3}. \tag{3.18}$$

Let's consider the range of loads from 1 nN to 1 μN (1 nN corresponds roughly to the strength of a single atomic bond, while 1 μN is just over the threshold for plastic deformation in this contact geometry):

load L (nN)	contact radius a (nm)	mean pressure p_{mean} (GPa)	vertical compression per contact Δz (nm)
1	1.9	0.089	0.0036
10	4.1	0.19	0.017
100	8.8	0.41	0.077
1000	19	0.89	0.36

From these results, we notice that for this small particle undergoing elastic deformation, the contact zone is always small, usually only a few nanometers across. The elastic compression of the particle is also small, no more than a few angstroms for the highest load. Since the particle radius is relatively small, the mean contact pressure is fairly high, and, for the largest load 1 μN, the pressure is comparable to the 1.5 GPa hardness value of steel, suggesting that, if we push much harder, the deformation will mainly be plastic rather than elastic and we will need a different analysis for the contact mechanics.

3.2.2.3 *Approximating an asperity contact as a flat punch*

While less applicable to as many situations as the sphere-on-flat geometry, another useful approximation of an asperity contact is a flat punch contacting a flat surface, as illustrated in Figure 3.9. A situation where a flat punch approximation would be better suited to describe the contact is that of an asperity worn flat by wear; while the sphere-on-flat approximation might be better for describing the initial contact of the asperity before wear occurs.

As illustrated in Figure 3.9, the geometry under consideration is a rigid, circular punch with a flat end with radius a being pushed against an elastic flat surface. The deformation and stress field for this situation have been solved by Sneddon (1946). Since the punch is rigid, the contact area is constant at πa^2 and the deformation Δz is given by

$$\Delta z = \left(1 - v_{substrate}^2\right) \frac{L}{2aE_{substrate}}, \tag{3.19}$$

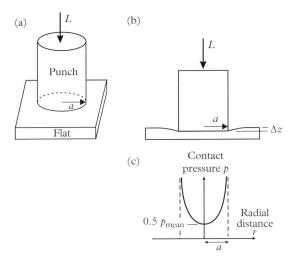

Figure 3.9 *(a) Three-dimensional (3D) view of a circular punch with a flat being pushed with a load L against a flat surface. (b) Cross-sectional view of the punch being pushed against the flat. (c) The radial distribution across the contact zone of the contact pressure acting normal to the flat surface.*

where L is the applied normal load. So, for this flat punch geometry, the deformation increases linearly with load L, rather than increasing nonlinearly as $L^{2/3}$ as it did for the sphere-on-flat geometry.

For the flat punch geometry, the contact pressure (normal stress) is given by

$$p(r) = p_{min}\left(1 - \frac{r^2}{a^2}\right)^{-1/2}, \tag{3.20}$$

where

$$p_{min} = \frac{L}{2\pi a^2} = \frac{p_{mean}}{2} \tag{3.21}$$

is the minimum value of the pressure, which occurs at the center of the punch ($r = 0$). The radial distribution of contact pressure across the contact zone is shown in Figure 3.9(c). This pressure distribution has quite a different behavior than the Hertzian pressure distribution for the sphere on flat geometry shown in Figure 3.6(c): for a sphere on flat, the peak pressure occurs at the center of the contact zone, and the pressure goes to zero at the edges of the contact zone; while, for a flat punch geometry, the contact pressure is minimum at the center of the contact zone and rises to a singularity at the edges of the contact zone.

In reality, when a flat punch is pushed up against another surface, the contact pressure does not go to infinity around the contact edges; instead some elastic deformation occurs around the edges of the punch, since it is not completely rigid, to relieve the stresses. In general, the peak stresses occur around the edges of any contact with a sharp corner, like that around the periphery of a flat punch, making these edges the most likely locations for the plastic deformation to initiate.

3.2.3   Elastic deformation from tangential loading

When one surface slides over another, this adds a tangential loading or traction force to the contacting interface. This tangential force generates additional stresses within the contacting materials, which can substantially alter the tribological behavior.

The simplest situation to analyze is when a sphere slides across a flat surface and where we assume that the coefficient of friction μ is the same everywhere within the contact zone; this means that traction stress q within the siding contact zone is given by

$$q(x, y) = \mu p(x, y). \tag{3.22}$$

Since the normal contact pressure p for a sphere sliding across a flat is given by eq. (3.11), the traction shear stress at the interface is

$$|q(r)| = \mu p_o \{1 - (r/a)^2\}^{1/2}$$
$$= \frac{3\mu L}{2\pi a^2} \{1 - (r/a)^2\}^{1/2}. \tag{3.23}$$

Solutions for the stresses before slip (including the phenomenon of partial slip that tends to occur around the contact edge) are discussed elsewhere (Johnson 1955, Popov 2010).

3.3   Plastic deformation

3.3.1   Basic relations

When a ductile material, such as the cylinder shown in Figure 3.10, is subjected to tensile stress high enough such that the applied force on the individual atoms exceeds the attractive restoring force, the atoms start to slide past one another along slip planes, as illustrated in Figures 3.2(d), 3.3, and 3.4. These regions of the material undergo *plastic deformation*. The tensile (or compressive) force per unit area needed to initiate plastic deformation is called the *yield stress Y*. One characteristic of plastic deformation is that the process is irreversible, that is, when the stress is removed, atoms that underwent plastic deformation are quite happy to stay in their new locations, so the cylinder on the right side of Figure 3.10 remains permanently elongated.

It can be shown that, if plastic deformation were to solely occur by the planes of atoms sliding past each other as shown in Figures 3.2(d) and 3.3, then the theoretical yield stress should be $Y_{\text{theoretical}} \sim E/10$ (Zhu and Li 2010). Most materials, however, plastically deform at stress values several orders of magnitude less than this theoretical limit. For ductile materials, the yield stress is lowered by the movement of dislocations as discussed in Section 3.1 and illustrated in Figure 3.4, while for brittle materials, failure is initiated at pre-existing cracks and flaws in the material that concentrate stresses by factors of 100 or far greater.

3.3.2   Yield criteria for plastic deformation

Since the yield strength Y of a material is almost always much less than the maximum theoretical value, Y is typically determined by applying a uniaxial stress as shown in Figure 3.10 to a test material. In practical situations, the applied forces are multi-axial, leading to multiple stress components acting on the infinitesimal volume elements within the material (Figures 3.5(c) and 3.10). Since it is not possible to test materials under all possible combinations of stress states, several yield criteria have been developed for correlating the onset of plastic deformation in an arbitrary stress field to the yield strength Y determined from uniaxial testing.

Since plastic deformation typically involves slippage of one atomic plane over another, we can intuitively see that the magnitude of the *shear stress* should determine whether plastic deformation occurs. Indeed, the Tresca criterion for plastic yield simply states that the yielding initiates when the maximum shear stress exceeds the yield shear stress τ_Y for that material. The yield shear stress is related to the yield strength Y measured in a uniaxial test by

$$\tau_Y = \frac{Y}{2}. \tag{3.24}$$

Another common yield criterion comes from the von Mises theory, which states that a ductile material will yield when the elastic strain energy that goes into distorting the material away from its original shape (as opposed to the elastic strain energy due to stresses that only a change its volume) exceeds a critical value. The distortional energy density is proportional to what's call the *von Mises stress*, which is defined as

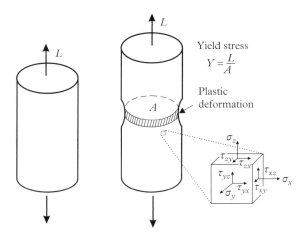

Figure 3.10 *(left) A cylinder subjected to a tensile load too weak to cause plastic deformation. (right) A cylinder of ductile material, such as a metal, subjected to a tensile load large enough to cause plastic deformation. Also shown on the right is an expanded view of an infinitesimally small volume element within the cylinder illustrating all the different components of normal stress σ and shear stress τ acting on the element.*

$$\sigma_{VM} = \sqrt{\frac{\left(\sigma_{xx} - \sigma_{yy}\right)^2 + \left(\sigma_{yy} - \sigma_{zz}\right)^2 + (\sigma_{zz} - \sigma_{xx})^2 + 6\left(\tau_{xy}^2 + \tau_{yz}^2 + \tau_{zx}^2\right)}{2}}. \qquad (3.25)$$

So, in the von Mises theory, the plastic deformation occurs at a location within a material when the von Mises stress σ_{VM} exceeds the yield strength Y obtained in uniaxial tensile test.

For a situation where only a shear stress τ exists, $\sigma_{VM} = \sqrt{3}\tau$ and plastic yielding occurs when τ reaches

$$\tau = \frac{Y}{\sqrt{3}} = 0.58Y. \qquad (3.26)$$

Comparing eqs. (3.26) and (3.24), we see that, for situations where only shear stress exists, the Tresca criterion is more conservative than the von Mises criterion, as the Tresca criterion predicts the onset of plastic deformation at a somewhat lower value of shear stress.

3.3.3   Plastic deformation in the sphere-on-flat geometry

To help visualize where plastic deformation is going to occur for a particular contact situation, it is useful to plot contours of either constant maximum shear stress or constant von Mises stress. Figure 3.11 shows such contour plots of the von Mises stress for the sphere-on-flat contact geometry with and without a tangential force applied to a sphere-on-flat geometry. As shown in Figure 3.11(a), when no friction exists between the sphere and the flat, the maximum shear stress occurs at a depth of $0.48a$ below the contact plane, and has a value of $0.31p_o$ or $0.47p_{mean}$ when $\nu = 0.3$ (Johnson 1985). Using the Tresca criterion, plastic deformation is initiated at this depth when the maximum shear stress exceeds $Y/2$, which for the sphere-on-flat geometry occurs when $p_o = 1.6Y$ or $p_{mean} = 1.1Y$. Using eq. (3.10), the load required to initiate plastic deformation for this frictionless sphere-on-flat geometry is

$$L_Y = \frac{1}{6}\left(\frac{R}{E_c}\right)^2 (1.6\pi Y)^3. \qquad (3.27)$$

Figure 3.11(b) shows the contours of constant von Mises stress for the case where a sphere is sliding across as smooth flat surface with a friction coefficient $\mu = 0.25$. The addition of shear stresses to the contacting interface from friction has several effects:

- It breaks the symmetry of the subsurface stress distribution, with the locations ahead of the contact line becoming more compressive and with those locations behind the contact line becoming less compressive and more tensile.
- It increases the magnitude of total stress at any given depth.
- It moves the location of the maximum stress towards the interface at $z = 0$.

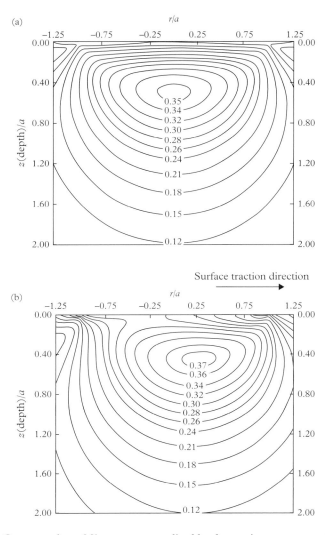

Figure 3.11 *(a) Contours of von Mises stress normalized by the maximum stress p_0, beneath a sphere-on-flat contact geometry in the $y = 0$ plane with a contact of radius a and with $v = 0.25$: (a) for a stationary contact and (b) for a sphere sliding across a flat surface with a friction coefficient of 0.25. Reproduced from Lee and Ren (1994) with permission. Copyright 1994 by Taylor and Francis.*

A consequence of the location of maximum stress shifting towards the sliding interface with increasing frictional coefficient is that eventually the maximum stress occurs at the surface rather than at a subsurface location when μ exceeds a critical value. For the sphere-on-flat geometry, this concentration of the maximum von Mises stress at the surface occurs when $\mu > 0.3$ and is located at the leading edge of the contact zone (Johnson 1985), and becomes the place where plastic yielding initiates.

For brittle materials, the situation is quite different in that these materials can handle a high amount of compressive stress without yielding, but are much weaker against tensile stresses. Since the shear stresses across the contact zone also act to concentrate these tensile stresses at the trailing edge of the contact zone, for brittle materials, failure starts with surface fractures and cracks forming at the trailing edge of the contact zone rather than with plastic deformation at the leading edge (Lawn 1992).

3.3.4 Hardness

When a spherical indenter is pushed against a flat so as to cause plastic deformation, the plastic zone is at first small (Figure 3.12(a)), and the elastic deformation of the surrounding material acts to contain the plastic deformation. As the load increases, the plastic zone expands to accommodate the increasing contact pressure. Eventually, it breaks out to the free surface, and material plastically flows around the sides of the

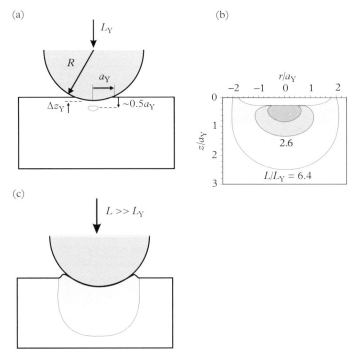

Figure 3.12 *Plastic deformation of a hard sphere indenting a flat surface of a softer material. (a) At the onset of plastic deformation, the plastic zone only occupies a small volume at the region of maximum pressure, where $a_Y = 2.5YR/E_c$ is the contact radius, Δz_Y is the penetration depth, and L_Y is the critical load when plastic deformation initiates. (b) The shaded areas show the extent of plastic deformation at several intermediate loads between the initial onset (L_Y) and complete break out of the plastic deformation at the surface; adapted with permission from Hardy et al. (1971), copyright 1971, John Wiley and Sons. (c) At higher loads, the plastic volume flows around the sides of the indenter.*

indenter, as illustrated in Figures 3.12(b) and (c). The load where plastic break out occurs in metals is typically 50–100 times greater than the load needed to initiate plastic flow below the surface (Johnson 1985). By the time full plasticity occurs, the mean contact pressure has risen to $p_{mean} = 3Y$. Once material starts to flow plastically around the indenter, the mean contact pressure remains fairly constant with increasing load; the area of plastic deformation simply increases proportionally to load. For this situation, we can define the hardness $H = load/area$ *of permanent indentation*. So, for the sphere-on-flat geometry where the material is fully plastic under the indenter Figure 3.12(c) (Tabor 1951):

$$H \simeq 3Y. \tag{3.28}$$

For non-spherical-on-flat contact geometries—such as for conical, pyramid, and flat end-shaped indenters—the ratio of indentation pressure and yield stress Y is also constant with the proportionality constant always close to 3, so eq. (3.28) still applies (Johnson 1985). The measured hardness does not depend so much on the shape of the apex of the indenter, but rather on the ability of the material to plastically flow around the sides of the indenter. Materials with different plastic behavior can have coefficients rather different than 3 in eq. (3.28) (Johnson 1970).

3.4   Real area of contact

Many tribological properties, such as friction and wear, tend to be proportional to the real area of contact. Since contacting surfaces touch only at the top of their asperities, the real area of contact is typically a small fraction of the apparent area of contact. As mentioned before, measuring the real contact area of the hidden contacting interface is usually extremely difficult. Instead, one typically has to resort to estimating the real area of contact by analyzing the contact mechanics of the contacting surfaces. As inputs to this analysis, one needs:

1. *Surface roughness.* As described in the previous chapter, an atomic force microscope (AFM) or 3D optical profilometer can be used to determine the surface topographies of the contacting surfaces and their roughness parameters. For rough surfaces, some commonly-used roughness parameters for a contact mechanics analysis are the *mean radius of curvature R* of the asperity summits and the *standard deviation of summit heights* σ_s. These values are then used to approximate the contact geometry by estimating the individual single asperities with spherical asperities all having the same radius of curvature R. Alternately, parameters extracted from the *power spectrum* of the surface topography can be used to analyze the roughness. This method is better suited for when asperities come in a wide range of lateral length scales and radii of curvatures.

2. *Deformation constants.* As discussed in the previous section, an indenter can be used to determine the *microhardness H* of the near surface region of materials. The same

indentation experiment can also be used to determine the *Young's modulus E* of the material (Doerner et al. 1986, Oliver and Pharr 1992, 2004, Pharr et al. 1992).

3. *Contact forces.* Contact forces are made up of the external forces applied both perpendicularly to the interface (the load L) and tangentially to the interface (the friction F), along with the internal adhesive forces (L_{adh}).

With these inputs, an finite element analysis (FEA) code can be used to obtain a reasonably accurate estimate for the real area of contact for a given contact situation. While an FEA can be fairly accurate, it is often difficult to see from such an analysis how the area of contact originates from the physical properties of the contacting interfaces. To better comprehend the interplay between the various roughness and deformation parameters, we next discuss some simple models for estimating the real area of contact.

3.4.1   Greenwood and Williamson model

In 1966 Greenwood and Williamson developed a model for elastic deformation of rough surfaces in contact that illustrates how contact area depends on the roughness parameters of summit curvature and distribution of summit heights together with the elastic modulus. In this model, Greenwood and Williamson (1966) extended the Hertzian theory of elastic contact for a sphere on a flat to the situation illustrated in Figure 3.13, with the following assumptions about the nature of the contacting interface:

1. A non-deformable flat surface contacts an elastically deformable rough surface.
2. The rough surface consists of spherically-shaped asperities, all with the same radius of curvature R.
3. The asperity heights vary randomly with either a Gaussian or exponential distribution of summit heights.

In the model, each asperity in contact (asperities with $z > d$, where z is the height of asperity when not in contact and d is the distance between the flat surface and the mean height of the rough surface) deforms elastically without interacting with its neighbors and supports a load (from eq. (3.12))

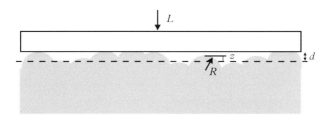

Figure 3.13 *Contact geometry assumed in the Greenwood and Williamson model.*

$$L_{\text{asperity}} = \frac{4}{3} E_{\text{c}} R^{1/2} (Z - d)^{3/2}. \tag{3.29}$$

The total load supported is given by integrating over the distribution of asperity heights $\phi(z)$ from $z = d$ to ∞:

$$L = \frac{4}{3} N E_{\text{c}} R^{1/2} \int_d^\infty (z - d)^{3/2} \phi(z) dz, \tag{3.30}$$

where N is the total of asperities.

As discussed in Chapter 2, the assumption of a Gaussian height distribution of spherically shaped summits does a reasonable job describing surfaces with an isotropic texture, such as produced by a grinding or random surface finishing process. Using a Gaussian distribution of summit heights, Greenwood and Williamson numerically calculated the real area contact for contact geometry in Figure 3.13. Using an exponential distribution of summit heights, they could derive analytical expressions for contact area and other contact parameters. We will discuss these analytical expressions in some detail in order to illustrate how the physical parameters influence contact.

An exponential distribution of summit heights can be expressed as

$$\phi(z) = \frac{C}{\sigma_{\text{s}}} \exp\left(\frac{-z}{\sigma_{\text{s}}}\right), \tag{3.31}$$

where z is the summit height relative to the surface mean and C is a constant. By choosing σ_{s} to be the standard deviation of the summit heights for a Gaussian distribution,[1] the results for the exponential distribution approximates those for the Gaussian distribution well. The approximation of an exponential distribution is most likely to be valid if contact is limited to the highest tenth of asperities. For elastic contact between a flat surface and an exponential distribution of summit heights, Greenwood and Williamson (1966) showed that:

- The number of asperities in contact is

$$n = \frac{L}{\pi^{1/2} E_{\text{c}} \sigma_{\text{s}}^{3/2} R^{1/2}}. \tag{3.32}$$

- The real area of contact is

$$A_{\text{r}} = \left(\frac{\pi R}{\sigma_{\text{s}}}\right)^{1/2} \frac{L}{E_{\text{c}}}. \tag{3.33}$$

---

[1] Note: the symbol σ does double duty in the chapter, serving as both the symbol for normal stress and the standard deviation. To help avoid confusion, we will use σ_{s} to stand for the standard deviation of summit heights and σ for normal stress.

- The mean area per asperity is

$$\frac{A_r}{n} = \pi R \sigma_s.\qquad(3.34)$$

- The mean contact radius is

$$a = \sqrt{R \sigma_s}.\qquad(3.35)$$

- The mean contact pressure is

$$p_{mean} = \frac{L}{A_r} = 0.56 E_c \left(\frac{\sigma_s}{R}\right)^{1/2}.\qquad(3.36)$$

From these analytical expressions, we learn some important aspects for the contact situation in Figure 3.13:

- The number of asperities and the real area of contact are proportional to load.
- The mean size of contact area per asperity is independent of load.
- The mean contact pressure is independent of load.

In this analysis, as the load increases, a sufficient number of new asperities are brought into contact so that these mean values of contact area, radius and pressure stay constant.

Similar results are also obtained if a Gaussian distribution is used for asperity heights. For example, with a Gaussian distribution, the Greenwood and Williamson model determines the mean contact pressure to be

$$p_{mean} = [0.3{-}0.4] \times E_c \left(\frac{\sigma_s}{R}\right)^{1/2}.\qquad(3.37)$$

This expression has the same dependence on E_c, σ_s, and R as that for an exponential distribution of heights, but the factor [0.3–0.4] now depends on the degree of loading and is slightly less than the 0.56 factor in eq. (3.36).

3.4.1.1   *Example: TiN contacts*

To see how the Greenwood and Williamson model works, we will use eqs. (3.32)–(3.36) to determine the contact parameters n, A_r, a, and P_{mean} for a rough TiN surface contacting a flat TiN surface with a load $L = 1$ N. (Titanium nitride coatings are commonly deposited onto metals and ceramics by physical vapor deposition in situations where a hard surface is desired.) The material parameters for TiN are $E = 600$ GPa, $\nu = 0.2$, $E_c = 313$ GPa, and $H = 17$ GPa (Bhushan 1999). We will assume the following roughness parameters for the rough surface: standard deviation of summit heights $\sigma_s = 20$ nm and mean radius of curvature of summits $R = 10 \ \mu m$;

$$n = \frac{L}{\pi^{1/2} E_{\mathrm{c}} \sigma_{\mathrm{s}}^{3/2} R^{1/2}} = \frac{1\mathrm{N}}{\pi^{1/2} \times 313 \text{ GPa} \times (20 \text{ nm})^{3/2} (10\mu\mathrm{m})^{1/2}} = 202; \quad (3.38)$$

$$A_{\mathrm{r}} = \left(\frac{\pi R}{\sigma_{\mathrm{s}}}\right)^{1/2} \frac{L}{E_{\mathrm{c}}} = \left(\frac{\pi \times 10 \text{ } \mu\mathrm{m}}{20 \text{ nm}}\right)^{1/2} \frac{1 \text{ N}}{313 \text{ GPa}} = 127 \text{ } \mu\mathrm{m}^2; \quad (3.39)$$

$$a = \sqrt{R\sigma_{\mathrm{s}}} = \sqrt{10 \text{ } \mu\mathrm{m} \times 20 \text{ nm}} = 0.45 \text{ } \mu\mathrm{m}; \quad (3.40)$$

$$p_{\mathrm{mean}} = 0.56 E_{\mathrm{c}} \left(\frac{\sigma_{\mathrm{s}}}{R}\right)^{1/2} = 0.56 \times 313 \text{ GPa} \times \left(\frac{20 \text{ nm}}{10 \text{ } \mu\mathrm{m}}\right)^{1/2} = 7.8 \text{ GPa}. \quad (3.41)$$

So, the Greenwood and Williamson model predicts that, even with the fairly substantial load of 1 N, only a relatively small number of asperities (202) will be in contact and that the average contact radius ($a = 0.45 \text{ } \mu\mathrm{m}$) is a very small fraction of the mean curvature of the asperities; thus the area of solid–solid contact is a small fraction of the apparent contact area. This turns out to be a fairly general result, that, for any contacting materials with moderate to high hardness and elastic modulus, the number of asperities in elastic contact will be fairly small and with fairly small contact zones. Since the load is concentrated on these small contact zones, the contact pressures are fairly high, in the GPa regime in this example. This approaches hardness of the TiN. If the TiN makes contact with a more compliant, softer material, then the stresses will be reduced since E_{c} will be less; however, the stresses can still be high enough to lead to plasticity or fracture. (Indeed, TiN is frequently used to coat cutting tool surfaces, so that the plastic deformation, which is the first step in the cutting process, is more likely to happen in the material being cut than in the material doing the cutting.)

3.4.1.2 *Real area of contact using the Greenwood and Williamson model*

In order to use the appropriate analysis for determining the real area of contact, one should first determine whether the deformation at the contacting asperities is predominately elastic or plastic. To achieve this, the Greenwood and Williamson model can be used to predict whether the average contact pressure is less than the hardness (indicating mostly elastic deformations) or greater than the hardness (indicating mostly plastic deformation). For this purpose, Greenwood and Williamson used their model to define a *plasticity index*, denoted ψ,

$$\psi = \left(\frac{\sigma_{\mathrm{s}}}{R}\right)^{1/2} \left(\frac{E_{\mathrm{c}}}{H}\right), \quad (3.42)$$

for characterizing contacts as elastic or plastic. If $\psi \geq 1$, the Greenwood and Williamson model predicts that the average contact pressure exceeds the hardness and the deformations are predominately plastic. If $\psi < 0.6$, the model predicts that the contact pressure is below the threshold for plasticity and the contacts deform elastically. From the definition of the plasticity index, we see that plastic deformation is more likely either when the material becomes softer (the hardness H decreases relative to composite elastic modulus E_c) or when the surface becomes rougher (σ increases relative to R). Indeed, comparing to eqs. (3.36) and (3.37), one sees that the plasticity index is directly related to the ratio of the mean pressure and the material hardness.

3.4.1.2.1   Plastic deformation   Under conditions for plastic deformation, the contacting asperities of the softer material will plastically flow so as to increase the contact area supporting the load until the mean contact pressure equals the hardness. So the contact geometry stabilizes when

$$p_{\text{mean}} = \frac{L}{A_r} = H, \tag{3.43}$$

which can be rearranged to give an expression for the real area of contact A_r; in the situations when plastic deformation dominates:

$$A_r \simeq \frac{L}{H}. \tag{3.44}$$

Note that no assumptions were made about the contact geometry in deriving this expression other than the condition of plasticity, so it is a fairly general expression for real contact area for plastic contacts.

3.4.1.2.2   Elastic deformation   Now for contacting situations that are primarily elastic, the simplest way to estimate the real area of contact is to use the Greenwood and Williamson model with a Gaussian distribution of summit heights to obtain the following expression from eq. (3.37):

$$A_r \simeq 3\left(\frac{R}{\sigma_s}\right)^{1/2}\frac{L}{E_c} \tag{3.45}$$

3.4.1.3   *Example: recording head on a laser textured disk surface*

For disk drives, during normal operations a recording head flies at a very low clearance over a rotating disk surface while it reads and writes information to a magnetic film on the disk surface. The surfaces of the disk and recording head are very smooth, which will result in high adhesive forces if the recording head lands on the disk when the disk stops spinning as the drive is powered off. Since about the mid-2000s, the recording heads have been unloaded to a ramp at the outer edge of the disk when the drive is powered off to avoid having the recording heads land on the disk as the disks are being spun down. Prior to mid-2000s, however, when recording heads flew over the disks at much higher clearances, it was practical to have the recording heads land near the ID hub of the on disk a deliberately textured landing zone (Johnson et al. 1996).

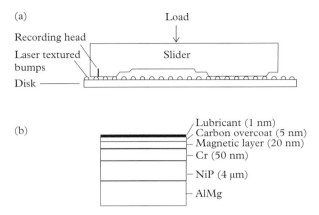

Figure 3.14 *(a) A recording head slider sitting on a laser textured landing zone of a disk surface inside a disk drive. (b) The different layers deposited onto an AlMg disk substrate, the type most commonly used for laser textured disks. Also shown are the typical layer thicknesses for laser textured disks in a disk drive from 2004.*

A common way to texture this ID disk landing zone was to use a pulsed laser to melt the disk substrate in localized spots that resolidified to form small bumps 5–40 nm in height on the disk surface (Baumgart et al. 1995, Johnson et al. 1996, Tam et al. 1996). By controlling the energy and time duration of the laser pulse, the height and shape of the bumps can be accurately controlled along with the spacing between bumps; hence, laser texturing allows for the control of the contact area between the head and the disk.

The schematic in Figure 3.14(a) shows a recording head slider resting on a laser textured landing zone. To a large extent this situation closely matches the idealized case of the Greenwood and Williamson model: An ideally smooth surface (the slider surface) contacts a rough surface (the laser textured disk surface) where the asperities have spherically shaped summits (or can be approximated as spherical; Gui et al. 1997) with nearly identical radii of curvatures and with a Gaussian distribution of summit heights. A major concern when designing this textured landing zone was to minimize the friction force between the slider and disk that needs to be overcome to start the drive spinning again when powered on, with the friction force being proportional to real area of contact.

So, for this example, we want to determine the real area of contact for the typical conditions for a slider parked on a laser texture zone: slider loading force $L = 20$ mN and laser textured bumps with spherical summits ($R = 200~\mu$m), with a standard deviation of bump heights $\sigma_s = 1.5$ nm, and with a density of 2000 bumps per mm^2.

First, we need to determine the composite elastic modulus E_c from the elastic modulus of the disk surface ($\sim$130 GPa) and from the elastic modulus of the Al$_2$O$_3$–TiC sintered material that makes up the slider body ($\sim$450 GPa) and assuming a Poisson ratio $\nu = 0.25$:

$$\frac{1}{E_c} = \frac{1-v_1^2}{E_1} + \frac{1-v_2^2}{E_2} = \frac{1-0.25^2}{130~\text{GPa}} + \frac{1-0.25^2}{450~\text{GPa}}. \tag{3.46}$$

$$E_c = 108~\text{GPa}$$

Next we need the hardness of the material where plastic deformation is most likely take place. Since the disk has a layered structure near its surface (Figure 3.14(b)), we need to determine the typical depth at which the shear stress responsible for initiating plastic deformation reaches a maximum (Section 3.3.3); for each asperity contact zone, this maximum occurs at a distance approximately $0.5a$ beneath the contact interface. Using eq. (3.35) to determine the mean contact radius a, the mean distance d from the contacting interface where plastic deformation initiates is

$$d \approx 0.5\sqrt{R\sigma_s} = 0.5\sqrt{(200 \ \mu m)(1.5 \ nm)} = 0.27 \mu m. \tag{3.47}$$

For the disk, this depth is below the magnetic and chromium layers and is in the NiP overlayer that coats the AlMg substrates typically used for laser textured disks (Figure 3.14(b)). The NiP layer has a hardness $H \sim 7$ GPa, which is much less than the hardness of the slider material ($H \sim 22$ GPa for sintered Al_2O_3–TiC).

Now, we are ready to determine the plasticity index ψ from eq. (3.42):

$$\psi = \left(\frac{\sigma_s}{R}\right)^{1/2}\left(\frac{E_c}{H}\right) = \left(\frac{1.5 \ nm}{200 \ \mu m}\right)^{1/2}\left(\frac{108 \ GPa}{7 \ GPa}\right) = 0.08. \tag{3.48}$$

Since $\psi < 0.6$, the deformation is primarily elastic, and we need to use eq. (3.45) for determining the real area of contact:

$$A_r \simeq 3\left(\frac{R}{\sigma_s}\right)^{1/2}\frac{L}{E_c} = 3\left(\frac{200 \ \mu m}{1.5 \ nm}\right)^{1/2}\frac{20 \ mN}{108 \ GPa} = 203 \ \mu m^2. \tag{3.49}$$

So, the actual contact area is a small fraction the apparent area of 1 mm² for a typical recording head slider sitting on a disk.

Equation (3.32) of the Greenwood and Williamson model also can be used to estimate the number of laser texture bumps in contact with the slider:

$$n = \frac{L}{\pi^{1/2}E_c\sigma_s^{3/2}R^{1/2}} = \frac{20 \ mN}{\pi^{1/2} \times 108 \ GPa \times (15 \ nm)^{3/2}(200 \ \mu m)^{1/2}} = 127. \tag{3.50}$$

So, underneath the 1 mm² area of the slider, only about 127 out of 2000 bumps contact the slider, which means the approximation of an exponential height distribution used for deriving eq. (3.32) should be valid.

In our estimate of the real area of contact, we assumed that the top of the laser textured bumps and the slider surfaces are ideally smooth. While very smooth, these surfaces still have a small amount of roughness, so potentially the real area of solid–solid contact may be less than that estimated by eq. (3.49). To re-estimate the contact area taking into account the roughness at the top of the of the laser textured bumps, we will assume, for the disk, a standard height deviation $\sigma_s = 0.9$ nm and a mean radius of curvature

$R = 80$ nm and, for the slider, an ideally smooth surface. In this case, the plasticity index ψ from eq. (3.42) is

$$\psi = \left(\frac{\sigma_s}{R}\right)^{1/2} \left(\frac{E_c}{H}\right) = \left(\frac{0.9 \text{ nm}}{80 \text{ nm}}\right)^{1/2} \left(\frac{108 \text{ GPa}}{5 \text{ GPa}}\right) = 2.3, \tag{3.51}$$

where we have used the hardness for the magnetic layer. Since most the deformation is plastic (since $\psi > 1$), we now need to use eq. (3.44) to estimate the real area of contact A_r:

$$A_r = \frac{L}{H} = \frac{20 \text{ mN}}{5 \text{ GPa}} = 4 \text{ }\mu\text{m}^2. \tag{3.52}$$

How do we account for this new estimate of the contact area being nearly two orders of magnitude smaller than the previous estimate from eq. (3.49)? Figure 3.15 illustrates what is going on. When a laser texture bump contacts the flat surface, the loading force elastically deforms the bump to create a contact zone with diameter $2a$, over which the nanoscale roughness makes contact. Since the nanoscale roughness is only a few nanometers in height, the gaps between the roughness peaks invariably fill with lubricant or mobile contaminants drawn into the narrow space by the molecular level attractive forces. So, the contact area estimated by eq. (3.49) (using the Greenwood–Williamson model with the radius of curvature of the laser texture bumps and ignoring their nanoscale roughness) includes both the solid–solid contact area at the summits of the nanoscale roughness and the area filled with lubricant/contaminant between the nanoscale summit contacts. The estimate from eq. (3.52) is then only for the solid–solid contact area at the summits of the nanoscale roughness, which is only a small fraction of the contact zone formed by the elastic deformation of the summit of the laser textured bump.

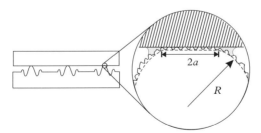

Figure 3.15 *Flat surface contacting three "sombrero" shaped laser textured bumps (left). Expanded view of the summit of a laser textured bump contacting the flat surface (right). The dashed line shows the macroscopic radius R, which is flattened at the summit due to elastic deformation to form a contact zone with diameter 2a. The nanoscale roughness is also shown with an exaggerated height scale. The gray shows the lubricant and contaminant that fill in the gaps between the nanoscale roughness and the flat.*

3.4.2   Persson theory of the contact mechanics of rough surfaces

While simple, several of the assumptions inherent within the Greenwood–Williamson model make it unsuitable outside of a narrow range of contact situations. Indeed, Greenwood himself has considered these shortcomings of his theory to be serious (Greenwood and Wu 2001). The main shortcomings of the Greenwood–Williamson model are as follows:

- The Greenwood–Williamson model assumes that the asperity summits can be described as spheres having a single, average radius curvature. For many surfaces, the heights, curvatures, and other topographical quantities can have multiscale behavior that is not well represented by a single radius of curvature.

- The Greenwood–Williamson model assumes the asperities heights are well described by a Gaussian distribution, particularly at the higher asperity heights. However, many surface finishing processes can skew this part of the distribution away from a Gaussian.

- The Greenwood–Williamson analysis is only legitimate for small interferences and contact areas, and quickly breaks down when the contact area becomes a significant fraction (>0.01%) of the apparent area of contact, which can easily be the case for compliant materials like rubber and other elastomers.

- Finally, the Greenwood–Williamson model ignores elastic coupling between nearby asperities. When one asperity is subjected to a load, the elastic displacement field near the loaded asperity extends for an appreciable distance, and nearby asperities will be displaced downward reducing the probability that they will make contact.

To overcome these limitations, B. N. J. Persson developed an alternative contact theory for rough surfaces that works well even when the roughness ranges over many lateral length scales, when the contact ranges from a small fraction of the apparent area to full contact, and when elastic coupling occurs between nearby asperities (Persson 2001, 2007, Yang and Persson 2008, Carbone and Bottiglione 2011). In the Persson theory, the distribution in contact pressure is determined via a parabolic partial differential equation with a diffusive term that contains the information of the surface roughness. The main inputs to Persson's model are the power spectrum of the surface heights and the elastic properties of the deforming surfaces.

As discussed in Chapter 2, the surface roughness power spectrum $C(\boldsymbol{q})$ is related to the two-dimensional Fourier transform $z(\boldsymbol{q})$ of the surface topography by

$$C(\boldsymbol{q}) = \frac{(2\pi)^2}{A} \left\langle \left| z(\boldsymbol{q})^2 \right| \right\rangle, \tag{3.53}$$

where q is the wavevector corresponding to each length scale λ under consideration ($q = 2\pi/\lambda$), A is the area over which $z(\boldsymbol{q})$ is calculated, and $z(\boldsymbol{q})$ is given by

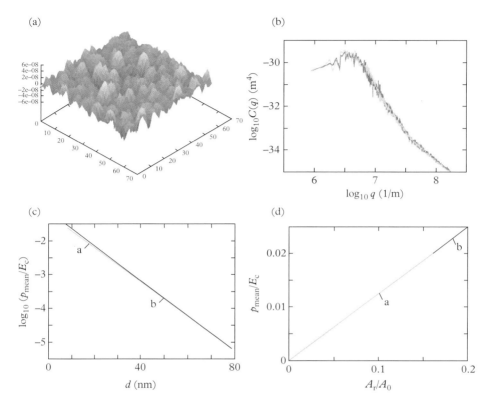

Figure 3.16 *(a) AFM topography image of a polymer surface with rms roughness $R_q = 14$ nm measured over a 10×10 μm^2 area. (b) The surface roughness power spectra C(q) of the polymer surface calculated from two topography images collected over different areas. (c) For this polymer surface, the predicted relationship between the squeezing pressure p_{mean} normalized by E_c and the mean separation distance d when squeezed against a rigid surface, as determined by (line a) the finite element calculation of Pei et al. (2005) and by (line b) the Persson model. (d) The relationship between the relative contact area A_r/A_0 and squeezing pressure p_{mean}/E_c, again determined by (line a) finite element analysis and by (line b) Persson's model. Reproduced with permission from Yang and Persson (2008). Copyright 2008 IOP Publishing.*

$$z(\boldsymbol{q}) = \frac{1}{(2\pi)^2} \int_A z(\boldsymbol{x}) e^{-i\boldsymbol{q}\cdot\boldsymbol{x}} d^2 x. \tag{3.54}$$

Figure 3.16 shows an example of a roughness power spectrum and the AFM image from which it was calculated.

In his model, Persson derives the following simple formula for the real area of contact A_r normalized by the apparent area of contact A_0 as a function of mean contact pressure $p_{mean} = L/A_0$:

$$\frac{A_r}{A_0} = \mathrm{erf}\left(\frac{p_{mean}}{E_c \left[\pi \int_{q_{min}}^{q_{max}} q^3 C(q) dq \right]^{1/2}} \right)$$

$$= \mathrm{erf}\left(\frac{p_{mean}}{E_c m_2^{1/2}} \right) \tag{3.55}$$

where it is assumed that the surface roughness is isotropic so that $C(q) = C(\mathbf{q})$ and where

$$m_2 = \pi \int_{q_{\min}}^{q_{\max}} q^3 C(q) dq \tag{3.56}$$

is the second moment of the roughness power spectrum. This equation for real contact area has the advantage over the eq. (3.45) from the Greenwood–Williamson model in that it is valid from very small contact areas up to full contact where $A_r/A_0 \simeq 1$.

In the limiting case of small loading forces and contact areas, eq. (3.55) can be approximated as

$$\frac{A_r}{A_0} \simeq \frac{2}{E_c} \left(\frac{1}{\pi m_2} \right)^{1/2} p_{\mathrm{mean}}. \tag{3.57}$$

This equation has a similar form as the Greenwood–Williamson eq. (3.45) for the real contact area, in that it predicts at small loading pressures that the real contact area to be linear to the loading force and inversely proportional to the composite elastic modulus.

Since the second moment m_2 is related to the rms slope measured over area A by

$$\sqrt{\langle (\nabla z)^2 \rangle} = \left[(2\pi)^2 \int_{q_{\min}}^{q_{\max}} q^3 C(q) dq \right]^{1/2},$$
$$= 2\sqrt{\pi m_2} \tag{3.58}$$

eq. (3.57) can be rewritten as

$$\frac{A_r}{A_0} \simeq \kappa \frac{p_{\mathrm{mean}}}{E_c \sqrt{\langle \nabla z^2 \rangle}}, \tag{3.59}$$

where κ is multiplicative factor that equals $\sqrt{8/\pi} \approx 1.6$ in Persson's theory. Interestingly, eq. (3.59) had been previously derived for small contact areas by Bush et al. (1975) using an updated version of the Greenwood–Williamson model, but in their equation $\kappa = \sqrt{2\pi} \approx 2.5$. Hyun et al. (2004) have done FEA of the contact area for a variety of self-affine surfaces and found in these simulations that eq. (3.59) is valid with $\kappa \sim 2 \pm 0.2$ and when $A_r/A_0 < 0.1$.

Persson's model can also be used for determining the average spacing d between the surface means as a function of mean contact or squeezing pressure p_{mean}. For small squeezing pressures, the model predicts

$$p_{\mathrm{mean}} = \beta E_c e^{-d/d_0}, \tag{3.60}$$

where d_0 is a characteristic length scale on the order of the rms roughness R_q. This prediction of an exponential relationship between the mean separation distance and the loading pressure is in good agreement with experiments and FEA of rough contacts.

Figure 3.16(c) shows an example of the p_{mean}/E_c calculated for a rigid flat squeezed against a polymer surface with the surface topography shown in Figure 3.16(a) and the roughness spectrum in Figure 3.16(b). In Figure 3.16(c) the Persson model prediction of $p_{mean} \propto \exp(-d/0.4R_q)$ is in good agreement with the FEA of Pei et al. (2005) for the same contact situation. Figure 3.16(d) shows the relationship and between p_{mean}/E_c and the contact area for $A_r/A_0 < 0.2$; again there is good agreement between the Persson model and the FEA of Pei et al.

Extensions of these models for plasticity (Pei et al. 2005) and adhesive surfaces have also been proposed (Mulakaluri and Persson 2011, Lorenz et al. 2013, Pastewka and Robbins 2014).

....

3.5 PROBLEMS

1. The surface stiffness or the normal contact stiffness k_N is defined as $k_N = \partial L(\Delta z)/\partial z$ where $L(\Delta z)$ is the normal loading force needed to elastically deform the surface a distance Δz along the z axis perpendicular to the surface.

   (a) Show that the normal contact stiffness is $k_N = 2aE_c$ for the Hertz model for a sphere on a flat.
   (b) Plot this normal contact stiffness for a 1 cm radius steel sphere pressed into contact with a rigid substrate as the load varies from 0 to 10 N.

2. Consider the following situation with two surfaces coming into contact: The lower surface is flat and rigid. The upper surface is flat but with a set of $n \times n$ protruding asperities all having the same height and with spherically shaped apexes. The surfaces span an area $100b \times 100b$ where b is a length. The asperities have a radius of curvature of R/n and are spaced apart by a linear distance of $100b/n$. The asperities also have modulus E and Poisson's ratio v. The two surfaces are pressed into contact with a loading force L. The asperities are spaced far enough apart that their stress and strain fields do not interact with each other. Calculate and plot on a log-linear plot the following quantities as a function of the number of asperities n^2 for $1 < n^2 < 10000$. Normalize your calculations to the value of the quantity for $n^2 = 1$:

   (a) The number of asperities per unit area
   (b) The load per asperity
   (c) The contact area for an asperity
   (d) The total contact area
   (e) The ratio of the total contact area to the apparent contact area ($100b \times 100b$)
   (f) The reduction in height d of an asperity from elastic deformation
   (g) The maximum contact stress at the interface and acting perpendicular to it at an asperity

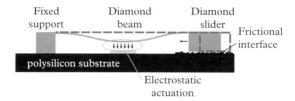

Figure 3.17 *Simplified schematic of a MEMS actuator developed by de Boer et al. (2004) for moving a diamond slider laterally across polysilicon substrate using electrostatic actuation. The dashed line corresponds to the beam and slider positions before electrostatic actuation.*

   (h) The absolute maximum shear stress (which is not necessarily at the interface!) in an asperity

Based on these plots, comment on the effect that splitting up a single asperity into an array of asperities has on the contact area and contact stresses.

3. A MEMS lateral actuator is fixed on one end and connected to a slider on the other, as shown in Figure 3.17. The slider has a nominal area of contact is $2 \times 200 \ \mu m^2$ with a rough polycrystalline silicon substrate on which rests. We wish to predict the friction force and tendency for wear to occur at the sliding interface. The following values are given:

- The applied load $= 160 \ \mu N$
- Standard deviation of distribution of summit heights of the silicon substrate $=$ 4.0 nm
- Average asperity radius of the silicon surface $= 80$ nm
- Asperity density of the silicon surface $= 200$ asperities per μm^2
- Young's modulus of the silicon $= 164$ GPa
- Poisson's ratio of the silicon $= 0.23$

Also assume that the adhesive force between the slider and silicon substrate is negligible because of the use of a low surface energy self-assembled monolayer (SAM) coating that passivates the surfaces, that the SAM layer has a negligible effect on the mechanical properties of the silicon surface, and that the diamond slider is smooth and rigid compared to the polysilicon substrate. Use the Greenwood–Williamson model of contact for analysis to determine the following:

   (a) Find the true area of contact between the slider and the substrate expressed both in μm^2 and as a percentage of the apparent area of contact. Also find the ratio of the area of contact to the load and find the number of asperities in contact.

   (b) Now let's change the geometry so that the apparent area of contact is now $30 \times 30 \ \mu m^2$. Given that all other parameters (load, surface properties, elastic properties) remain the same, re-calculate the quantities specified in part (a).

(c) Based on this elastic analysis, what fraction of the asperities has stresses that exceed the 6 GPa yield stress of silicon? What is the plasticity index?

4. Consider the fractured rock sample described in Section 2.4.4.1 and assume that it makes contact across a fault. For simplicity, assume the opposite surface is perfectly flat and that the composite elastic modulus for the rock is $E_c = 90$ GPa. The AFM data for the rock surface shown in Figure 2.16 were obtained from topographic information collected from lateral lengths scales ranging from 10 μm down to 60 nm.

(a) Calculate the rms slope ∇z_{rms} for this data set.

(b) From this, calculate the mean real contact pressure (L/A_r) the rock experiences in contact.

(c) Compare this to the theoretical yield stress $Y_{theoretical} \sim E/10$ expected for the rock. For more discussion on this point see Thom et al. (2017).

. . . .

3.6 REFERENCES

Baumgart, P., D. J. Krajnovich, T. A. Nguyen and A. C. Tam (1995). "A new laser texturing technique for high-performance magnetic disk drives." *IEEE Transactions on Magnetics* **31**(6): 2946–51.

Bhushan, B. (1999). *Principles and applications of tribology*. New York: John Wiley & Sons.

Bush, A. W., R. D. Gibson and T. R. Thomas (1975). "The elastic contact of a rough surface." *Wear* **35**(1): 87–111.

Carbone, G. and F. Bottiglione (2011). "Contact mechanics of rough surfaces: a comparison between theories." *Meccanica* **46**(3): 557–65.

de Boer, M. P., D. L. Luck, W. R. Ashurst, R. Maboudian, A. D. Corwin, J. A. Walraven and J. M. Redmond (2004). "High-performance surface-micromachined inchworm actuator." *Journal of Microelectromechanical Systems* **13**(1): 63–74.

Doerner, M. F., D. S. Gardner and W. D. Nix (1986). "Plastic properties of thin films on substrates as measured by submicron indentation hardness and substrate curvature techniques." *Journal of Materials Research* **1**(6): 845–51.

Greenwood, J. A. and J. B. P. Williamson (1966). "Contact of nominally flat surfaces." *Proceedings of the Royal Society of London, Series A: Mathematical and Physical Sciences* **295**(1442): 300–19.

Greenwood, J. A. and J. J. Wu (2001). "Surface roughness and contact: an apology." *Meccanica* **36**(6): 617–30.

Gui, J., D. Kuo, B. Marchon and G. C. Rauch (1997). "Stiction model for a head–disc interface: experimental." *IEEE Transactions on Magnetics* **33**(1): 932–7.

Haasen, P. (1996). *Physical metallurgy* (3rd ed.). Cambridge: Cambridge University Press.

Hamilton, G. M. (1983). "Explicit equations for the stresses beneath a sliding spherical contact." *Proceedings of the Institution of Mechanical Engineers Part C: Journal of Mechanical Engineering Science* **197**: 53–9.

Hardy, C., C. N. Baronet and G. V. Tordion (1971). "The elasto plastic indentation of a half-space by a rigid sphere." *International Journal for Numerical Methods in Engineering* **3**: 451–62.

Hertz, H. (1882). "On the contact of elastic solids." *Journal für die reine und angewandte Mathematik* **92**: 156–71.

Hyun, S., L. Pei, J.-F. Molinari and M. O. Robbins (2004). "Finite-element analysis of contact between elastic self-affine surfaces." *Physical Review E* **70**(2): 026117.

Jacobs, T. D. B., C. M. Mate, K. T. Turner and R. W. Carpick (2014). "Understanding the tip–sample contact: an overview of contact mechanics from the macro- to the nanoscale." In: *Scanning probe microscopy in industrial applications: nanomechanical characterizatio*n, D. Yablon, Ed. Hoboken, NJ: John Wiley & Sons.

Johnson, K. E., C. M. Mate, J. A. Merz, R. L. White and A. W. Wu (1996). "Thin-film media—current and future technology." *IBM Journal of Research and Development* **40**(5): 511–36.

Johnson, K. L. (1955). "Surface interaction between elastically loaded bodies under tangential forces." *Proceedings of the Royal Society of London, Series A: Mathematical and Physical Sciences* 230(1183): 531–48.

Johnson, K. L. (1970). "The correlation of indentation experiments." *Journal of the Mechanics and Physics of Solids*18(2): 115–26.

Johnson, K. L. (1985). *Contact mechanics*. Cambridge: Cambridge University Press.

Lawn, B. (1992). "Friction processes in brittle fracture." In: *Fundamentals of friction: macroscopic and microscopic processes*, I. L. Singer and H. M. Pollock, Eds. Springer Netherlands, pp.137–65.

Lee, S. C. and N. Ren (1994). "The subsurface stress field created by three-dimensionally rough bodies in contact with traction." *Tribology Transactions* **37**(3): 615–21.

Lorenz, B., B. A. Krick, N. Mulakaluri, M. Smolyakova, S. Dieluweit, W. G. Sawyer and B. N. J. Persson (2013). "Adhesion: role of bulk viscoelasticity and surface roughness." *Journal of Physics: Condensed Matter* **25**(22): 225004.

Mulakaluri, N. and B. N. J. Persson (2011). "Adhesion between elastic solids with randomly rough surfaces: comparison of analytical theory with molecular-dynamics simulations." *Europhysics Letters* **96**(6): 66003.

Oliver, W. C. and G. M. Pharr (1992). "An improved technique for determining hardness and elastic modulus using load and displacement sensing indentation experiments." *Journal of Materials Research* **7**(6): 1564–583.

Oliver, W. C. and G. M. Pharr (2004). "Measurement of hardness and elastic modulus by instrumented indentation: advances in understanding and refinements to methodology." *Journal of Materials Research* **19**(1): 3–20.

Pastewka, L. and M. O. Robbins (2014). "Contact between rough surfaces and a criterion for macroscopic adhesion." *Proceedings of the National Academy of Sciences* **111**(9): 3298–303.

Pei, L., S. Hyun, J. Molinari and M. O. Robbins (2005). "Finite element modeling of elasto-plastic contact between rough surfaces." *Journal of the Mechanics and Physics of Solids* **53**(11): 2385–409.

Persson, B. N. J. (2001). "Theory of rubber friction and contact mechanics." *The Journal of Chemical Physics* **115**(8): 3840–61.

Persson, B. N. J. (2007). "Relation between interfacial separation and load: a general theory of contact mechanics." *Physical Review Letters* **99**(12): 125502.

Pharr, G. M., W. C. Oliver and F. R. Brotzen (1992). "On the generality of the relationship among contact stiffness, contact area, and elastic-modulus during indentation." *Journal of Materials Research* **7**(3): 613–17.

Popov, V. L. (2010). *Contact mechanics and friction: physical principles and applications*. Berlin: Springer Science & Business Media.

Sackfield, A. and D. A. Hills (1983). "Some useful results in the tangentially loaded hertz contact problem." *Journal of Strain Analysis for Engineering Design* **18**(2): 107–10.

Sneddon, I. N. (1946). "Boussinesq's problem for a flat-ended cylinder." *Mathematical Proceedings of the Cambridge Philosophical Society* **42**(1): 29–39.

Tabor, D. (1951). *The hardness of metals*. Oxford: Clarendon Press.

Tam, A. C., I. K. Pour, T. Nguyen, D. Krajnovich, P. Baumgart, T. Bennett and C. Grigoropoulos (1996). "Experimental and theoretical studies of bump formation during laser texturing of Ni-P disk substrates." *IEEE Transactions on Magnetics* **32**(5): 3771–3.

Thom, C. A., E. E. Brodsky, R. W. Carpick, G. M. Pharr, W. C. Oliver and D. L. Goldsby (2017). "Nanoscale roughness of natural fault surfaces controlled by scale-dependent yield strength." *Geophysical Research Letters* **44**(18): 9299–307.

Williams, J. A. and R. S. Dwyer-Joyce (2001). "Contact between solid surfaces." In: *Modern tribology handbook*, B. Bhushan, Ed. Boca Raton, FL: CRC Press, pp.121–62.

Yang, C. and B. N. J. Persson (2008). "Contact mechanics: Contact area and interfacial separation from small contact to full contact." *Journal of Physics: Condensed Matter* **20**(21): 215214.

Zhu, T. and J. Li (2010). "Ultra-strength materials." *Progress in Materials Science* **55**(7): 710–57.

4

Friction

In the last chapter we considered contact between materials where a loading force acts perpendicularly to the contacting surface. In this chapter we consider what happens when a tangential force is applied to slide one material over another. When materials move against each other, *frictional* forces opposing this motion are generated in the contact regions. Consider, for example, the forces acting on a stationary block sitting on the inclined plane shown in Figure 4.1(a), where the incline angle φ can be adjusted by using a lab jack. Figure 4.1(b) shows the three distinct forces acting on the block to keep it stationary:

1. the weight of the block due to gravity, $L = mg$,
2. a reactive force, $L \cos \varphi$, that the incline exerts perpendicularly to the block's surface in response to the weight, and
3. a static friction force, $L \sin \varphi$, that the incline exerts tangentially to the block's surface.

For small φ, the reactive and friction forces adjust as the angle φ changes so that their vector sum balances the weight of the block, keeping the block stationary. As φ increases, the static friction force increases proportionally to ϕ until reaching a maximum value at some critical angle φ_s. Above this angle, the static friction force is constant, so the tangential component of the block's weight exceeds the opposing friction force, and the block slides down the incline.

(a)　　　　　　　　　　　　　　　　(b)

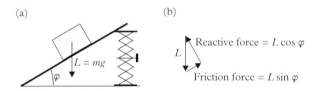

Figure 4.1 *(a) Block of mass m sitting on an inclined plane whose angle φ can be adjusted with a lab jack. (b) Diagram of the forces acting on the block to keep it stationary.*

Tribology on the Small Scale: A Modern Textbook on Friction, Lubrication and Wear. Second edition. C. Mathew Mate and Robert W. Carpick. © Oxford University Press 2019. Published in 2019 by Oxford University Press.
DOI: 10.1093/oso/ 9780199609802.001.0001

In this chapter, we examine how a friction force originates at the microscale contacts between solid surfaces and begin to examine the atomic origins of this force. In later chapters, we examine the friction of lubricated surfaces and delve more deeply into the atomic and molecular origins of friction.

4.1 Amontons' and Coulomb's laws of friction

In many situations, the friction force during sliding can be described by the following "laws:"

First law of friction   The friction force is proportional to the normal load.

Second law of friction   The friction force is independent of the apparent area of contact.

The first law can be expressed by the simple equation:

$$F = \mu L \tag{4.1}$$

where μ is called the *coefficient of friction*. This expression also implies the second friction law that friction is independent of the apparent area of contact. For the block sliding on the incline in Figure 4.1, $\mu = \tan \varphi_s$, where φ_s is the incline angle where the block starts to slide.

While the friction force usually has some dependence on sliding velocity, this dependence is typically fairly small, leading to a third law of friction:

Third law of friction   Kinetic friction is independent of sliding velocity.

This law is never strictly true, so it is best thought of as an approximation for the weak dependence of friction during sliding on velocity rather than a law. Once sliding stops, however, the force needed to initiate sliding (static friction) is usually greater than the force needed to sustain sliding (kinetic friction). In terms of the friction coefficient, the coefficients of static and kinetic friction are labeled μ_s and μ_k, respectively, and obey the relationship $\mu_s \geq \mu_k$.

The first two laws are referred to as Amontons' laws of friction after Guillaume Amontons (1663–1705) who published them in 1699 in the *Proceedings of the French Royal Academy of Sciences* (Dowson 1978). The third law, that kinetic friction has a weak dependence on velocity, was provided by Coulomb (1736–1806) eighty years later and is often referred to as Coulomb's law of friction. The first two friction laws had actually been deduced previously by Leonardo da Vinci (1452–1519) in the middle of the fifteenth century, who described them in his notebooks that have only recently been discovered. Since Amontons published his results while da Vinci didn't, the glory of having his name associated with the laws of friction belongs to Amontons. (Recent experiments have confirmed his observations, using a faithfully rebuilt version of the equipment da Vinci's (Pitenis et al. 2014).)

4.1.1   Coefficients of friction

Since Amontons' law is empirical, values of the coefficient of friction cannot be calculated (as of yet) from fundamental principles and consequently need to be determined from experimental measurement. Within Amontons' law, the coefficient of friction μ simply represents the empirical proportionality constant between the friction force and the normal load (eq. (4.1)). Over the years, friction coefficients have been measured for a wide variety of situations and material pairs, some of which are summarized in Figure 4.2.

When discussing friction coefficients, it is useful to distinguish between "dry" and "wet" contacts, where the distinguishing feature of wet contacts is that they are separated by a fluid film that provides hydrodynamic lubrication. For wet contacts with good hydrodynamic lubrication, the apparent coefficient of friction can be very low (typically $\mu_k = 0.001–0.05$). As discussed in Chapter 9, the friction force for hydrodynamic lubrication depends not so much on the materials doing the sliding, but rather on the properties of the fluid that separates them and the geometry of the hydrodynamic bearing. Consequently, for wet contacts the friction force depends strongly on additional factors to the loading force, such as the sliding speed, fluid film thickness, and the dependence of the fluid's viscosity on pressure and temperature; therefore, using a coefficient of friction may be an incomplete way to characterize a wet contact.

For dry sliding contacts, many common materials exhibit friction coefficients in the range 0.2–0.6. As indicated in Figure 4.2, however, many other materials exhibit friction

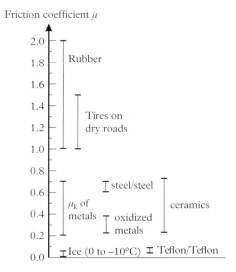

Figure 4.2 *Range of friction coefficient values for some materials under dry sliding conditions.*

coefficients outside of this range. For example, Teflon is known for have having a very low friction coefficient, typically less 0.1, while rubber can have a high coefficient, with $\mu > 1$.

While the type of material in contact is a major factor in determining the friction coefficient, other factors are also important; consequently it is better to think of a friction coefficient as being a *system property* rather than a *material property*. A few system factors that influence the values of friction coefficients:

- *Surface roughness*—As mentioned in Section 2.1, rougher surfaces tend to have higher values of coefficients of friction, and smoother surfaces lower values; but, making the surface too smooth makes the adhesion and the real area of contact increase dramatically, leading to a dramatic increase in friction.

- *Molecular films*—Adsorbed molecules from either contamination from the environment or deliberated deposited molecular films can substantially lower the friction coefficient by lowering the shear strength of the sliding interface (Section 10.2).

- *Oxide films*—Metal surfaces exposed to oxidative or corrosive environments typically form oxide layers that have substantially lower coefficient of friction. At sufficiently high loading force, however, the oxide films are worn off and the higher friction coefficient of the bare metal returns (this is observed, for example, for sliding contacts involving aluminum).

- *Sliding speed*—For example, clean metal surfaces typically have coefficients of friction around 0.6 at low sliding speeds, but this drops to around 0.2 when the sliding speed becomes high enough for frictional heating to induce melting of the metal surface.

- *Wear damage*—The friction coefficients tend to rise if wear increases the roughness of the sliding surfaces, but decrease if wear generates particles that roll between the sliding surfaces.

- *Relative humidity*—On many materials, a molecular layer of water adsorbs at higher relative humidity that lowers friction by lubricating the surface. For human skin, however, higher relative humidity tends to hydrate and soften the skin, which results in a higher coefficient of friction (Section 4.2.1.1).

4.2 Physical origins of dry friction

Amontons' first law that friction is proportional to load was readily accepted by the French Royal Academy, as it fitted with the everyday experience of the members that heavy objects are more difficult to slide than light objects. The second law, that friction is independent of apparent area of contact, however, was hotly disputed as many thought that friction should somehow also scale with contact area. We now know that friction does indeed scale with area but with the real area of contact, which is usually proportional to load, rather than the apparent area. This dispute prompted De la Hire to repeat

Amontons' experiments, and he confirmed Amontons' conclusions, further solidifying support within the Academy for Amontons' laws.

At that time, another major concern (which continues to the present day) was that these laws are not strict physical laws derived from fundamental principles, but rather empirical laws based on experimental observations. Since these laws apply to many practical situations, they are a good starting point for most analyses of friction. Many situations, however, deviate from these laws. In order to know when these laws can be applied and when they cannot, it helps to understand their physical origins.

Amontons, like many people who have studied friction, realized that the surfaces that he worked with were not completely smooth, and he suspected that roughness is somehow responsible for friction. He proposed that friction originated from the surface roughness in two ways: (1) For rigid asperities, friction was due to the force needed to pull the weight up the slopes of the surface roughness, as shown in Figure 4.3(a). (2) For deformable asperities, the asperities act as flexible springs that are bent over during sliding so that friction increases with the amount of deflection, as illustrated in Figure 4.3(b). For both these explanations, friction depends only on load and not the size of the contacting objects. De la Hire also believed that friction originated from roughness, but added the concept that friction also involved the work that comes from asperities breaking off during the sliding process, as illustrated in Figure 4.3(c). In his studies of friction, Coulomb found that the static friction force was substantially higher than kinetic friction, and would increase the longer the two surfaces stayed in stationary contact. Since the difference between static and kinetic friction was higher for wood than for metal surfaces, Coulomb proposed that the wood surfaces were covered by fibers

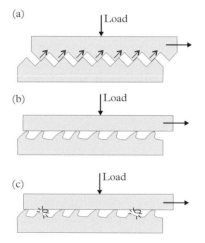

Figure 4.3 *Early concepts of how friction originated from roughness. (a) Amontons' concept for rigid asperities that friction originates from the force needed to raise the surfaces up the slopes of the roughness. (b) Amontons' concept for flexible asperities that the friction originates from the force bending the asperities. (c) De la Hire's concept that some of the friction comes from breaking off the asperity summits.*

oriented like bristles on a brush. When two surfaces were pressed into contact the bristles slowly penetrate each other, increasing the static friction over time. Once the surfaces start to slide, Coulomb proposed that the bristles folded back on themselves to reduce friction.

These initial ideas involving how friction originates from roughness have now evolved over the intervening 300 years to where friction between two solid surfaces in dry contact is thought to originate via two mechanisms:

1. an *adhesion force* needed to shear the contacting junctions where adhesion occurs;
2. a *plowing force* needed to plow the asperities of the harder surface through the softer surface.

4.2.1 Adhesive friction

While numerous researchers had pointed out that adhesion occurs between two materials when they are brought into contact and that these interatomic forces should contribute to friction, it was Bowden and Tabor who developed in the 1930s and 1940s a model to describe the connection between adhesive forces and friction (Bowden and Tabor 1950).

The model is fairly simple. When two surfaces touch, the contacting asperities undergo elastic and plastic deformation, as analyzed in the previous chapter and illustrated in Figure 4.4. At each contacting junction, labeled i, a small contact area A_i is generated. Within A_i, the surface atoms are in intimate contact generating attractive and repulsive interatomic forces between the atoms as discussed in Section 3.1. When a tangential force is applied to slide one object over the other, shear stresses develop over the junction interfaces to resist this force. At low shear stresses, the interatomic forces between the atoms are strong enough to prevent the atoms from sliding over each other, and the material around the contact junctions deforms elastically to counter the applied tangential force. At some critical shear stress, the applied force acting on the individual atoms exceeds the sum of all the interatomic forces that are trying to keep the atoms in their positions, and they start to slide over each other. The shear stress needed to start and sustain sliding is called the shear strength s, and the force needed to shear the ith junction is $F_i = A_i s$.

If we assume that all junctions have the same shear strength s, the adhesive friction force F_{adh} to shear all the junctions and slide the object is given by

$$F_{adh} = A_r s, \tag{4.2}$$

where A_r is the total real area of contact $(A_r = \sum A_i)$. In reality the shear strength is likely to vary from junction to junction, so s more appropriately corresponds to the average shear strength of the junctions. In this model, the adhesive friction is proportional to load (Amontons' law) if the real area of contact is proportional to load and the shear strength of contact junctions is independent of contact pressure. A key advantage of this adhesive friction model is that the expression $F_{adh} = A_r s$ also applies in situations where A_r is not proportional to load—situations where Amontons' law is not valid—such as

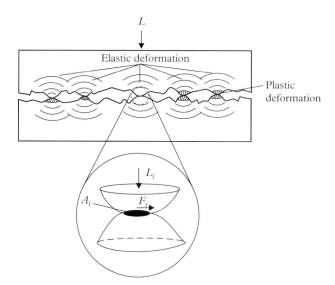

Figure 4.4 *Contact between two rough surfaces. Regions surrounding the contact junctions undergo elastic deformation, while plastic deformation occurs at some of the junctions. The expanded view shows the contact area A_i formed at the ith contact zone between two of the contacting asperities.*

when two atomically smooth surfaces are brought together so that real area of contact becomes similar to the apparent area of contact.

As discussed in Chapter 3, the real area of contact between rough surfaces is generally proportional to load. In particular, the Greenwood and Williamson model shows that, if the summit heights of the surface roughness have a Gaussian distribution and the deformation at the contacting asperities is primarily elastic, A_r can be approximated (eq. (3.45)) as

$$A_r \simeq 3 \left(\frac{R}{\sigma_s} \right)^{1/2} \frac{L}{E_c},$$

where R is the average radius of curvature of the summits, σ_s the standard deviation of summit heights, L the loading force, and E_c the composite elastic modulus. If instead the contact regions mainly deform plastically, A_r can be approximated by eq. (3.44):

$$A_r \simeq \frac{L}{H},$$

where H is the hardness of the softer material. For both cases, the real area of contact varies linearly with the loading force L pushing the two surfaces together, so the friction force $F_{adh} = A_r s$ is also proportional to L. If the shear strength is independent of contact pressure, F_{adh} is also independent of the apparent area of contact.

For example, metal surfaces finished by grinding or polishing, even though smooth to the touch, usually still have asperities sharp enough such that most asperities plastically deform during initial contact ($\psi > 1$). For this situation, $A_r = L/H$ and $F_{adh} = sL/H$, leading to

$$\mu_{adh} = \frac{s}{H} \qquad (4.3)$$

for shearing junctions of contacting metals or other contacting materials where most of the deformation is plastic. If the strong adhesion occurs at the contacting surfaces, the junctions rupture during shear within the weaker of the two materials, and s equals the shear strength of the softer material. From the Tresca yield criterion discussed in Section 3.3.2, $s = Y/2$, where Y is the yield stress of the softer material. Since $H \simeq 3Y$ (eq. (3.28)),

$$H \sim 6s \qquad (4.4)$$

and

$$\mu_{adh} \simeq \frac{s}{H} \sim \frac{1}{6}. \qquad (4.5)$$

From this equation, we see that friction is directly related to shear strength and hardness of the softer material in those situations where strong adhesion and plastic deformation occur at the asperity contacts.

4.2.1.1 Human skin friction

The friction coefficient of skin depends on multiple factors, but the most sensitive factor is the amount of moisture present, with friction increasing as the amount of moisture increases. So, the friction of our clothes against our skin is lower when them off before a shower or bath than when we put them back on afterwards (Dowson 2009).

In general, friction coefficients for dry skin are found experimentally to range from $\mu = 0.2$ to $\mu = 1$, while for wet skin $\mu > 1$ (Derler and Gerhardt 2012). This increase in skin friction with moisture is attributed to water becoming adsorbed into the skin, where it has a plasticizing effect on the molecular structure that greatly reduces the skin's elastic modulus. The lower elastic modulus then leads to a much higher real contact area at a given load, which from eq. (4.2) should result in increased friction (Adams et al. 2007, Derler and Gerhardt 2012).

Adams et al. (2007) have proposed that skin friction can be modeled as the sum of two non-interacting terms:

$$F = F_{adh} + F_{def}, \qquad (4.6)$$

where F_{adh} is the adhesive friction term and F_{def} is the deformation term. The deformation component comes from the energy dissipated within the skin due to its viscoelastic deformation during the sliding process. Since the deformation term contributes only $\mu_{\text{def}} \sim 0.04$ to the skin's coefficient of friction, or an order magnitude less than the measured values for μ, it can be neglected compared to the adhesive term.

For the interfacial shear strength s of skin, Adams et al. propose that it has the same linear dependence on the mean contact pressure p as previously found for organic thin films (Briscoe and Tabor 1975):

$$s = s_0 + \alpha p. \tag{4.7}$$

Neglecting the deformation component of friction, we can then write

$$\mu_{\text{skin}} = \frac{F_{\text{adh}}}{L} = \frac{s_0}{p} + \alpha = \frac{s_0 A_{\text{r}}}{L} + \alpha. \tag{4.8}$$

For dry skin, its roughness results in the real contact area A_{r} being only a small fraction of the apparent contact area A. In this situation A_{r} increases linearly with the loading force L, and μ_{skin} is independent of load. As the skin becomes hydrated at a given load, either with increasing humidity or with the application of water, the fraction of the apparent area in contact grows both due to the skin becoming suppler with increased moisture and due to the increase adhesion from water menisci forming around the contact points. At sufficiently wet conditions, the real area of contact approaches the apparent area ($A_{\text{r}} \to A$), and the friction force saturates.

Figure 4.5 shows measurements of the shear strength of wet skin under conditions where $A_{\text{r}} \sim A$ (Adams et al. 2007). The measured values of the skin shear strength at low loads ($s_0 \sim$ 4–6 kPa) are actually fairly low when compared either to the interfacial shear strength of glassy organic polymers (typically a few MPa; Briscoe and Tabor 1975) or to metals (typically a few GPa). So, when we slide a finger across a surface with a light

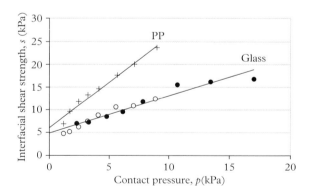

Figure 4.5 *The dependence of the interfacial shear strength of skin on contact pressure when either a polypropylene (PP) or a glass probe is rubbed against wet skin on a forearm. The solid lines show the fits to eq. (4.7) with $s_0 = 4.8$ kPa and $\alpha = 0.8$ for glass and $s_0 = 6.1$ kPa and $\alpha = 2.0$ for PP. Reproduced with permission from Adams et al. (2007). Copyright Springer Science + Business Media, LLC 2007.*

touch, the low contact area combined with the low shear strength provides for smooth sliding. But in wet conditions, even though the shear strength is low, since $A_r \sim A$, the total friction can be relatively high (on the order of 1 N/cm^2).

One interesting aspect to notice regarding the results in Figure 4.5: even though the shear strengths are similar for the polypropylene and glass probes at low contact pressures, the hydrophobic polypropylene has a much higher value of α than the hydrophilic glass. Presumably, the hydrophilic nature of the glass surface enables it to attract extra water into the sliding interface to help lubricate it.

4.2.2　Plowing friction

The contribution to the friction force from plowing hard asperities through a softer surface is called the *plowing friction* or F_{plow}. A simple estimate of F_{plow} can be made by multiplying the total projected area A_0 in the direction of motion of the contacting asperities by the pressure that must be exerted on the softer material in order to induce plastic flow (approximately, its hardness H):

$$F_{\text{plow}} \approx A_0 H. \tag{4.9}$$

This is illustrated in Figure 4.6 for a single, cone shaped asperity plowing through a soft material. A cone shaped asperity is used to keep the geometry simple (Figure 4.6(b)). The single asperity in Figure 4.6 exerts a plowing force F_{plow}, to displace the softer material, that is H times the cross-sectional area of the groove $ax = x^2 \tan \varphi$:

$$F_{\text{plow}} = Hx^2 \tan \varphi. \tag{4.10}$$

As the asperity plows through the softer material, the loading force L is supported by the contact pressure H of the softer material acting over an area $\pi a^2/2$ underneath the cone:

$$L = \frac{1}{2} H \pi a^2 = \frac{1}{2} H \pi x^2 \tan^2 \varphi. \tag{4.11}$$

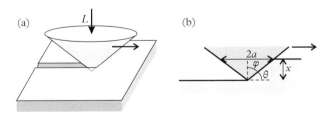

Figure 4.6 *(a) A rigid cone shaped asperity plowing a groove through a softer material. (b) Schematic showing the cone's semi-angle φ and the groove depth x.*

This leads to a coefficient of friction for plowing of

$$\mu_{\text{plow}} = \frac{F_{\text{plow}}}{L} = \frac{2}{\pi} \cot \varphi. \tag{4.12}$$

As the slopes of most surfaces of practical interest are less than $10°$ ($\varphi > 80°$), we should expect μ_{plow} to be less than 0.1.

While the adhesive and plowing contributions to friction are not completely independent, it is convenient to treat them separately, as we have been doing, and express the total friction force F as the sum

$$F = F_{\text{adh}} + F_{\text{plow}}. \tag{4.13}$$

For the particular case where one of the contacting materials is much softer than the other and most contact junctions deform plastically, so that both eqs. (4.5) and (4.12) are valid, the total friction arising from the combination of adhesion and plowing can be expressed by the coefficient of friction:

$$\mu = \frac{S}{H} + \frac{2}{\pi} \cot \varphi$$
$$\mu \simeq \frac{S}{H} + \tan \theta \tag{4.14}$$

where $\theta = 90° - \varphi$ represents a typical slope of the plowing asperities. Since $s/H \sim 0.17$ (eq. (4.5)), eq. (4.14) predicts that μ should not exceed 0.2, for contacting metals with similar hardness (so that the plowing contribution is negligible), and μ should not exceed 0.3 for a hard metal sliding on a softer one. Experimentally, however, the friction coefficients for metals are found to be several times these values, indicating that other mechanisms are contributing to friction. The two main ones for sliding metals are *work hardening* and *junction growth*.

4.2.3   Work hardening

The previous analyses assumed that the yield strength of materials remains constant during the sliding process. Many ductile materials, however, undergo a process of work hardening or strain hardening where the material's yield stress, hardness, and shear strength increase as the amount of plastic deformation increases.

At the atomic scale, work hardening occurs as a result of the formation of slip planes and the movement of dislocations during plastic deformation (Section 3.1 and Section 12.6.2.1). Eventually, the slip planes and dislocations run into obstacles, such as impurities or grain boundaries, or other dislocations, pinning them and making further deformation more difficult. Since a higher stress has to be applied to continue the plastic flow, the material now has a higher hardness and shear strength. As asperities sheared during sliding are more plastically deformed near the surfaces of the contact zones than regions far away where the load is supported, the surface shear strength s increases faster than the near surface hardness, leading to higher values of μ_{adh} than expected from eq. (4.5) or (4.14).

4.2.4 Junction growth

A more important contributor to higher values of μ_{adh} is junction growth (Tabor 1959, Ovcharenko et al. 2008). In junction growth, the combination of normal and tangential stresses acting on adhering junctions cause them to plastically deform in a manner that increases the area of contact being sheared. For clean metals, without any contamination layer that might lower the shear strength below the bulk metal value, junction growth leads to fairly high coefficients of friction ($\mu > 1$) and even seizure. Seizure or cold welding is particularly acute if the clean metals are rubbed against each other in high vacuum conditions where the contamination layers cannot reform.

When we derived the contact area expression for plastically deforming asperities ($A_r = L/H$, eq. (3.44)), we assumed that only normal forces were acting on the contacting asperities. If tangential forces act simultaneously on these asperities, plastic deformation occurs at a lower value of the normal contact pressure. To illustrate this, consider the idealized asperity shown in Figure 4.7. First, a loading force L is applied so that the contact pressure p_y in the asperity is at the onset of yield. Then, a tangential force F is applied to the asperity to generate a shear stress τ within the asperity. The extra stress from τ causes plastic flow of the asperity until the normal pressure is reduced to p_1. The relationship between p_1, τ, and p_y is determined by the particular shape of the asperity and the yield criterion. For the cubic shaped asperity in Figure 4.7,

$$p_1^2 + 3\tau^2 = p_y^2, \tag{4.15}$$

which is a statement of the von Mises criterion for yield (eq. (3.25)). Converting pressure and stresses into the normal and tangential forces ($p_1 = L/A$, $\tau = F/A$), we obtain

$$L^2 + 3F^2 = A^2 p_y^2. \tag{4.16}$$

Since p_y, the plastic yield stress, is a material property that remains constant, the contact area A will grow as the applied tangential force F is increased.

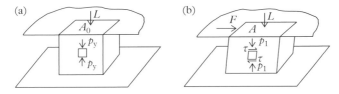

Figure 4.7 *Model for plastic deformation an idealized, cubic shaped asperity pressed against a counterface. The stresses are shown acting on an internal element: (a) with only a normal force L and (b) with both a normal force L and an applied tangential force F.*

The physical origin of this effect is simple: the von Mises criterion assumes that the distortional strain energy (the total strain energy excluding hydrostatic contributions) is constant as a material yields. Shear stresses and normal stresses both contribute this energy. If we add shear stress by applying the tangential force F, the normal stress must decrease to keep the total constant. Growth of the junction's area at constant load accomplishes this by reducing the normal stress. Another way to understand the effect is to view the addition of shear stress as leading to increased plastic flow, thus the asperity is flattened and its contact area grows.

In principle, the junction area growth could increase indefinitely but is usually limited by two factors: (1) the reduced ductility of the asperity by work hardening (increase of p_y) and (2) the presence of a weak interfacial film that has a lower shear strength s_i than the bulk shear strength s_0. Lubricant films are examples of weak interfacial films that are deliberately applied to the surfaces to provide a low shear strength.

It is straightforward to extend the model to include the effect of a weak interfacial film: As the tangential force F increases, the junction area A grows until it reaches the value

$$A_{max} = \frac{F_{max}}{s_i} \tag{4.17}$$

when sliding occurs at the interface film with a low shear strength s_i. The von Mises yield criterion can be used to relate the bulk shear strength to the bulk yield stress (eq. (3.26)):

$$p_y = \sqrt{3} s_0. \tag{4.18}$$

From eq. (4.16), the maximum contact area A_{max} of the junction before the onset of sliding is given by

$$A_{max} = \frac{L}{\sqrt{3 \left(s_0^2 - s_i^2\right)}}. \tag{4.19}$$

Combining eqs. (4.16)–(4.19), we derive the maximum adhesive friction coefficient in the presence of junction growth:

$$\mu_{adh} = \frac{F_{max}}{L} = \frac{1}{\sqrt{3 \left[(s_0/s_i)^2 - 1\right]}}. \tag{4.20}$$

From eqs. (4.17) and (4.20), we can see that, as s_i tends to the bulk value s_0, both the maximum contact area and the friction of the junction tend to infinity. We also see that surface films with low interface shear strengths greatly reduce friction by inhibiting junction growth. In the limit where the interface film has a very low shear strength and $s_i \ll s_0$, no junction growth occurs: $F = As_i$ and $L = Ap_y$, leading to

$$\mu_{adh} = \frac{F_{max}}{L} = \frac{s_i}{p_y},$$ (4.21)

which is similar to eq. (4.3). This equation shows what we already know from intuition: that a practical way to reduce friction between two sliding surfaces is to interpose a film, such as a lubricant film, with a low shear strength.

4.3 Static friction

From our everyday experience in dealing with friction, most of us have learnt that the force required to start an object sliding is greater than the force needed to maintain sliding, that is, static friction is greater than kinetic friction. What is less obvious is that the value of static friction typically increases the longer the two surfaces stay in stationary contact as illustrated in Figure 4.8(a) and that the kinetic friction typically decreases as the sliding velocity increases (Figure 4.8(b)). Generally, static friction asymptotes so quickly to its final value that we tend to think of static friction as having a constant value. Similarly, the drop in kinetic friction to a lower value asymptote, as the sliding velocity ramps up, happens over such a small range of velocities near zero that kinetic friction is typically approximated as being independent of velocity (remember the third law of friction).

The increase in static friction with longer rest times can be understood at the atomic and molecular level as follows: When two surfaces are brought into contact with an applied load L at time $t = 0$, a contact area $A_c(t = 0)$ initially forms, and the lateral force needed to overcome the adhesive friction is $F_{adh} = sA_c(0)$. At $t = 0$, the atoms in each contact zone start experiencing the atomic forces acting across the contact, with the long range adhesive forces pulling atoms and molecules into the junction, and the short range repulsive forces pushing atoms out from the center of the contact junctions (Figure 4.9(a)). This causes the atoms to move over time to energetically more favorable locations at the junction edges (Figure 4.9(b)). As the solid–solid contact area $A_c(t)$ grows with time, the adhesive friction F_{adh} grows.

In addition to the movement of solid atoms, molecules from the surrounding environment are drawn into the crevices around the contacting asperities (as illustrated in

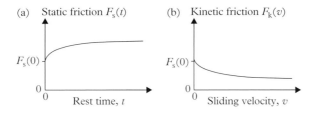

Figure 4.8 *Typical plots of (a) static friction as function of rest time and (b) kinetic friction as a function of sliding velocity.*

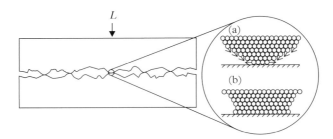

Figure 4.9 *(a) Arrangement of atoms on an asperity at t = 0. The arrows indicate the directions that the atoms move to reach their new equilibrium positions in (b).*

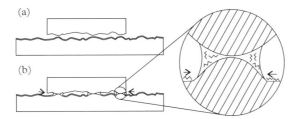

Figure 4.10 *(a) Two surfaces before they touch. The bottom surface has a thin film of mobile molecules, such as a thin film of liquid lubricant or of adsorbed molecules from the air, such as adsorbed water vapor. (b) The inset illustrates how, after the two surfaces touch, capillary action pulls molecules in around the asperity contacts to form menisci.*

the inset of Figure 4.10) by the attractive interatomic forces. As these molecules collect around the contacting asperities, they form liquid menisci that exert an adhesive force on each asperity due to the capillary pressure inside each meniscus. The net adhesive force L_{adh} from the sum of all the menisci forces adds to the applied load L. In accordance with Amontons' law, the static friction force $F_s(t)$ increases as the net normal force $(L + L_{adh}(t))$ increases:

$$F_s(t) = \mu\,(L + L_{adh}(t))\,. \tag{4.22}$$

A common example of this: When the humidity is high, people complain about it feeling "sticky" as the water in their sweat, which is less likely to evaporate at high humidity, exerts meniscus forces on contacting parts of the body. Similarly, at high humidity, water vapor condenses around the contact points between objects with hydrophilic surfaces, such as dust particles on glass windows, causing them to stick more strongly.

Another example: a thin film of lubricating oil applied to reduce kinetic friction, such as illustrated in Figure 4.10, often makes static friction worse. This is caused by lubricant being drawn into and accumulating in the narrow gaps between the contacting surfaces, when the surfaces stay in stationary contact, leading to a higher meniscus adhesion force and higher friction.

These effects also occur during sliding, when contact junctions are constantly being formed and ruptured. During the brief periods of contact, the atoms are pushed and pulled toward new equilibrium positions, increasing the contact area of the junction. So, the mechanism that results in kinetic friction of solids decreasing with increasing sliding speed (Figure 4.8(b)) has a similar physical origin as the increase in static friction with rest time: for slower sliding speeds, the contact junctions exist for longer, allowing for more time for their contact areas to grow, leading to higher friction. As discussed in a later chapter on lubrication, when a liquid film is sheared between two solid surfaces, however, the opposite trend is observed: *friction increases with velocity*, due to the frictional forces being generated by viscous dissipation within the liquid.

4.3.1 Stick-slip

One consequence of static friction's being higher than kinetic friction and of kinetic friction's decreasing with increasing sliding speed is the phenomenon of *stick-slip* where sliding occurs either as a sequence of sticking and slipping or as an oscillation at a resonance frequency of the system, rather than sliding at a constant speed. Stick-slip can manifest itself as an unpleasant squealing and chattering noise, such as a squeaky door hinge or the squeak of chalk on a blackboard. It is also what occurs during an earthquake: the sudden, unsteady sliding of the earth's tectonic plates against each other after decades of static loading.

Stick-slip is also frequently a sign of a malfunctioning device such as when your car brakes start to squeal. In another example, as surfaces of windshield wipers degrade when exposed to sunlight, their static friction increases relative to the kinetic friction, and eventually they stick-slip ineffectively across the windshield. In a few cases, stick-slip may actually be desirable, for example, in the action of a bow rubbing across the strings of a violin, resulting in the string's producing music.

4.3.1.1 *Velocity-controlled stick-slip*

Stick-slip arises from the interplay of friction with the dynamics of the mechanical system. To better see how stick-slip comes about, we consider the mechanical system shown in Figure 4.11, where a block with mass m sits on a belt moving with velocity V. A spring element, with stiffness k and dashpot damper η, is attached to the fixed support on the left, restraining the motion of the block. A load L (the weight of the block mg plus the externally applied load and adhesive forces) pushes the block in the normal direction against the belt, and a friction force $F = \mu L$ acts in the tangential direction to oppose relative motion between the block and belt. When the displacement x of the block away from the spring's equilibrium position is sufficiently small so that $F < \mu_s L$, the block moves with the same velocity as the belt ($\dot{x} = V$). In this case, the tangential forces acting on the block are

$$|\eta V + kx = F| < \mu_s L. \tag{4.23}$$

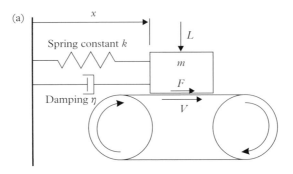

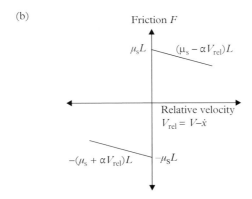

Figure 4.11 *(a) A block sitting on a belt moving with velocity V and restrained by a spring element and dashpot. (b) How friction force F acting on the block depends on relative velocity between the block and the belt.*

As displacements become larger, the spring elongates and eventually exerts a force exceeding the static friction $\mu_s L$, at which point the block starts to slide against the belt with a relative velocity $V_{rel} = V - \dot{x}$. When the block slides over the belt, its equation of motion is

$$m\ddot{x} + \eta\dot{x} + kx = F. \qquad (4.24)$$

Here we consider *velocity-controlled stick-slip*, which originates from a friction force decreasing with increasing velocity. To simplify the analysis, we assume that friction decreases linearly with increasing relative velocity:

$$
\begin{aligned}
F &= (\mu_s - \alpha V_{rel})\,L \quad &\text{if } V_{rel} > 0 \\
F &= -(\mu_s + \alpha V_{rel})\,L \quad &\text{if } V_{rel} < 0
\end{aligned}
, \qquad (4.25)
$$

where α is the slope of friction as a function of velocity. The assumption that kinetic friction is linear with velocity can be valid for a sufficiently small range of the friction vs. velocity curve in Figure 4.8(b); for example, Figure 4.11(b) shows the friction for a small range of sliding speeds near zero velocity.

Combining eqs. (4.24) and (4.25) leads to

$$m\ddot{x} + (\eta - \alpha L)\,\dot{x} + kx = (\mu_s - \alpha V)\,L \quad \text{if } V_{rel} > 0$$
$$m\ddot{x} + (\eta - \alpha L)\,\dot{x} + kx = (-\mu_s - \alpha V)\,L \quad \text{if } V_{rel} < 0 \tag{4.26}$$

An interesting feature of eq. (4.26) is the term $(\eta - \alpha L)$ on the left side of the equations, which represents the effective damping coefficient of the mechanical system. When this term is negative, $(\eta - \alpha L) < 0$, the energy being supplied by the friction force to the mechanical system is greater than the energy being dissipated in the dashpot damper, driving an oscillatory or stick-slip motion. This instability occurs at high load L or at high α, the slope of the $\mu_k - V_{rel}$ curve. Otherwise, for $(\eta - \alpha L) > 0$, the effective damping term is positive, indicating that any oscillation in the spring system is eventually damped out and no steady-state oscillation or stick-slip motion occurs.

Figure 4.12 shows solutions for the motion of the block initially sitting on the belt at $t = 0$, $x = 0$, and $\dot{x}(0) = V = 1\text{m/s}$ using the differential equation (4.26) for when $V_{rel} \neq 0$ and using the condition (4.23) to determine when $V_{rel} = 0$. At first, the block moves with the belt until the static friction is exceeded.

- For the low load case, where the effective damping coefficient is positive, once the static friction is exceeded, the displacement of the spring quickly settles to a constant value corresponding to a constant friction force acting on the block as it slides smoothly against the belt.
- If the load L or the slope α is increased, the effective damping coefficient becomes less positive. For the special case where $(\eta - \alpha L) = 0$, once the static friction is exceeded the block undergoes a sinusoidal motion about the displacement corresponding to the average friction during sliding. During this oscillation, the

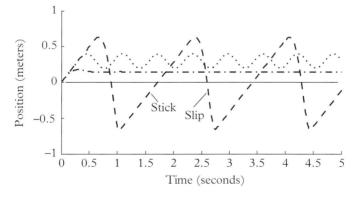

Figure 4.12 *Motion of the block on the moving belt illustrated in Figure 4.11. For this analysis, $m = 1$ kg, $k = 100$ N/m, $\mu_s = 0.5$, $\alpha = 0.2$ s/m, $V = 1$ m/s, $\eta/2m = 10$ s^{-1} (η chosen to provide critical damping of the mass–spring system). Different loads are used to generate different values of the effective damping coefficient $(\eta - \alpha L)$: dot-dash line, $L = 50$ N, $(\eta - \alpha L) = +10$ N s/m; dotted line, $L = 100$ N, $(\eta - \alpha L) = 0$ N s/m; and dashed line, $L = 150$ N, $(\eta - \alpha L) = -10$ N s/m.*

energy received by the block from friction while traveling in the same direction of the belt is lower than the energy dissipated through friction when moving in the opposite direction of the belt, as the friction is lower for the higher relative velocity. The extra energy from the difference in friction drives the oscillation through the cycle before being dissipated in the spring damping.

- At higher loads, where the effective damping coefficient is negative, stick-slip instabilities occur. The kinetic friction is high enough that the block stops moving relative to the belt or "sticks" to the belt for part of the oscillation; once the static friction is exceeded, the block "slips" to the next part of the oscillation where it sticks again.

4.3.1.2 Time-controlled stick-slip

In the previous section, we assumed that the static friction force did not increase significantly as the stick time increased. As discussed earlier, static friction tends to rise the longer two surfaces are in stationary contact, as illustrated in Figure 4.8(a). This phenomenon, in the context of geology and earthquake faults, is known as "aging." If the increase in static friction during a stick phase of the oscillation is greater than the variation in kinetic friction during the slip phase of the oscillation, the stick-slip process is said to be *time-controlled* rather than *velocity-controlled*. To understand how a higher value of static friction leads to stick-slip, we will follow the approach originally developed by Rabinowicz (1958). We again consider the block on the moving belt in Figure 4.11(a), but this time with the friction increasing with stick time as shown in Figure 4.13. If the block is initially at rest at $t = 0$, the force exerted by the spring on the block during

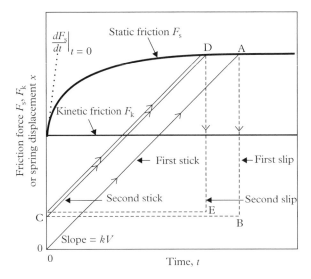

Figure 4.13 *Static friction increasing with time for generating time-controlled stick-slip. Reproduced from Rabinowicz (1958).*

the first stick phase increases as kVt. At the point A, the spring force exceeds the static friction, and the block slips across the belt surface until coming to rest at point B. Since kinetic friction F_k is assumed constant (represented by the straight line in Figure 4.13), the friction force at point B is as much below F_k as point A is above F_k. The block undergoes a second stick phase from C to D, followed by a second slip to E before settling into a steady state stick-slip oscillation. From the diagram in Figure 4.13, we see that increasing the spring constant k or the belt velocity V increases the slope kV of the stick phase, leading to a smaller amplitude of the stick-slip oscillation. If $kV > (dF/dt)_{t=0}$, smooth sliding rather occurs than stick-slip.

4.3.1.3   *Displacement-controlled stick-slip*

In addition to being velocity-controlled and time-controlled, stick-slip can also be *displacement-controlled*, where stick-slip is caused by the friction force varying as function of position over the sliding surface. For example, if the belt in Figure 4.11(a) has a non-uniform surface so that the friction force F varies along the belt, the spring displacement x varies according to $x = F/k$ for smooth sliding. If the block encounters a portion of the belt surface where the friction drops suddenly so that the gradient ∇F is greater than the spring constant k, a slip occurs as the pull from the spring is momentarily greater than the slowing force from friction.

Displacement stick-slip occurs not only for macroscale systems like the block on the belt in Figure 4.11, but also at the atomic scale. This type of atomic scale slick-slip process can be readily observed in an atomic force microscope (AFM) such as the one shown in Figure 4.14(a), which has a cantilever that flexes easily in the direction of the friction

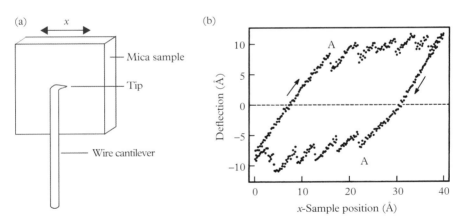

Figure 4.14 *(a) Geometry of an atomic force microscope (AFM) tip contacting a mica surface. The tip was made by etching the end of a tungsten wire to a sharp point (tip radius $R \sim 300$ nm) and bending it toward the sample so that the cantilever spring constant is the same perpendicularly to the surface and in the x-direction. (b) Friction loop for an AFM tip sliding over a cleaved mica surface. The deflection of the AFM cantilever parallel to surface is plotted vs. the sample position in the weak direction of the cantilever, with a spring constant $= 100$ N/m and a load $L = 10^{-6}$ N. Reprinted with permission from Erlandsson et al. (1988). Copyright 1988, American Institute of Physics.*

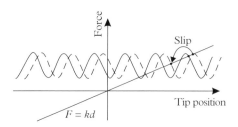

Figure 4.15 *The friction force for two sample positions (solid and dashed lines) as the sample moves from left to right, and the negative force from the cantilever deflection ($-F = kd$, where $k =$ cantilever spring constant and $d =$ wire deflection). The tip position on the surface is stable when the forces on the tip balance, as indicated by the intersection of the plots. As the surface moves underneath the tip, a slip occurs when, at the intersection point, the derivative of the frictional force equals the spring constant. Reprinted with permission from Mate et al. (1987). Copyright 1987 by the American Physical Society.*

force. Figure 4.14(b) shows a "friction loop" generated by such an AFM tip sliding back and forth over a freshly cleaved mica surface. When the direction of sample motion changes, the tip initially moves with the sample, and the wire cantilever deflection is proportional to the static friction force. At point A, the cantilever wire exerts enough lateral force on the tip to overcome the static friction, and the tip starts to slide across the surface. The sliding process is not uniform, but rather occurs as a series of slips with each having a distance corresponding to the 5.2 Å unit cell size of the hexagonal SiO_4 layer of the cleavage plane of muscovite mica.

This displacement-controlled stick-slip process occurs when the spring constant of the AFM cantilever parallel to the surface is less than the spatial derivative of the friction force that the AFM tip experiences as it slides over the surface. When the tip slides over a crystal surface with a periodic arrangement of surface atoms, the tip experiences a friction force with the same periodicity as the crystal lattice. When the slope of the decreasing portion of the periodic force is greater in magnitude than the cantilever spring constant, the tip slips to the next stable position. Figure 4.15 illustrates how this occurs. This is discussed further in Chapter 11.

4.4 PROBLEMS

1. Consider the MEMS lateral actuator in Problem 3 of Chapter 3. Assuming that AFM measurements determined an interfacial shear strength of $s = 200$ MPa and that the adhesive friction mechanisms account for the friction force, what is the lateral force needed to displace the slider parallel to the substrate?

2. In discussing Figure 4.13, we state that "the friction force at point B is as much below F_k as point A is above F_k." Prove this statement. *Hint:* Using an energy argument is a relatively simple approach that will work.

4.5 REFERENCES

Adams, M. J., B. J. Briscoe and S. A. Johnson (2007). "Friction and lubrication of human skin." *Tribology Letters* **26**(3): 239–53.

Bowden, F. P. and D. Tabor (1950). *The friction and lubrication of solids*. Oxford: Clarendon Press.

Briscoe, B. J. and D. Tabor (1975). "The effect of pressure on the frictional properties of polymers." *Wear* **34**(1): 29–38.

Derler, S. and L. C. Gerhardt (2012). "Tribology of skin: review and analysis of experimental results for the friction coefficient of human skin." *Tribology Letters* **45**(1): 1–27.

Dowson, D. (1978). *History of tribology*. London: Longman.

Dowson, D. (2009). "A tribological day." *Proceedings of the Institution of Mechanical Engineers Part J: Journal of Engineering Tribology* **223**(J3): 261–73.

Erlandsson, R., G. Hadziioannou, C. M. Mate, G. M. McClelland and S. Chiang (1988). "Atomic scale friction between the muscovite mica cleavage plane and a tungsten tip." *Journal of Chemical Physics* **89**(8): 5190–3.

Mate, C. M., G. M. McClelland, R. Erlandsson and S. Chiang (1987). "Atomic-scale friction of a tungsten tip on a graphite surface." *Physical Review Letters* **59**(17): 1942–5.

Ovcharenko, A., G. Halperin and I. Etsion (2008). "In situ and real-time optical investigation of junction growth in spherical elastic–plastic contact." *Wear* **264**(11): 1043–50.

Pitenis, A. A., D. Dowson and W. Gregory Sawyer (2014). "Leonardo da Vinci's friction experiments: an old story acknowledged and repeated." *Tribology Letters* **56**(3): 509–15.

Rabinowicz, E. (1958). "The intrinsic variables affecting the stick-slip process." *Proceedings of the Physical Society* **71**: 668–75.

Tabor, D. (1959). "Junction growth in metallic friction – the role of combined stresses and surface contamination." *Proceedings of the Royal Society of London, Series A: Mathematical and Physical Sciences* **251**(1266): 378–93.

5

Surface Energy and Capillary Pressure

To most people, it is not initially obvious that surfaces should have an energy associated with them. Not only are significant amounts of energies associated with the surfaces and interfaces of solids and liquids, these surface energies are responsible for driving many surface phenomena. For example, surface energies account for how detergents clean our clothes, how paints spread on surfaces, and how insects are able to walk on water. Consequently, the concept of surface energy underlies much of our understanding of the surface phenomena that make up tribology. In this chapter, we introduce surface energy and several related concepts: work of adhesion, capillary pressure, and adhesion hysteresis. The following chapter builds on these concepts to understand the forces between surfaces and the relationships between surface energy, lubrication, and friction.

First, let's cover a few points of nomenclature. The extra energy per unit area needed to form a material surface exposed to gas or vacuum is called the *surface free energy*, the *excess surface free energy*, or the *surface tension*. In this book, we use the terms *surface energy* when describing solids and *surface tension* when describing liquids. For a surface that forms an interface between two condensed phases, the extra energy per unit area associated with this interface is called the *interfacial energy*. The energy associated with surfaces or interfaces is represented by the symbol γ and has units of energy per unit area (mJ/m^2 = ergs/cm^2) or, equivalently, force per unit distance (mN/m = dynes/cm, equivalent to ergs/cm^2, which are the units for surface tension).

The terms *surface* and *interface* are often used interchangeably to describe the boundary between two material media or phases. The term *surface*, however, is generally used when a condensed (solid or liquid) phase is on one side and a rarified (gas or vacuum) phase is on the other side of the interface. In contrast, the term *interface* is used when condensed phases (either one or both of which may be solid or liquid) are present on both sides of the interface.

5.1 Liquid surface tension

You have probably noticed that when water droplets form they have a spherical shape, rather than a cubic or irregular shape. The reason for this spherical shape is that when a droplet forms extra energy is needed to form the droplet's surface; to minimize this

Tribology on the Small Scale: A Modern Textbook on Friction, Lubrication and Wear. Second edition. C. Mathew Mate and Robert W. Carpick. © Oxford University Press 2019. Published in 2019 by Oxford University Press.
DOI: 10.1093/oso/ 9780199609802.001.0001

surface energy, the liquid rearranges to have the minimum surface area, which occurs with the spherical shape.

Figure 5.1 illustrates the forces acting on molecules in a liquid that give rise to surface energy. A liquid molecule in the bulk feels the attractive cohesive forces from its neighbors uniformly distributed around it, while the cohesive forces acting on a molecule at the surface are unbalanced since this surface molecule lacks neighbors on the surface side. So, if a molecule starts to move away from the surface, the attractive cohesive forces pull it back into the liquid, creating a tension that pulls the liquid surface flat, hence preference for the name *surface tension* when discussing liquid surfaces. Thermodynamically, the surface tension γ is the reversible work W required to create a unit area A of a liquid surface:

$$\gamma = \frac{dW}{dA}. \tag{5.1}$$

In other words, the surface tension is the work done per unit area against the cohesive forces to bring molecules from the bulk to the surface. Surface tensions of typical liquids in air at room temperature range from 20–40 mN/m for hydrocarbons (e.g., $\gamma = 25$ mN/m for paraffin wax), to 73 mN/m for water, and up to 485 mN/m for mercury. The increasing surface tension from hydrocarbons to water to mercury indicates increased bond strength (higher cohesive energy) between the atoms in liquid mercury compared to the bond strength between the liquid hydrocarbon molecules.

An easy way to visualize that tension really exists at liquid surfaces is to sprinkle some pepper on the surface of water in your kitchen sink and then dip a bar of soap into the water. The pepper floating on the water moves rapidly away from the location where the soap touches the water. What is happening? As the water surface becomes covered with a monolayer of soap, the surface tension drops from that of bare water (73 mN/m) down to that of soap (30 mN/m). When the boundary of the spreading soap film encounters the grains of pepper, the 42 mN/m difference in surface tension pulls the pepper across

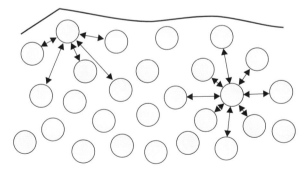

Figure 5.1 *Schematic of molecules near a liquid surface. The molecule at the surface on the left feels attractive cohesive forces from predominantly one side, pulling it back into the liquid, while the bulk molecule on the right experiences cohesive forces from all sides so that the average net force is zero.*

the surface in front of the spreading soap, since, at the boundary between the covered and bare water surfaces, the soap surface molecules exert smaller cohesive forces on the pepper grains than the water surface molecules.

A simple way of measuring the surface tension is to use a Wilhelmy plate setup, as illustrated in Figure 5.2, which converts the surface tension into a measurable force. In this measurement, a plate, with width b and thickness Δb, is lowered into the liquid, whose surface tension γ is being measured. The material of the plate is chosen so that the liquid wets the plate, that is, the meniscus contact angle $= 0°$. As discussed later in Section 5.4.1, this condition occurs when the work of adhesion between the liquid and the solid is greater than the liquid's internal work of cohesion; that way, when the plate touches the liquid surface, the attraction of the plate material for the liquid pulls it up to form a meniscus with a zero contact angle, enabling the liquid surface tension to act parallel to the plate surface, that is, in the vertical direction. At the contact line at the top of the meniscus, the liquid exerts on the plate the same tension as occurs on the liquid surface. The total force F_{men} exerted on the plate by the meniscus equals the contact line perimeter times the liquid surface tension γ:

$$F_{\text{men}} = 2\,(b + \Delta b)\,\gamma. \tag{5.2}$$

By measuring force acting on the plate in excess of the weight of the plate ($F_{\text{men}} = W_{\text{total}} - W_{\text{plate}}$), the liquid surface tension γ can be determined by solving eq. (5.2) for γ.

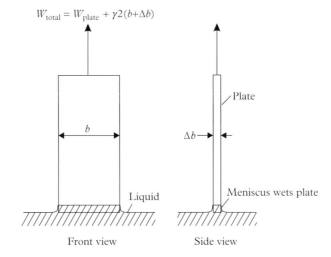

Figure 5.2 *Wilhelmy plate technique for measuring the surface tension γ of a liquid. The bottom edge of the plate is kept level with the liquid surface to minimize buoyancy forces. The material comprising the plate surface needs to be chosen so that the liquid has a contact angle equal to zero with it. Weight of the plate $= W_{plate}$.*

Numerous other methods exist for measuring liquid surface tensions which are detailed in the book by Adamson and Gast (1997).

5.2 Capillary pressure

As said earlier, the liquid rearranges inside a water droplet to form a sphere as this shape minimizes the surface energy. If the droplet is initially non-spherical, the liquid needs to move around within the droplet to achieve the low energy spherical configuration; this movement of liquid is driven by pressure differences between regions of the surface that have different curvature; these pressure differences originate from the surface tension pushing and pulling the liquid toward the minimum energy geometry.

The pressure inside a liquid associated with a curved liquid surface is called the *capillary* or *Laplace pressure* and is described by the equation developed by Young and Laplace in 1805. A simple derivation of this equation (Adamson and Gast 1997) can be obtained by considering a soap bubble with an equilibrium radius r and pressure difference ΔP between air inside and outside the bubble (Figure 5.3); the goal of the derivation is to determine the relationship between r and ΔP. For this soap bubble, the total surface free energy is $4\pi r^2 \gamma$, where γ in this case refers to the combined surface tension from the two sides of the soap film. If the bubble shrinks by an amount dr, the total surface energy becomes

$$4\pi (r - dr)^2 \gamma = 4\pi r^2 \gamma - 8\pi r dr\gamma + 4\pi dr^2 \gamma \simeq 4\pi r^2 \gamma - 8\pi r\gamma dr. \tag{5.3}$$

The decrease in surface energy $8\pi r\gamma\, dr$ is converted into the work $4\pi r^2 \Delta P dr$ compressing the air inside the bubble:

$$8\pi r\gamma\, dr = 4\pi r^2 \Delta P dr. \tag{5.4}$$

Rearranging eq. (5.4), we arrive at the equation for the equilibrium pressure difference between the inside and outside of the bubble:

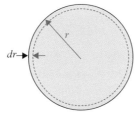

Figure 5.3 *Soap bubble with an initial radius r that shrinks by an amount dr.*

$$\Delta P = \frac{2\gamma}{r}. \tag{5.5}$$

This derivation for the pressure difference can be extended to the more general case of an arbitrarily curved liquid surface. This pressure difference ΔP, the *Laplace pressure* or the *capillary pressure* (P_{cap}), is given by the equation of Young and Laplace:

$$P_{cap} = \gamma \left(\frac{1}{r_1} + \frac{1}{r_2} \right), \tag{5.6}$$

where r_1 and r_2 are the two principal radii of curvature that define the curved surface. If the surface is spherical in shape, $r_1 = r_2 = r$ and capillary pressure is given by eq. (5.5).

The sign of r_1 or r_2 depends on which way the surface is curved: If the liquid meniscus curves away from you when looking from the outside of the liquid, $r > 0$, as for the droplet illustrated in Figure 5.4(a). If the surface curves towards you when looking from the outside, then $r < 0$, as for the meniscus that wets the inside of a narrow capillary tube like that illustrated in Figure 5.4(b). From the equations for capillary pressure (eqs. (5.5) and (5.6)), we see that the signs of the radii of curvature govern whether the pressure in the liquid is greater than or less than the surrounding environment. For example, for the liquid droplet shown in Figure 5.4(a), the pressure in the liquid is higher than that of the air outside the droplet. In Figure 5.4(b), having $r < 0$ leads to the liquid rising up the capillary tube. The "capillary rise" occurs since the pressure of the liquid around the meniscus is less than that of the outside air; the liquid is therefore pulled up the

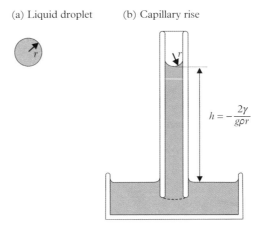

(a) Liquid droplet   (b) Capillary rise

$$h = -\frac{2\gamma}{g\rho r}$$

Figure 5.4 *(a) Liquid droplet with radius $r > 0$, resulting in the pressure inside the droplet being higher than in the air outside. (b) The capillary rise of a liquid in a narrow cylindrical tube where the liquid wets the wall of the tube, for example, water in a glass tube with a hydrophilic surfaces. The meniscus has a radius $r < 0$, resulting in the liquid pressure around the meniscus being lower than the air outside, which pulls the liquid up the tube a distance h.*

tube until an equilibrium height h is reached where the weight of the liquid in the tube exerts a pressure equal and opposite to the capillary pressure: equilibrium occurs when $hg\rho = -2\gamma/r$, where g is the acceleration due to gravity and ρ is the liquid density. This capillary rise to the same height h occurs even if the air is removed so that a vacuum (pressure equal to zero) exists outside the liquid, indicating that the absolute liquid pressure can be less than zero.

One might be tempted to think that, since the pressure is higher inside the liquid droplet than outside it, the droplet should expand until the pressure difference becomes zero. Likewise, for the capillary rise geometry in Figure 5.4(b), one might think that, since the pressure is higher above the meniscus than below it, the meniscus would be pushed down below the liquid reservoir surface, in the same manner that the high pressure behind a piston pushes it towards the low pressure end of a tube. Two analogies to other physical systems can help clarify one's understanding. (1) Within the Earth, the weight of the Earth's outer layers pushes down on the interior generating very high pressures; this positive pressure, however, does not cause the Earth's surface to move toward outer space as this pressure is counterbalanced by the gravitational forces. (2) A balloon has significantly higher pressure inside than out, but does not expand as the air pressure pushing against the balloon's interior surface is counterbalanced by the elastic stress generated by stretching the balloon. Similarly, a liquid droplet does not swell due to the higher pressure inside, as this internal pressure is counterbalanced by the stress generated from the surface tension that pulls on the curved liquid surface. In fact, the droplet's internal pressure is this physical consequence of the energy and tension associated with the liquid surface. One can think of a droplet as a pressurized balloon, where the bulk liquid interior is equivalent to the gas, and the liquid surface itself is equivalent to the rubber balloon. Like the rubber membrane of the balloon, the surface of the liquid is experiencing in-plane tensile stress.

5.2.1   Capillary pressure in confined places

In tribology, we are frequently interested in what happens in confined spaces, such as in the narrow gap between two solids when they are brought into contact. In these confined spaces, liquids are pulled in by capillary action or condensation and, once there, can have a tremendous influence on adhesion, friction, and lubrication. In a later chapter, we discuss directly how the capillary pressure influences adhesion forces; in this section, we outline some of the possible ways that liquids distribute themselves within confined geometries, and the associated capillary pressures.

First let's consider the example where a drop of water is placed on a glass microscope slide, and then another slide is placed on top of it. As the two slides come together, the droplet spreads out within the gap between the slides to form a pillbox-shaped meniscus (Figure 5.5) with diameter R.

If the slides have clean, hydrophilic surfaces (Figure 5.5(a)), the water contact angle θ is typically small, certainly $<90°$, so the meniscus has a concave surface with $r < 0$ at the water–air interface. Using eq. (5.6) with $r_1 = r$ and $r_2 = R$, we see that

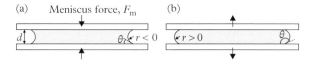

Figure 5.5 *Drop of water between two microscope slides with either (a) hydrophilic or (b) hydrophobic surfaces. For both cases, $P_{cap} \sim \gamma/r$, but it is negative for the hydrophilic case (a) and positive for the hydrophobic case (b).*

$$P_{cap} = \gamma \left(\frac{1}{R} - \frac{\cos\theta}{d/2} \right) \approx -\frac{2\gamma\cos\theta}{d}. \tag{5.7}$$

The negative value of the capillary pressure P_{cap}, means that the meniscus force $F_{men} = P_{cap} \times$ (mensicus area) acts to pull the two slides together. Since the capillary pressure increases as the separation d decreases, the two slides will continue to be pulled together by the meniscus force until either trapped particles or surface roughness prevent them from coming any closer, or the meniscus reaches the edges of the glass microscope slides. If the meniscus is still within the edges of the slides, the final capillary pressure can be extremely high. For example, if the final separation distance is $d = 1$ µm and $\theta \sim 0°$, then

$$P_{cap} = -2\gamma/d$$
$$= -2 \times (73 \text{ mN/m}) /10^{-6}\text{m} \tag{5.8}$$
$$= -146 \text{ kPa} = -14.6 \text{ N/cm}^2.$$

So, for every square centimeter of the gap filled with the water, a force of 14.6 N in the normal direction would be needed to separate the slides. (The glass usually breaks when this much force is applied.) An easier way to separate slides is to place them underwater so that the gap completely fills with water, eliminating the air–water meniscus surface and the force derived from its capillary pressure.

Figure 5.5(b) shows the case where the surfaces of the microscope slides are hydrophobic, which can be achieved by coating the slides with wax or bonding organic monolayers to their surfaces (Fadeev and McCarthy 2000). When the contact angle θ is greater than $90°$, $r > 0$ and $P_{cap} > 0$; since the capillary pressure is positive, the meniscus force F_m acts to repel the two slides, rather than to pull them together as for the hydrophilic case, and the final separation distance d occurs when the meniscus force balances the weight of the top slide. Another way to observe a repulsive meniscus is to use a hydrophobic slide as the plate in the Wilhelmy plate experiment shown in Figure 5.2 with water as the liquid; since the water contact angle $> 90°$, the meniscus will bend down rather than up, so the normal force on the hydrophobic slide/plate acts to push it out of the water rather than pull it in as for the hydrophilic slide/plate. Water strider insects provide a more elegant example of this effect: the high contact angle of water against legs of these insects generates a positive capillary pressure that provides an upward force sufficient to lift the insect body above the water surface thus enabling them to walk on water. Water pollution, like a thin oil film that lowers the surface tension, can reduce the insect's buoyancy to the point where drowning rather than walking on water occurs.

Figure 5.6 shows examples of liquid menisci in a few other confined geometries along with the associated capillary pressures (P_{cap}):

- Figure 5.6(a) shows a liquid meniscus inside a narrow capillary tube or a cylindrically shaped pore with a circular cross section. The geometry is the same as for the capillary rise geometry shown in Figure 5.4(b): a hemispherical meniscus with radii of curvature $r_1 = r_2 = r$, where r equals the inner tube radius when the liquid wets the walls of the tube (contact angle $\theta = 0°$). From eq. (5.6), the capillary pressure $P_{cap} = 2\gamma/r$ acts to pull the liquid along the capillary.
- The left side of Figure 5.6(b) shows a liquid film coating the walls of a capillary tube or pore. For this geometry, $r_1 = r$, $r_2 = \infty$, so $P_{cap} = \gamma/r$. As the liquid film becomes thicker, for example, by increasing the vapor partial pressure to increase condensation, the liquid film becomes unstable, tending to jump the gap across the pore and form a meniscus as shown on the right side of Figure 5.6(b). The capillary pressure after the jump is now $P_{cap} = 2\gamma/r$, and the meniscus pulls in liquid that was coating the pore surfaces to leave a much thinner film.
- Figure 5.6(c) shows a spherical particle sitting on a surface covered with a thin liquid film. The liquid from the film wicks up around the sphere to make a meniscus with radii of curvature r_1 and r_2, as defined in Figure 5.6(c). (Note: for this geometry, the radii are opposite in sign: $r_1 < 0$ and $r_2 > 0$.) If r_1 and r_2 are comparable in size, the capillary pressure is given by eq. (5.6). For the more common situation where $|r_2| \gg |r_1|$, the capillary pressure can be approximated $P_{cap} \simeq \gamma/r_1$. This negative pressure pulls liquid into the meniscus until the capillary pressure equals the pressure in the film. The pressure in the film, which originates from the molecular interactions between the solid surface and the liquid, is called the *disjoining pressure* and is discussed in detail in a later chapter.

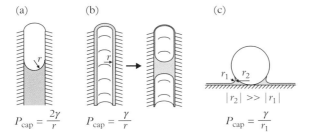

Figure 5.6 *(a) Liquid meniscus in a capillary or pore. (b) Liquid film coating the walls of a capillary tube (left). For thicker films, the liquid surfaces become unstable and condense to form a meniscus (right). (c) Spherical particle sitting on a surface with liquid film that wicks up around the particle to form a meniscus.*

5.2.2 The Kelvin equation and capillary condensation

One consequence of the pressure difference across a curved liquid surface is that the vapor pressure over a liquid surface depends on the degree of curvature of the surface. This leads to the phenomenon of *capillary condensation*, where vapors condense into small cracks and pores at vapor pressures significantly less than the saturation vapor pressure, as illustrated in Figure 5.7. Capillary condensation frequently has a major impact on the tribology of contacting surfaces since, even at fairly low vapor pressures, water and organic vapors will condense into the nanometer size gaps and crevices between contacting surfaces, greatly influencing their adhesion and lubrication characteristics (Binggeli and Mate 1994, 1995).

The expression describing how vapor pressure depends on curvature is called the *Kelvin equation* and is easily derived as follows.

For the liquid, the excess Gibbs free energy of a curved liquid surface relative to a flat liquid surface is given by

$$\Delta G = \int V_{\mathrm{m}} dP$$
$$= 2\gamma V_{\mathrm{m}}/r,$$

(5.9)

where V_{m} is the molar volume and using $P_{\mathrm{cap}} = 2\gamma/r$.

For the vapor, the excess Gibbs free energy of the vapor pressure P over the curved liquid surface relative to the saturation vapor pressure P_{s} over the flat surface is given by

$$\Delta G = RT \ \ln \left(\frac{P}{P_{\mathrm{s}}} \right).$$

(5.10)

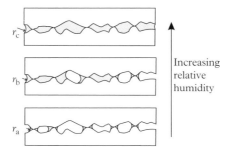

Figure 5.7 *Capillary condensation of water vapor between two hydrophilic rough contacting surfaces. At low humidity (bottom case), small water menisci (gray), with a small meniscus radius r_a given by the Kelvin equation, form around the contacting asperities. As the humidity increases, the meniscus radius increases, causing the menisci to grow and to fill in some of the spaces between the contacting asperities (middle case). At sufficiently high humidity (top case), the meniscus radius r_c is sufficiently large that the condensed water floods all the gaps between the contacting surfaces.*

Equating eqs. (5.9) and (5.10) yields the *Kelvin equation*:

$$2\left(\frac{1}{r_1} + \frac{1}{r_2}\right)^{-1} = r_k = \frac{2\gamma V_m}{RT \ln (P/P_s)}, \qquad (5.11)$$

where r_k is called the *Kelvin radius* and corresponds to the mean radius of curvature of the condensed meniscus. Equation (5.11) has been verified to hold for r_k as small as 4 nm for condensed organic liquids (Fisher and Israelachvili 1981) and for r_k as small as 5 nm for condensed water (Kohonen and Christenson 2000). The Kelvin equation shows that the saturation vapor pressure P over liquid surfaces that curve inward (like those shown in Figures 5.4(b), 5.5(a), and 5.6) is lower than the bulk saturation vapor pressure P_s; for those that are curved outward (Figures 5.4(a) and 5.5(b)), condensation occurs for vapor pressures higher than P_s. Physically, this can be understood very simply: at a surface with negative overall local curvature (like the meniscus shown in Figure 5.6(c)), the average density of atoms or molecules is larger than for a flat surface. This provides increased bonding and thus stability, allowing the liquid to form (i.e., for liquid–vapor equilibrium to be established) at a lower vapor pressure than for a flat surface.

5.2.2.1   *Example: capillary condensation of water in a nano-size pore*

Consider water condensing in a narrow pore, such as shown in Figure 5.6(a), to form a spherical meniscus that wets the walls of the pore and has a radius of curvature r. For water $\gamma = 73$ mN/m, $V_m = 18$ cm^3, and the Kelvin equation predicts $r = \infty$ at $P/P_s = 1$, $r = -10$ nm at $P/P_s = 0.9$, and $r = -1$ nm at $P/P_s = 0.34$. So, for relative humidities in the range 34–90%, condensation of water vapor only forms a meniscus in nanometer sized pores, that is, in those with diameters ranging from 2 to 20 nm.

5.2.2.2   *Example: capillary condensation of an organic vapor at a sphere on flat geometry*

On hydrophobic surfaces or at low relative humidities, capillary condensation of water becomes energetically unfavorable. Organic vapors, which always exist to a certain extent in the atmosphere, more readily condense onto surfaces due to their lower surface tensions. Here, we examine the condensation of the organic vapor, hexane, around the sphere-flat contact geometry (e.g., the particle on flat illustrated in Figure 5.6(c) or a single asperity contact in Figure 5.7). For hexane $\gamma = 18$ mN/m and $V_m = 130$ cm^3, and assuming $|r_2| \gg |r_1|$, then the Kelvin equation predicts $r = -11$ nm at $P/P_s = 0.9$ and $r = -1$ nm at $P/P_s = 0.3$. So, just as for water condensation in pores, hexane and other organic vapors condense only in nanometer sized crevices at contacting interfaces for partial pressures much below saturation.

5.3   Interfacial energy and work of adhesion

Before discussing solid surface energies and how they differ from liquid surface tensions, it is convenient to first introduce the concepts of interfacial energy and work of adhesion.

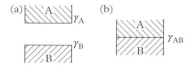

Figure 5.8 *(a) Two materials A and B when separated have surface energies γ_A and γ_B. (b) When joined together, they have an interfacial energy γ_{AB}.*

When two materials contact each other, like materials A and B shown in Figure 5.8, the interface between them has an energy γ_{AB} associated with it. To better understand this energy, we first consider the *work of adhesion* for this interface, labeled W_{AB}. The work of adhesion is defined to be the change in free energy or the thermodynamically reversible work done per unit area to separate the two contacting materials to infinity in vacuum. For separating at a hypothetical interface within a single component material (A = B), the interfacial energy is defined to be zero ($\gamma_{AA} = 0$); in this case, the work of adhesion W_{AA} is called the *work of cohesion*. Since the energy associated with creating a unit of surface area of material exposed to vacuum is γ_A or γ_B, the work of cohesion of material A can be related to its surface energy by

$$\gamma_A = \frac{1}{2} W_{AA}. \qquad (5.12)$$

For dissimilar materials contacting across an interface, the energy needed to create this interface can be divided into two steps: the energy first needed to create two surfaces A and B separated in vacuum (Figure 5.8(a)), then the energy gained when A and B are brought into contact (Figure 5.8(b)). Therefore the change in total free energy γ_{AB} to create this interface is

$$\gamma_{AB} = \frac{1}{2} W_{AA} + \frac{1}{2} W_{BB} - W_{AB} = \gamma_A + \gamma_B - W_{AB}, \qquad (5.13)$$

which is called the *Dupré equation*. The Dupré equation can be extended to the separation of materials A and B in a third medium, say a vapor or liquid C:

$$\gamma_{AB} = W_{AB} + W_{CC} - W_{AC} - W_{BC} = \gamma_{AC} + \gamma_{BC} - W_{ACB}, \qquad (5.14)$$

where W_{ACB} is the work of adhesion between materials A and B in medium C. Equations (5.13) and (5.14) are fairly general, where materials A and B can be either solids or liquids.

Next, we consider the special case where one is solid, which we label S, and the other is a liquid, which we label L. For the case of a solid–liquid interface in the presence of a vapor V, the Dupré equation (5.14) is written as

$$\gamma_{SL} = \gamma_{SV} + \gamma_{LV} - W_{SVL}. \qquad (5.15)$$

For future convenience, we will drop the V subscripts so that eq. (5.15) becomes

$$\gamma_{SL} = \gamma_S + \gamma_L - W_{SL}. \qquad (5.16)$$

The vapor is assumed still to be present and can have a major impact on these surface energies, as, over time, vapor molecules like water and hydrocarbons may absorb onto otherwise clean surfaces, lowering their surface energies.

5.4 Contact angles

When a liquid is placed on a solid surface with lower surface energy than the liquid, the liquid will bead up to form droplets with finite contact angle θ, as illustrated in Figure 5.9. In addition to the difference between the liquid and solid surface energies, the interfacial energy between the liquid and the solid is also an important factor as to determining how a liquid wets a surface. Consequently, studying contact angles of liquids on solids provides useful ways of studying surface energies and work of adhesions.

The relationships between a contact angle θ and the surface and interfacial energies and the work of adhesion are defined by the *Young's equation* and *Young–Dupré equation*. Let's start by considering the edge of a liquid droplet shown in Figure 5.9 advancing over a flat solid surface in the presence of its own vapor. When the droplet moves a distance *ds*, it covers, per unit length of the droplet edge, an area *ds* on the solid and creates an area *ds* cos θ of fresh liquid surface. As the droplet edge advances, the contact angle θ decreases, due to the finite volume of the droplet, and the edge stops advancing when θ is such that the surface energy lost at the solid–vapor interface equals the energy expended to create the new areas at the liquid–vapor and liquid–solid interfaces. This corresponds to an equilibrium contact angle θ satisfying *Young's equation*:

$$\gamma_{SL}\, ds + \gamma_L\, ds \cos\theta - \gamma_S\, ds = 0 \qquad (5.17)$$

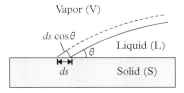

Figure 5.9 *Edge of liquid droplet on a solid surface, with a contact angle θ. An advance of the droplet by the distance ds creates an area ds cos θ of fresh liquid surface per unit length of the droplet edge.*

or

$$\cos\theta = \frac{\gamma_S - \gamma_{SL}}{\gamma_L}. \qquad (5.18)$$

So, if the liquid surface tension γ_L is higher than the difference between the solid surface energy and the interfacial energy ($\gamma_L > \gamma_S - \gamma_{SL}$), the liquid droplet has a finite contact angle; while, if $\gamma_L < \gamma_S - \gamma_{SL}$, the liquid spreads over the solid surface. Rearranging eq. (5.18) yields an expression for the solid–vapor surface energy:

$$\gamma_S = \gamma_L \cos\theta + \gamma_{SL}. \qquad (5.19)$$

Combining eq. (5.19) with the Dupré equation (5.15) results in the *Young–Dupré equation*:

$$W_{SL} = \gamma_L (1 + \cos\theta), \qquad (5.20)$$

which relates the contact angle to the work of adhesion between the solid and liquid. Therefore, if $0 < W_{SL} < 2\gamma_L$, the liquid has a finite contact angle on the solid, and the solid-liquid work of adhesion W_{SL} can be obtained by measuring the contact angle. On the other hand, if $W_{SL} \geq 2\gamma_L$, $\theta = 0$ and the liquid wets the solid.

An important consequence of eq. (5.19) is that γ_S can be determined from the equation if the three quantities γ_L, θ, and γ_{SL} are known. The liquid surface tension γ_L can be determined using the Wilhelmy plate method. The contact angle θ can be measured by placing a droplet on the solid surface as described below. The challenge comes in determining or estimating the solid-liquid interface energy γ_{SL}, which is discussed further in Section 5.5.2.

5.4.1 Types of wetting

As indicated by eq. (5.20), the work of adhesion between a solid and a liquid determines what type of wetting occurs. When a liquid droplet is placed on a surface, as shown in Figure 5.10, three types of wetting can be defined:

- *Wetting*—the contact angle θ between the liquid and solid is zero or so close to zero that the liquid spreads over the solid. Spreading occurs when the work of adhesion between the liquid and solid is equal to or greater than the liquid's internal work of cohesion.
- *Partial wetting*—the contact angle $\theta > 0$.
- *Non-wetting*—a subset of partial wetting where the contact angle $\theta > 90°$. While, thermodynamically, a contact angle θ of 90° has no particular significance, above this angle, liquids tend to ball up and easily roll off the surface if the contact angle hysteresis is low. Also, for $\theta > 90°$, $\gamma_{SL} > \gamma_S$ indicating that it takes more energy to create a solid–liquid interface than is gained from the loss of solid-vapor interface.

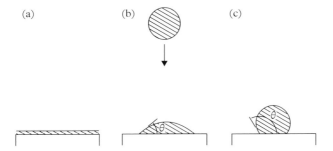

Figure 5.10 *When a droplet (above) is placed on a surface three types of wetting can occur: (a) wetting, (b) partial wetting, and (c) non-wetting—a subcategory of partial wetting where* θ > 90°.

Surfaces may be described as:

- *Hydrophilic*—the surface has an affinity for water and a low water contact angle.
- *Hydrophobic*—the surface has a repulsion for water and a high water contact angle.
- *Superhydrophobic*—a surface is considered superhydrophobic if the contact angle of water exceeds 150°.

5.4.1.1 *Superhydrophobicity*

Superhydrophobic surfaces have the remarkable property that, when a water droplet is dropped on it, the droplet rebounds almost elastically and easily rolls off the surface. This phenomenon is often dubbed the "lotus effect" as lotus leaves are covered with small hairs made of a wax-like material that lead to superhydrophobic behavior where rain droplets tend to bounce on impact and then easily roll off the leaves.

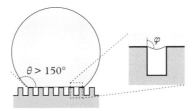

Figure 5.11 *Water droplet sitting on superhydrophobic surface with an apparent contact angle* θ > 150°. *A superhydrophobic surface consists of roughness features made of a hydrophobic material with a water contact angle* φ > 90°; *these rough features are close enough together to make it difficult for the water to penetrate the gaps between them.*

Superhydrophobic surfaces derive their extreme anti-wetting character by combining a hydrophobic material together with a high degree of roughness (Figure 5.11). In particular, superhydrophobic surfaces have roughness features that are close enough together so that water has difficulty penetrating into the gaps between the roughness features. As a consequence, a water droplet sitting on a superhydrophobic surface only makes contacts with the summits of the surface roughness, resulting in a very small contact area and a small value for total water–solid adhesion energy; this low adhesion energy (and correspondingly low adhesion hysteresis) is what lets water easily roll these surfaces.

Another example of superhydrophobic surfaces in nature is that of the legs of the water strider and other water insects (Feng et al. 2007). The water striders' legs are covered with specialized "hairpiles" or needle-shaped microsetae with elaborate nanogrooves. Since water has difficulty penetrating the nanogrooves and these hairpiles, this enables water striders to use the water's surface tension to provide a sufficient lifting force to walk on water, as shown in Figure 5.12.

5.4.2   *Measuring contact angles*

The most common way to measure a contact angle θ is to examine the edge of a droplet resting on a flat surface (Figure 5.13). Typically, a syringe is used to place the droplet on the surface, and a short working distance telescope is used to provide a magnified view of the droplet edge. The contact angle is measured either by visual examination through the telescope eyepiece with a goniometer scale or, more commonly nowadays, by capturing the image with a video camera and determining θ with computer software. Sometimes the droplet is enclosed so that the air around the droplet becomes saturated with the liquid's vapor; this ensures that the droplet is in equilibrium with the vapor, preventing evaporation, and giving an equilibrium value of γ_S.

For the droplet method in Figure 5.13, the simplest way of placing the droplet on the surface is to let it drop a small height from the syringe. Another common approach is to position the syringe with its end in contact with the droplet, as illustrated in Figure 5.14; this has the advantage that both the *advancing* and *receding* contact angles can be measured by, respectively, swelling and shrinking the droplet on the surface. When measuring contact angles, most researchers prefer to use the advancing angle rather than

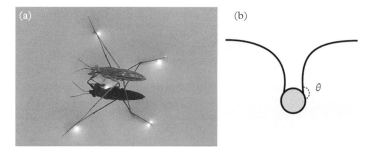

Figure 5.12 *(a) Water strider in a pond, courtesy of Schnobby. Creative Commons 3.0. https://creativecommons.org/licenses/by/3.0/legalcode. (b) Cross-sectional schematic of one of the strider's legs illustrating how the superhydrophobic nature of this leg allows it to bend the water surface without breaking it, so that the water's surface tension provides enough lifting force for the strider to walk on water. Adapted with permission from Feng et al. (2007). Copyright 2007 American Chemical Society.*

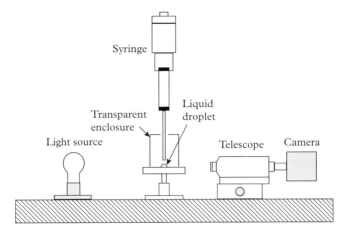

Figure 5.13 *Experimental setup for measuring contact angle.*

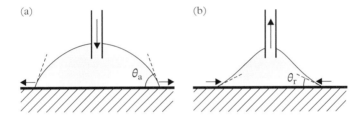

Figure 5.14 *A liquid droplet sitting on a flat surface with the end of a syringe immersed in the liquid. (a) Liquid entering the droplet via the syringe to generate an advancing contact angle θ_a. (b) Liquid being withdrawn via the syringe to generate a receding contact angle θ_r.*

the receding angle for determining solid surface energies, as the solid surface for this case is not yet been altered by having liquid flowing over it.

5.4.2.1 Contact angle hysteresis

If the droplet was able to be in true thermodynamic equilibrium with the solid surface, the advancing and receding contact angles should have the same value given by Young's equation (eq. (5.18)). Typically, however, the advancing angle is significantly larger than the receding angle, a phenomenon known as *contact angle hysteresis*. For most situations, the underlying causes for this hysteresis are not known, though possible mechanisms have been extensively debated in the scientific literature (Miller and Neogi 1985, Adamson and Gast 1997, Israelachvili 2011). Frequently, contact angle hysteresis is thought to originate from the chemical nature of the solid–liquid interface changing once the surface is covered with liquid. Examples of chemical changes include the slow reorientation of molecular groups in the presence of the liquid, the reaction of surface functional groups with chemical species dissolved in the liquid, the dissolution of contaminants or chemical species adsorbed on the surface into the liquid, and the deposition of chemical species dissolved in the liquid onto the surface.

One cause of contact angle hysteresis that is fairly well understood is surface roughness: for advancing liquids, slopes that provide large contact angles are favored, while, for receding liquids, slopes that provide small angles are favored (Miller and Neogi 1985, Chen et al. 1999, Israelachvili 2011). Another cause that is well understood is chemical heterogeneity: the advancing droplet edge preferentially seeks regions with low work of adhesion, while a receding edge preferentially remains in contact with regions of high work of adhesion.

A practical consequence of contact angle hysteresis is that droplets are pinned to the surface until a sufficient shear force is applied to the droplet to initiate motion. For example, rain droplets remain pinned on a car's windshield due to the contact angle hysteresis until a sufficient shear force is generated by the wind to move the droplet across the windshield (at most safe speeds, windshield wipers are still needed to remove droplets, particularly the smaller ones).

5.5   Surface energy of solids

5.5.1   Why solids are not like liquids

For solid surfaces, the concept of surface energy presents some difficulties. First, solid surfaces are rarely uniform or equilibrated to their lowest energy state. Second, when a force is applied to change a solid's shape then released, plastic deformation or viscous forces often prevent it from rapidly reverting back into its original shape.

For liquids, the concept of surface energy is a fairly straightforward: When the surface area of a liquid increases, a certain amount of work has to be expended to move liquid atoms or molecules from the bulk to their new equilibrium positions on the surface; the amount of work divided by the newly created surface area equals the surface energy. This work is reversible: when the liquid surface area is reduced back to its original area, the liquid does work on its surroundings equal to the energy gained from the loss of surface area; the energy comes from the liquid atoms or molecules moving from the higher energy surface sites to lower energy bulk configurations. For most liquids, the movement back and forth between the surface and bulk positions happens over a very short distance, a few atomic or molecular diameters; consequently, for low viscosity liquids, the expenditure or release of surface energy happens quickly. Therefore, the surface tension of liquids does not change as the surface area changes ($\partial\gamma/\partial A = 0$).

For solids, the situation is quite different, since, by definition, in a solid the atoms do not flow when a stress is applied unless the stress exceeds the threshold for plastic yield. So, if we apply a tensile stress to a solid bar, the atomic bonds stretch, the surface area becomes slightly larger. If we stay in the elastic regime, the atoms stay in their positions relative to each other, and the work expended stretching the atomic bonds is recovered when the atoms move back to their original positions when the stress is released (the work is reversible). If the applied stress exceeds the solid's plastic threshold, the solid atoms rearrange through the movement of dislocations and the sliding of slip planes as described in Section 3.1. During plastic flow, some of the energy applied to deforming the solid goes to moving atoms to and from the surface, changing the surface area. For solids, unlike liquids, the surface energy does not necessarily remain constant as we change surface area $(\partial \gamma / \partial A \neq 0)$.

In general, the energy needed to increase the surface area of a solid depends greatly on how it was created and the history of the surface. Consider trying to pull a perfect, defect-free crystalline solid apart into two pieces using tension. For a perfectly brittle crystal, it can be cleaved to produce two perfect crystal planes as surfaces. The energy per unit surface area required to do this is much higher if the crystal is ductile, since plastic flow will occur. For the cleaved crystal, the energy expended simply breaks the bonds across the cleaved crystal plane so that the surface atom positions terminate the bulk crystal structure. However, when plastic flow occurs (e.g., when a metal bar is stretched past its yield point, as illustrated in Figure 3.10), the movement of atoms takes place mainly through the movement of dislocations. These dislocations first have to be created, which requires energy to move the atoms out of their energetically favorable positions in the crystal lattice. In addition to the energy expended breaking bonds to generate surface atoms, energy has also to be expended to create and move the dislocations responsible for plastic flow. A surface formed by plastic deformation has a significant number of dislocations, steps, and other defects left over from the deformation process, which have higher formation energy than the surface atoms that simply terminate the bulk crystal structure. The fraction of the surface covered with these higher energy sites depends on the history of the deformation process, so the surface energy also depends on this history.

Once a solid surface has been created by cleavage, plastic deformation, or some other process, it is not necessarily stable and may relax to a lower energy configuration, if one exists, as follows. For cleaved surfaces, the termination of the bulk structure is not the lowest energy configuration for the surface atoms (Somorjai 1994). Instead, the spacing in the first and second layer is often found to be significantly smaller than the bulk spacing, as the loss of atomic bonding on one side of these surface atoms leads to stronger bonding and shorter bond lengths for the atoms on the other side. These same forces often further reconstruct by moving surface atoms laterally relative to the bulk positions. This is especially true for covalently bonded solids like semiconductors, as the dangling bonds created when the surface is cleaved cannot be easily satisfied except through a drastic rearrangement of the surface atoms.

For dislocations and defects, these high energy surface sites often anneal out at high temperatures through the diffusion of surface and bulk atoms. For two rough surfaces pressed together, the solid–solid contact area will slowly grow as atoms in defect sites diffuse to new low energy sites (discussed in Section 4.3 and illustrated in Figure 4.9).

The surface energy also changes due to the absorption of molecules from the environment. Solid surfaces bare of any adsorbed molecules occur only in very specialized situations, such as inside an ultra-high vacuum chamber or in outer space, where a low probability exists for molecules striking the surface. Most surfaces are exposed to gas or liquid environments where molecules are impinging all the time. Such atoms and molecules can adsorb to the surface if doing so lowers the local surface energy through the formation of a physical or chemical bond.

Experimental determination of a solid's surface energy is problematic. Since, as discussed above, the surface energy depends on the history of the surface and can potentially change over time, one cannot simply measure it once and log it in a table for future reference, as for liquid surfaces. Due to the possibility of plastic deformation, surface energy measurements involving stretching of the surface (which work well for liquids, e.g., the Wilhelmy plate method) are difficult to apply to solids, unless done at high enough temperatures to promote rapid atomic diffusion, or by stretching the surface area at an exceptionally slow rate to relieve the bulk stress.

Experimental values for solid surface energies should be viewed as more qualitative than quantitative, given all the difficulties in achieving equilibrated, non-hysteretic results, and since surface energies of solids can change over time.

5.5.2   Estimating solid surface energies from contact angles

As mentioned previously, by depositing liquid droplets onto a solid surface and measuring their contact angles, the surface energy of that solid can be determined using the Young's equation (5.19) if the three quantities γ_L, θ, and γ_{SL} are known. Contact angle measurements have become the most commonly employed technique for estimating the surface energy of materials. This is in large part due to the relative ease of the measuring contact angles, and it is now possible to obtain highly reproducible measurements of contact angles with modern equipment, automated to deposit axisymmetric droplets on surfaces with a slowly advancing contact angles (Kwok and Neumann 1999, Tavana and Neumann 2007).

As mentioned previously, the main difficulty to using the Young's equation to determine the solid surface energy from contact angles is that γ_{SL} is generally not known. Several methods have developed for estimating the value of γ_{SL}, which are discussed below.

5.5.2.1   *Equation of state method*

One common approach for determining γ_{SL} is to assume that an equation of state exists between $\gamma_L \cos \theta$, γ_L, and γ_S (Kwok and Neumann 1999), that is,

$$\gamma_L \cos \theta = f(\gamma_L, \gamma_S), \tag{5.21}$$

and that another equation of state exists relating γ_{SL} to γ_L and γ_S, that is, the Young's equation (5.18). This assumption of an equation of state (eq. (5.21)) derives from the fact that, when the contact angles of a series of liquids deposited on the same surface are plotted versus γ_L (Figure 5.15), they yield a smooth line, and that different solids yield offset parallel lines.

One particular method for determining γ_{SL} was developed by Girifalco and Good (1957). Using combining rules developed for molecular interactions (Israelachvili 2011),

Surface energy of solids **129**

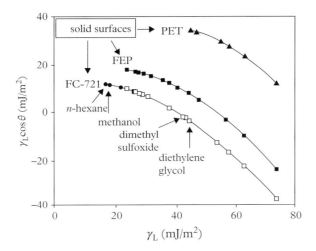

Figure 5.15 *Plots of $\gamma_L \cos\theta$ versus γ_L for a variety of liquids with different molecular properties and a range of surface tensions γ_L, the highest being that of water with $\gamma_{water} = 73$ mN/m. That the contact angles lie on a smooth line for a given material indicates that an equation of state exists between $\gamma_L \cos\theta$, γ_{SL}, γ_L, and γ_S. These measurements were done on the surfaces of three types of polymers: a fluorochemical polymer (Fluorad FC-721), fluorinated ethylene propylene (FEP), and poly(ethylene terephthalate) (PET). The smooth curves show the best fits to eq. (5.25), which yield the following values for γ_S and β: FC-721 ($\gamma_S = 11.78$ mJ/m^2 and $\beta = 0.000121$ (mJ/m^2)$^{-2}$), FEP ($\gamma_S = 17.85$ mJ/m^2 and $\beta = 0.000134$ (mJ/m^2)$^{-2}$), and PET ($\gamma_S = 35.22$ mJ/m^2 and $\beta = 0.000111$ (mJ/m^2)$^{-2}$). Reproduced with permission from Tavana and Neumann (2007). Copyright 2007, Elsevier.*

the solid–liquid work of adhesion W_{SL} is estimated in terms of γ_L and γ_S as a geometric mean of γ_L and γ_S multiplied by a factor Φ:

$$W_{SL} = 2\Phi\sqrt{\gamma_L\gamma_S}, \tag{5.22}$$

which leads to the expression

$$\gamma_{SL} = \gamma_L + \gamma_S - 2\Phi\sqrt{\gamma_L\gamma_S}. \tag{5.23}$$

Various forms have been proposed for Φ, and the one that we will highlight here is $\Phi = \exp(-\beta(\gamma_S - \gamma_L)^2)$, which was proposed by Li and Neumann (1990) and which has the characteristics that $\Phi \sim 1$ when $\gamma_S \sim \gamma_L$ and that Φ decreases rapidly as the difference between γ_S and γ_L increases. Incorporating this into eq. (5.23) leads to the following expression for equation of state for interfacial tension:

$$\gamma_{SL} = \gamma_L + \gamma_S - 2\sqrt{\gamma_L\gamma_S}e^{-\beta(\gamma_L - \gamma_S)^2}. \tag{5.24}$$

Combining eq. (5.24) with Young's equation (5.18) yields

$$\cos\theta = -1 + 2\sqrt{\frac{\gamma_S}{\gamma_L}}e^{-\beta(\gamma_L-\gamma_S)^2}. \tag{5.25}$$

Figure 5.15 shows that this equation of state works well in predicting contact angles: in this figure the smooth lines show the best fits of the contact angle data eq. (5.25), where γ_S and β are treated as the adjustable parameters for the fit.

5.5.2.2 Surface tension components method

An alternative approach to the equation of state method for estimating interfacial energies, developed by Fowkes (1964), involves first assuming that the surface energy or surface tension can be expressed as a sum of components corresponding to specific types of molecular interactions:

$$\gamma = \gamma^d + \gamma^p + \gamma^i + \dots, \tag{5.26}$$

where d, p, and i superscripts stand for, respectively, dispersion, polar, and induction components, and the dots represents other less significant terms. Since dispersion and polar terms typically dominate over the other surface energy terms, the work of adhesion can be expressed (using eq. (5.22) with $\Phi = 1$) as

$$W_{SL} \simeq 2\sqrt{\gamma_L^d\gamma_S^d} + 2\sqrt{\gamma_L^p\gamma_S^p}. \tag{5.27}$$

The procedure for using the Fowkes analysis to determine a solid's surface energy γ_S is to first determine its dispersive component γ_S^d (Waltman et al. 2003). This is done by using a liquid, such as an alkane, that interacts with other materials predominately via the dispersive van der Waals forces ($\gamma_{alkane} \simeq \gamma_{alkane}^d$) to measure the contact angle θ_{alkane} of that liquid against the solid in question. This contact angle and surface tension are then used in a combination of eqs. (5.20) and (5.27) to yield γ_S^d:

$$\gamma_S^d = \frac{\gamma_{alkane}^d(1 + \cos\theta_{alkane})^2}{4}. \tag{5.28}$$

Next, the contact angle is measured for a liquid with known values of γ_L^d and γ_L^p where γ_L^p is a substantial fraction of the total liquid surface tension, such as for water where $\gamma_L^d = 20$ mN/m and $\gamma_L^p = 53$ mN/m. From the contact angle for this polar liquid, γ_S^p is first determined by combining eqs. (5.20) and (5.27), then γ_S is determined by substituting into the Young-Dupré equation the following expression for the solid-liquid interfacial energy (a combination of eqs. (5.20) and (5.27)):

$$\gamma_{SL} = \gamma_S + \gamma_L - 2\sqrt{\gamma_S^d\gamma_L^d} - 2\sqrt{\gamma_S^p\gamma_L^p} \tag{5.29}$$

and solving for γ_S.

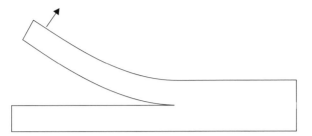

Figure 5.16 *Cleaving a solid crystal to measure the work of adhesion.*

5.5.3　Other methods for estimating solid surface energies

Since using contact angles to estimate solid surface energies is limited to those materials with surface energies less than the more readily available liquids (with the surface tension of water $\gamma_{water} = 73$ mN/m representing the high end of this range), other methods have to be used for higher surface energy surfaces. A few of the other methods developed for experimentally determining the surface energies of solids are (Bikerman 1970, Overbeek 1974, Adamson and Gast 1997):

- *Cleavage of brittle crystals*—used for brittle materials. The value of γ is determined using eq. (5.12) by measuring the work done cleaving a single crystal to expose the two crystal faces on opposite sides of the cleavage plane. One way of measuring the work of adhesion is illustrated in Figure 5.16. A more sophisticated way of doing this is to use the surface force apparatus (SFA), discussed in Chapter 8.

  o For example, the surface free energy of muscovite mica surfaces has been measured by cleavage and by SFA to be a few thousand ergs/cm^2 in vacuum (Obreimoff 1930), 413 ergs/cm^2 in dry nitrogen, and 185 ergs/cm^2 in laboratory air (Frantz et al. 1997, Frantz and Salmeron 1998), which shows the effect that airborne contaminations, such as water (Christenson 1993), have on lowering surface energies. Further physical reasons for why the surface energies of mica measured by cleavage vary over such a wide range have been reviewed by Christenson and Thomson (2016).

- *Calorimetry*—used for powders. When a crystal dissolves, the heat of solution is determined from the temperature rise. When a powder of small crystallites is dissolved, an additional heat is given off from the loss of surface energy. The surface energy is determined by dividing the extra heat by the powder surface area, which has to be determined by a separate method.

5.6　Adhesion hysteresis

When we introduced the concept of work of adhesion in Section 5.3, a basic assumption was that the contact process occurs under ideal equilibrium and thermodynamically reversible conditions, meaning that the work done to bring two surfaces together equals the work needed to separate them. When real surfaces are brought together, however, equilibrium conditions are rarely achieved initially; instead, the atoms and molecules at the contacting interfaces slowly diffuse and reorient themselves in an effort to eventually achieve an equilibrium configuration. As a consequence, the work expended to separate two surfaces under most realistic conditions is greater than the work gained when bringing them together.

According to Israelachvili (Chen et al. 1991, Israelachvili 2011), the underlying mechanisms for adhesion hysteresis can be divided into two categories: *mechanical hysteresis* that arises from mechanical instabilities and *chemical hysteresis* that arises from either the movement of atoms and molecules at the contacting interfaces or the formation and breaking of chemical bonds at the interface.

5.6.1   Mechanical adhesion hysteresis

As pointed out by Israelachvili (Chen et al. 1991), mechanical instabilities during the measurement of adhesion can be a major source of hysteresis in experimental values for the work of adhesion. Figure 5.17 illustrates a typical experiment where the work of adhesion is determined by integrating the force F acting perpendicularly to the surfaces over the separation distance D:

$$W = \int FdD, \qquad (5.30)$$

where the integral is over the path that the atoms take during approach and separation. Since the force F is never measured at the surfaces themselves (S in Figure 5.17), but rather some distance away (S′), the force is transmitted through the elastic elements (springs, connectors, etc.) in between points S and S′, with effective spring constant K_s (Figure 5.17 (middle, left)). As the two surfaces approach along the force vs. distance curve shown in Figure 5.17 (bottom), at some separation distance D_A, the surfaces suddenly jump into contact when the gradient of the attractive force is greater than the spring constant K_s. As one tries to separate the surfaces, one has to apply a force large enough to overcome the attractive adhesive force. Again the elasticity creates an instability on separation at D_s, and the surfaces jump apart out from D_s to D_R.

From the force vs. distance curve in Figure 5.17 (bottom), we see that the surfaces do not follow the same path when they separate as when they approach: the area enclosed in the cycle line through D_R–D'–D_A–D_s–D_R corresponds to extra work that needs to be performed to complete the cycle and to separate the surfaces, that is, the adhesion hysteresis ΔW.

This type of mechanical hysteresis occurs not only for systems that have a mechanically compliant spring element for measuring adhesion force, but can also occur due to the elasticity of the material itself. All materials have some degree of elasticity, and so some degree of mechanical adhesion hysteresis can occur for many contacts. Indeed, it can occur even at the atomic level, where the sudden separation of two surfaces

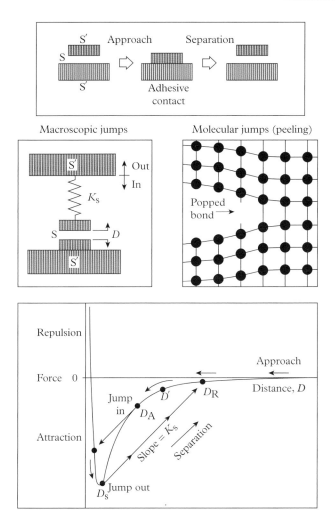

Figure 5.17 *Origin of mechanical adhesion hysteresis during approach and separation for two solid surfaces. The force is not measured at the surfaces S, but rather at some distance back, at S′ with the elasticity of the elements in between providing an effective spring constant K_s (top and middle, left). Bonds breaking at the atomic and molecular level during separation (middle, right). Force versus separation distance during approach and separation of two attracting surfaces (bottom). On approach, the rate of increase in the attractive force overwhelms the spring constant at point $D = D_A$, and the two surfaces jump into contact. A similar instability occurs at $D = D_S$ when the surfaces are pulled apart. Reprinted with permission from Chen et al. (1991). Copyright 1991 American Chemical Society.*

involves the spontaneous breaking of bonds, as illustrated in Figure 5.17 (middle, right). The criterion for whether this atomic- and microscale adhesion results macroscopic observable adhesion and adhesion hysteresis is elaborated on in Section 6.2.4.

Other situations that promote mechanical adhesion hysteresis include:

- *Chemically homogeneous surfaces supported by elastic materials.* This type of adhesion hysteresis contributes to "rolling" friction and elastoplastic adhesive contacts during loading–unloading cycles (Bowden and Tabor 1967, Greenwood and Johnson 1981, Maugis 1985, Michel and Shanahan 1990).

- *Surface roughness.* As explained in Section 3.4, as the roughness increases, the real area of contact decreases. Also, when the real area of is a small fraction of the apparent area (as it is for rough surfaces), the effective stiffness of the contact is much less than the composite modulus. This leads to rougher surfaces exhibiting a higher degree of adhesion hysteresis than smooth surface (Wei et al. 2010).

5.6.2   Chemical adhesion hysteresis

Chemical induced adhesion hysteresis comes from the movement of atoms and molecules at contacting interfaces. This can occurring a variety of ways: diffusion of along the interface or across the interface, interdigitation of polymers across the interface, reorientation of molecules at the interface, the exchange of chemical species across the interface, and the formation of bonds across the interface. What distinguishes chemical hysteresis from mechanical is that the chemical nature of the surfaces is different after separation from during contact.

Figure 5.18 illustrates two types of potential movement of atoms and molecules at a contacting asperity between two rough surfaces: (1) solid atoms diffusing under the influence of the atomic level attractive and repulsive forces, and (2) capillary condensation of water or organic vapors (as discussed in Section 5.2.2). By the time the surfaces are pulled apart, the atoms and molecules have diffused or condensed to energetically more favorable sites, requiring more work to disrupt the new bonding geometries formed during contact, increasing the work of adhesion. In Section 4.3, we discussed how this same process leads to the friction force increasing with the contact time between two surfaces (aging). A third mechanism that can increase adhesion, not shown, is the formation of chemical bonds across the interface between atoms that are originally unbonded (Vigil et al. 1994).

For surfaces covered with polymer or surfactant films, interdigitation of molecules on the opposing surfaces is proposed to be a major cause of the adhesion hysteresis observed for these surfaces, even if they are nominally smooth and chemically homogeneous. For example, Israelachvili and coworkers have attributed adhesion hysteresis between thin layers of vertically-oriented surfactant molecules on atomically smooth mica surfaces brought together in a SFA to interdigitation of the molecules (Chen et al. 1991, Yoshizawa et al. 1993a, 1993b, Israelachvili et al. 1994, Yoshizawa and Israelachvili 1994).

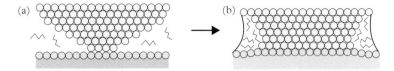

Figure 5.18 *When rough surfaces contact each other, atoms and molecules diffuse to grow the contact area and to form menisci around the contacting asperities, leading to higher adhesive forces and thus to adhesion hysteresis. (a) An initially sharp asperity right before it makes contact. The squiggly lines represent organic molecules in vapor form. (b) The asperity after a short period of contact, after which the atoms within the initially sharp asperity have moved under the influence of the attractive and repulsive atomic level forces to new equilibrium positions. Also, the vapor organic molecules have now undergone capillary condensation to form a meniscus around the contacting asperity.*

5.7 PROBLEMS

1. Prove the Young–Laplace equation (eq. (5.6)) by referring to force equilibrium shown in the free body diagram in Figure 5.19. The diagram depicts an infinitesimal piece of a curved liquid surface, where $\hat{n}$ is the unit vector perpendicular to the surface. Note that the free body diagram does not show the force due to the capillary pressure. Assume that a capillary pressure P_{cap} acts on the concave inside of the curved membrane (i.e., below the liquid surface as drawn), which is experiencing a surface tension γ that produces the forces F and curvature radius R shown.

2. Steel is denser than water, yet a steel paper clip will remain at the surface of water if placed (carefully) on top, as shown in Figure 5.20, due to the paper clip surface being somewhat hydrophobic.

    (a) To explain why this happens, prove explicitly that this is feasible by using equations that compare the relevant forces and by assuming that the contact angle between water and steel is greater than 90°. Consider a single cross-section of a 1.0 mm diameter steel wire for this calculation. To simplify your proof, ignore

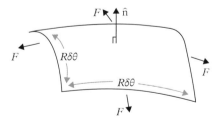

Figure 5.19 *An infinitesimal square section of a curved liquid surface with radius of curvature R that subtends an infinitesimal angle δθ. The surface tension produces as a force F acting on each edge with length Rδθ.*

Figure 5.20 *A steel paper clip floating on water.*

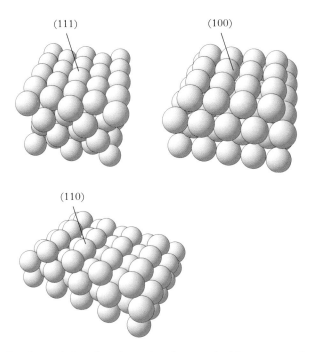

Figure 5.21 *The three lowest index plane faces of an fcc crystal with a one atom basis.*

buoyancy and just consider gravity acting on the paper clip; also assume that the contact line is below the horizontal mid-plane of the paper clip.

(b) What will happen if a drop of soap is deposited on the water? Explain physically why this happens in a few sentences, with specific reference to relevant surface and/or interfacial energies. Use additional diagrams as necessary.

3. Consider the (111), (100), and (110) surfaces of a face-centered cubic (fcc) crystal with a one atom basis and with no reconstruction, as shown in Figure 5.21. The coordination number (number of nearest neighbors) of the topmost atoms is 9, 8, and 7, respectively, for these three surfaces. Considering the favorable nature of surface close packing, which of these three fcc low index planes do you expect to have the lowest surface energy and why? *Hint*: Be sure to not just consider the role of the coordination number of the surface atoms, but also the atomic surface density.

4. (a) Which do you expect is larger: the work of adhesion between FEP (a Teflon^{TM}-like material) and water, or the work of adhesion between PET and water? Explain why. Then, check your answer by using the information provided in Figure 5.15 and the associated equations.

(b) The contact angle of a certain liquid is measured on FEP and on PET. It wets PET with a contact angle θ ($< 90°$), and happens to form a contact angle on FEP of $180° - \theta$. Estimate the surface tension of this liquid, and then estimate θ. Note: the exact value can be solved from eq. (5.25) but is cumbersome to do. However, an estimate can be garnered more easily by considering the form of eq. (5.25) and then using Figure 5.15 graphically.

5.8 REFERENCES

Adamson, A. W. and A. P. Gast (1997). *Physical chemistry of surfaces* (6th ed.). New York: John Wiley & Sons.

Bikerman, J. J. (1970). *Physical surfaces*. New York: Academic Press.

Binggeli, M. and C. M. Mate (1994). "Influence of capillary condensation of water on nanotribology studied by force microscopy." *Applied Physics Letters* **65**(4): 415–17.

Binggeli, M. and C. M. Mate (1995). "Influence of water vapor on nanotribology studied by friction force microscopy." *Journal of Vacuum Science & Technology B* **13**(3): 1312–15.

Bowden, F. P. and D. Tabor (1967). *Friction and lubrication* (2nd ed.). London: Methuen.

Chen, W., A. Y. Fadeev, M. C. Hsieh, D. Oner, J. Youngblood and T. J. McCarthy (1999). "Ultrahydrophobic and ultralyophobic surfaces: some comments and examples." *Langmuir* **15**(10): 3395–9.

Chen, Y. L., C. A. Helm and J. N. Israelachvili (1991). "Molecular mechanisms associated with adhesion and contact-angle hysteresis of monolayer surfaces." *Journal of Physical Chemistry* **95**(26): 10736–47.

Christenson, H. K. (1993). "Adhesion and surface-energy of mica in air and water." *Journal of Physical Chemistry* **97**(46): 12034–41.

Christenson, H. K. and N. H. Thomson (2016). "The nature of the air-cleaved mica surface." *Surface Science Reports* **71**(2): 367–90.

Fadeev, A. Y. and T. J. McCarthy (2000). "Self-assembly is not the only reaction possible between alkyltrichlorosilanes and surfaces: monomolecular and oligomeric covalently attached layers of dichloro- and trichloroalkylsilanes on silicon." *Langmuir* **16**(18): 7268–74.

Feng, X.-Q., X. Gao, Z. Wu, L. Jiang and Q.-S. Zheng (2007). "Superior water repellency of water strider legs with hierarchical structures: experiments and analysis." *Langmuir* **23**(9): 4892–6.

Fisher, L. R. and J. N. Israelachvili (1981). "Experimental studies on the applicability of the Kelvin equation to highly curved concave menisci." *Journal of Colloid and Interface Science* **80**(2): 528–41.

Fowkes, F. M. (1964). "Attractive forces at interfaces." *Industrial & Engineering Chemistry* **56**(12): 40–52.

Frantz, P. and M. Salmeron (1998). "Preparation of mica surfaces for enhanced resolution and cleanliness in the surface forces apparatus." *Tribology Letters* **5**(2–3): 151–3.

Frantz, P., A. Artsyukhovich, R. W. Carpick and M. Salmeron (1997). "Use of capacitance to measure surface forces. 2. Application to the study of contact mechanics." *Langmuir* **13**(22): 5957–61.

Girifalco, L. A. and R. J. Good (1957). "A theory for the estimation of surface and interfacial energies." *Journal of Physical Chemistry* **61**: 904–9.

Greenwood, J. A. and K. L. Johnson (1981). "The mechanics of adhesion of viscoelastic solids." *Philosophical Magazine A: Physics of Condensed Matter Structure Defects and Mechanical Properties* **43**(3): 697–711.

Israelachvili, J. N. (2011). *Intermolecular and surface forces* (3rd ed.). Burlington, MA: Academic Press.

Israelachvili, J. N., Y. L. Chen and H. Yoshizawa (1994). "Relationship between adhesion and friction forces." *Journal of Adhesion Science and Technology* **8**(11): 1231–49.

Kohonen, M. M. and H. K. Christenson (2000). "Capillary condensation of water between rinsed mica surfaces." *Langmuir* **16**(18): 7285–8.

Kwok, D. Y. and A. W. Neumann (1999). "Contact angle measurement and contact angle interpretation." *Advances in Colloid and Interface Science* **81**(3): 167–249.

Li, D. and A. Neumann (1990). "A reformulation of the equation of state for interfacial tensions." *Journal of Colloid and Interface Science* **137**(1): 304–7.

Maugis, D. (1985). "Subcritical crack-growth, surface-energy, fracture-toughness, stick slip and embrittlement." *Journal of Materials Science* **20**(9): 3041–73.

Michel, F. and M. E. R. Shanahan (1990). "Kinetics of the JKR experiment." *Comptes Rendus de l'Academie des Sciences Series II* **310**(1): 17–20.

Miller, C. A. and P. Neogi (1985). *Interfacial phenomena: equilibrium and dynamic effects.* New York: Marcel Dekker.

Obreimoff, J. W. (1930). "The splitting of strength of mica." *Proceeding of the Royal Society of London, Series A* **127**: 290–7.

Overbeek, J. T. G. (1974). *Colloid and surface chemistry: a self-study course.* Cambridge: Department of Chemical Engineering and Center for Advanced Engineering Study, University of Cambridge.

Somorjai, G. A. (1994). *Introduction to surface chemistry and catalysis.* New York: John Wiley & Sons.

Tavana, H. and A. W. Neumann (2007). "Recent progress in the determination of solid surface tensions from contact angles." *Advances in Colloid and Interface Science* **132**(1): 1–32.

Vigil, G., Z. Xu, S. Steinberg and J. Israelachvili (1994). "Interactions of silica surfaces." *Journal of Colloid and Interface Science* **165**(2): 367–85.

Waltman, R. J., D. J. Pocker, H. Deng, N. Kobayashi, Y. Fujii, T. Akada, K. Hirasawa and G. W. Tyndall (2003). "Investigation of a new cyclotriphosphazene-terminated perfluoropolyether lubricant. Properties and interactions with a carbon surface." *Chemistry of Materials* **15**(12): 2362–75.

Wei, Z., M.-F. He and Y.-P. Zhao (2010). "The effects of roughness on adhesion hysteresis." *Journal of Adhesion Science and Technology* **24**(6): 1045–54.

Yoshizawa, H. and J. Israelachvili (1994). "Relation between adhesion and friction forces across thin-films." *Thin Solid Films* **246**(1–2): 71–6.

Yoshizawa, H., Y. L. Chen and J. Israelachvili (1993a). "Fundamental mechanisms of interfacial friction. 1. Relation between adhesion and friction." *Journal of Physical Chemistry* **97**(16): 4128–40.

Yoshizawa, H., Y. L. Chen and J. Israelachvili (1993b). "Recent advances in molecular-level understanding of adhesion, friction and lubrication." *Wear* **168**(1–2): 161–6.

6

Surface Forces Derived from Surface Energies

While surface energy is a valuable concept for understanding surfaces, in tribology it is often more important to know the forces between contacting surfaces rather than the interaction energy. For macroscopic objects, forces like gravity tend to be more significant than the attractive forces between contacting surfaces arising from molecular level interactions. At the micro- or nanoscale, however, the situation becomes reversed, as the attractive surface forces scale with the radius of curvature, while gravity and inertia forces scale with the object's volume. So, these attractive surface forces, which may seem inconsequential in our everyday experience, dominate phenomena occurring at the small scale. This has enabled evolution to endow insects and other small creatures with the ability to adhere and climb up vertical surfaces with ease, while much larger humans can only wish that they had this ability (though researchers are hoping to eventually develop a "spiderman suit" to enable humans to climb walls; Pugno 2007).

Since force and energy are often related, it is natural to expect that forces at surfaces will be related to surface energies. In the last chapter, we discussed briefly several examples of forces related to surface energy and surface tension:

- Wilhelmy plate technique (Figure 5.2);
- capillary force (Figure 5.5);
- work of adhesion (Figure 5.8).

In this chapter, we go into greater detail on these surface forces that are derived from surface energies.

6.1 The Derjaguin approximation

The *Derjaguin approximation* provides an elegant way to relate the force between surfaces to the interaction energy. It can also be used to relate the forces encountered in one contact geometry to other types of contact geometries. In the Derjaguin approximation,

Tribology on the Small Scale: A Modern Textbook on Friction, Lubrication and Wear. Second edition. C. Mathew Mate and Robert W. Carpick. © Oxford University Press 2019. Published in 2019 by Oxford University Press.
DOI: 10.1093/oso/ 9780199609802.001.0001

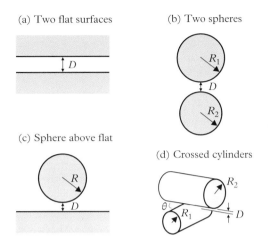

Figure 6.1 *The Derjaguin approximation relates the interaction energy per unit area W(D) between two parallel flat surfaces separated by a distance D (as illustrated in (a)) to the force law F(D) and between two curved surfaces, such as between (b) two spheres, (c) a sphere above a flat, and (d) two crossed cylinders. In order for the Derjaguin approximation to be valid, the separation distance D needs to be much smaller than the radii of curvature, that is, D << R, R_1, and R_2.*

the force law $F(D)$ between two curved surfaces is expressed in terms of the interaction energy per unit area $W(D)$ of two bodies with flat surfaces oriented parallel to each other. Figure 6.1 illustrates the geometry of two flat surfaces as well as the geometries of various types of curved surfaces.

The approximation made in the Derjaguin approximation is that the separation distance D and the range of the force interaction are both much less than the radii of curvature of the two surfaces. Within this approximation, it is straightforward to show (Derjaguin 1934, Israelachvili 2011: pp. 215–16) for two spheres with radii R_1 and R_2 (Figure 6.1(b)) and for any interaction law that is a function of separation distance that:

$$F(D)_{\text{sphere/sphere}} = 2\pi \left(\frac{R_1 R_2}{R_1 + R_2} \right) W(D). \tag{6.1}$$

So, if we know the interaction energy per unit area W in a *planar* geometry with particular separation D, either from a calculation or a measurement of energy, we can determine the force F between two *spheres* with radii R_1 and R_2 with the same separation distance D.

A special case of eq. (6.1) is the sphere-on-flat geometry ($R_1 = R$, $R_2 = \infty$; Figure 6.1(c)), where $F(D)$ becomes

$$F(D)_{\text{sphere–flat}} = 2\pi R W(D). \tag{6.2}$$

Similarly, for two cylinders with radii R_1 and R_2 and crossed at an angle θ to each other (Figure 6.1(d)):

$$F(D)_{\text{cylinder−cylinder}} = 2\pi \sqrt{R_1 R_2}\, W(D)/\sin\theta. \qquad (6.3)$$

Equation (6.3) is frequently used with the surface force apparatus (SFA) technique to determine, as a function of their separation distance, the interfacial energy $W(D)$ between two surfaces from the measured force $F(D)$. In a typical SFA experiment, two cylindrically shaped surfaces are crossed at a right angle ($\theta = 90°$) and $R_1 \sim R_2 \sim R$, so, in this situation, eq. (6.3) reduces to eq. (6.2), indicating that the SFA cross cylinder geometry is equivalent to the sphere-on-flat geometry.

The Derjaguin approximation greatly facilitates the comparison of theories of the interaction energy between parallel surfaces with experiments involving interaction forces between curved surfaces. While a geometry with two flat, parallel surfaces is convenient for theoretical analysis, such a geometry is generally impractical from an experimental point of view as it is extremely difficult to align two surfaces so that they are perfectly parallel. Instead, it is far easier to measure forces between curved surfaces, such as between two spheres, a sphere and a flat, or crossed cylinders, all of which generate a well-defined *point of contact*, and then to use the Derjaguin approximation to relate the experimental forces to the theoretical energies for parallel surfaces.

6.1.1   Derivation of the Derjaguin approximation

Here we present a simple derivation for the Derjaguin approximation of eq. (6.1) for two spheres with radii R_1 and R_2 and separated by distance D, as illustrated in Figure 6.2 (Israelachvili 2011: pp. 215–16). If we divide the surfaces of the spheres up into pairs of circular rings that each have radius r, width dr, and area $2\pi r dr$ and that are separated by a distance $Z = D + z_1 + z_2$ (Figure 6.2(b)), then the net force acting between the spheres is obtained by integrating the force per unit area $f(Z)$ acting along the z-direction between each pair of rings over the surfaces of the two spheres:

$$F(D) = \int_{Z=D}^{Z=\infty} 2\pi r\, dr\, f(Z), \qquad (6.4)$$

where an approximation has been made of setting the upper integration limit to infinity, which is valid since $f(Z) \sim 0$ for $Z > R_1, R_2$ and $D << R_1, R_2$. Applying the Pythagorean theorem to the triangle defined by r, R_1, and $R_1 − z_1$ in Figure 6.2(a),

$$R_1^2 = (R_1 − z_1)^2 + r^2, \qquad (6.5)$$

then expanding and taking advantage of $z_1 << R_1$, one can obtain $r^2 \approx 2R_1 z_1$ (and similarly $r^2 \approx 2R_2 z_2$). This leads to

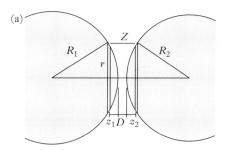

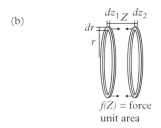

Figure 6.2 *(a) The geometry considered for the derivation of the Derjaguin approximation, consisting of two spheres of identical material, with radii R_1 and R_2, and separated by a distance D. (b) An isolated three-dimensional view of the integration element used to determine the net force acting between the spheres, which consists of a pair of circular rings separated by a distance $Z = z_1 + D + z_2$.*

$$Z = D + z_1 + z_2 \approx D + \frac{r^2}{2}\left(\frac{1}{R_1} + \frac{1}{R_2}\right). \tag{6.6}$$

and

$$dZ = \left(\frac{1}{R_1} + \frac{1}{R_2}\right) r\, dr. \tag{6.7}$$

Upon substitution into eq. (6.4), the net force becomes eq. (6.1):

$$F(D) = \int_D^\infty 2\pi \left(\frac{R_1 R_2}{R_1 + R_2}\right) f(Z)\, dZ = 2\pi \left(\frac{R_1 R_2}{R_1 + R_2}\right) W(D), \tag{6.8}$$

where we have used the following relationship between the interaction energy $W(D)$ per unit area between two bodies with planar surfaces separated by a distance D and the force $f(Z)\, dZ$ acting per unit volume along the z-direction between the bodies:

$$W(D) = \int_D^\infty f(Z)\, dZ. \tag{6.9}$$

6.2   Dry environment

6.2.1   Force between a sphere and a flat

As an example of the use of the Derjaguin approximation to determine a force from surface energy, let's consider a rigid sphere of material A sitting on a rigid flat surface of material B, as shown in Figure 6.3(a). For this case, the sphere and flat are in a dry environment (no liquids are present to form a meniscus around the contact point). If we assume that the surface roughness is negligible (i.e., $D \sim$ an atomic distance), then $W(D)$ becomes the work of adhesion W_{AB}. Using the Derjaguin approximation as expressed in eq. (6.2), the adhesion force L_{adh} acting between the sphere and flat is related to the work of adhesion by

$$L_{adh} = 2\pi RW_{AB}. \tag{6.10}$$

The symbol L_{adh} is used here for the adhesion force as L_{adh} acts like a loading force pushing the sphere toward the flat. If the sphere and flat have the same surface energy ($\gamma_S = \gamma_A = \gamma_B$), then $W_{AB} = 2\gamma_S$ and the adhesion force becomes

$$L_{adh} = 4\pi R\gamma_S. \tag{6.11}$$

6.2.1.1   *Example: adhesion force between two polystyrene spheres*

Let's calculate the adhesion force between two 1 μm diameter polystyrene spheres contacting each other. Using the Derjaguin approximation, eq. (6.1) the force between two spheres becomes, using $R_1 = R_2 = R$ and $W = 2\gamma_{polystyrene}$,

$$L_{adh} = 2\pi R\gamma_{polystyrene}. \tag{6.12}$$

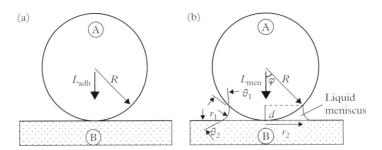

Figure 6.3 *Rigid sphere of material A sitting on a rigid flat of material B. (a) With dry surfaces. (b) With a liquid meniscus.*

so the sphere–sphere adhesion is half that for a sphere-on-flat. With $\gamma_{polystyrene} = 33$ mN/m,

$$L_{adh} = 2\pi \, (0.5 \ \mu m) \times (33 \ mN/m)$$
$$= 0.1 \ \mu N. \tag{6.13}$$

6.2.1.2   *Example: adhesion force between a polystyrene sphere and a PTFE flat*

Next, let's examine the adhesion force for a sphere-on-flat when they are made of dissimilar materials, by considering the special case of a 1 µm diameter polystyrene sphere sticking to a surface of polytetrafluoroethylene (more commonly called PTFE or Teflon), where $\gamma_{PTFE} = 18$ mN/m. To determine the work of adhesion between the dissimilar materials, we will use eq. (5.22) with $\Phi = 1$:

$$W_{PTFE-polystyrene} = 2\sqrt{\gamma_{PTFE}\gamma_{polystyrene}}. \tag{6.14}$$

The adhesion force for this sphere-on-flat situation is

$$L_{adh} = 4\pi R\sqrt{\gamma_{PTFE}\gamma_{polystyrene}}$$
$$= 4\pi \, (0.5 \ \mu m) \sqrt{(18 \ mN/m)(33 \ mN/m)}$$
$$= 0.15\mu N. \tag{6.15}$$

Even though PTFE or Teflon is often called "non-stick" because of its very low surface energy of 18 mN/m compared to other solid surfaces, the adhesive force is still not negligible. For this example of a 1 µm diameter spherical particle, the adhesive force is much greater than the force from gravity ($\sim 5 \times 10^{-15}$ N) acting on the sphere.

So how big does the polystyrene particle have to be for the force of gravity to equal the adhesive force? In this case,

$$\frac{4}{3}\pi g\rho R^3 = 4\pi R\gamma_S$$
$$R = \sqrt{\frac{3\gamma_S}{g\rho}} \tag{6.16}$$

where g is the acceleration due to gravity and ρ is the sphere density. For the case of a polystyrene sphere of PTFE flat, $\gamma_S = (\gamma_{PTFE}\gamma_{polystyrene})^{1/2} = 24$ mN/m. So, if the polystyrene sphere has a radius less than

$$R = \sqrt{\frac{3 \times 24 \ mN/m}{(98 \ m/s^2)(1050 \ kg/m^3)}}$$
$$= 2.7 \ mm, \tag{6.17}$$

the adhesive force from surface energy will be greater than the force of gravity acting on the sphere. Equation (6.17) predicts that a fairly large ~0.5 cm plastic particle should stick to a nominally low energy surface like Teflon, indicating that the contribution of surface energy to the adhesion force can be quite strong. In practice, it is rare for such large particles to stick as microroughness tends to reduce the surface energy contribution. (We will come back to this point shortly.)

6.2.1.3   *Example: adhesion force for an atomically sharp asperity*

As a final example, we look at the adhesive force acting on a single asperity contacting a flat surface, where the radius of curvature is so small that the apex is made up of only a few atoms or molecules. Such sharp asperities are desirable in atomic force microscopes (AFMs) and scanning tunneling microscopes (STMs) when one wants the tip to resolve atomic and molecular surface features. A major problem, however, is keeping these tips sufficiently sharp to achieve atomic resolution, since adhesive forces may pull off those few atoms at the apex responsible for the atomic resolution.

Let's consider the example of a tip with a 1 nm radius of curvature, which is small enough to achieve atomic resolution, but large enough that it is still reasonable (just barely) to apply the concept of surface energy. To take advantage of PTFE's low surface energy to reduce the adhesive force, one might be tempted to use a PTFE tip. For a PTFE tip scanning over polystyrene, we can calculate the adhesive force by substituting $R = 1$ nm into eq. (6.15) resulting in $L_{adh} = 0.3$ nN. While this is indeed a low adhesive force, if we assume that it is acting over an area with a diameter the same as the chain diameter of a PTFE molecule (~6 Å), then this would correspond to an applied tensile stress of 1 GPa, a hundred times higher than the yield strength of bulk PTFE (9 MPa). Consequently, we would expect that adhesive forces would plastically deform an initially atomically sharp PTFE tip, increasing its contact area to the point where the adhesive stress is below the plasticity threshold. So, while AFM tips are sometimes coated with PFTE to reduce meniscus adhesive forces, the tip atoms that do the atomic imaging need be of a much harder material.

One such hard material commonly used for AFM tips is silicon–nitride (hardness = 20 GPa). Using a surface energy of 70 mN/m for silicon nitride in eq. (6.15), $L_{adh} = 0.6$ nN for a 1 nm tip radius in contact with a polystyrene surface. Again assuming that this adhesive force acts over an area of 6 Å in diameter, the stress from this adhesive force corresponds to 2 GPa, a factor of ten less than the hardness of silicon nitride.

6.2.2   Adhesion-induced deformation at a sphere-on-flat contact

6.2.2.1   *The Johnson–Kendal–Roberts (JKR) theory*

When we first considered the sphere-on-flat geometry in Chapter 3, we used the Hertz analysis to describe how an external loading force elastically deforms the contacting geometry. The Hertz analysis assumes that no adhesive forces act between the bodies, which leads to the consideration of only compressive stresses within the solids. In a major advance, Johnson et al. (1971) extended the Hertz analysis to include the elastic

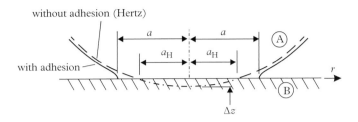

Figure 6.4 *Deformation and contact radius of an elastic sphere contacting a rigid flat. Solid line: Johnson–Kendal–Roberts (JKR) contact with adhesion. Dashed line: Hertzian contact without adhesion.*

deformations arising from the adhesive forces associated with surface energy. Within the Johnson–Kendal–Roberts (JKR) theory, a short-range adhesive force results in a tensile stress being added to the Hertzian compressive stress, so the total normal stress $p(r)$ at the interface becomes

$$p(r) = p_0 \left(1 - r^2/a^2 \right)^{1/2} + p'_0 \left(1 - r^2/a^2 \right)^{-1/2},$$
(6.18)

where a is the contact radius and r is the distance from the center of the contact (Figure 6.4). The first term in eq. (6.18), which is positive, is the compressive stress from eq. (3.11), while the second term, which has a negative value of p'_0, is the tensile stress induced by the adhesion. Equation (6.18) indicates that the net stress is compressive towards the center of the contact zone and tensile toward the outer edges of the contact zone. As illustrated in Figure 6.4, the tensile stress elastically deforms the edges of the contact zone, increasing its size so that the actual contact radius a becomes larger than the a_H predicted by the Hertz analysis.

While eq. (6.18) predicts that the stress should go to infinity at the edge of the contact zone, this theoretical divergence only comes about as JKR theory implicitly assumes that the adhesion forces act over an infinitesimally small distance exactly at the sphere–flat interface. Once the theory is modified to have an attractive force with a finite range, such as by using a Lennard-Jones potential (Muller et al. 1980, 1983, Fogden and White 1990), the infinite tensile stress disappears. An infinite stress would of course never occur in actual solids, as once the stress exceeds the threshold for plastic flow, the atoms move to more stable bonding configurations.

The basis of the JKR theory (Johnson 1985: pp. 125–9) is the simple concept that the total surface energy is reduced by extending the contact area through elastic deformation, which is opposed by the increase in strain energy due to the deformation; the equilibrium contact radius a corresponds to the balance of the two energies. The total energy U_T of the sphere-on-flat geometry is the sum of surface energy U_S lost due to the formation of a finite contact area plus the energy U_E stored in the elastic deformations:

$$U_T = U_S + U_E.$$
(6.19)

If the overall elastic deformation Δz of the two objects in the normal direction is kept constant, the equilibrium contact radius a occurs when

$$\left[\frac{\partial U_T}{\partial a}\right]_{\Delta z} = \left[\frac{\partial U_E}{\partial a}\right]_{\Delta z} + \left[\frac{\partial U_S}{\partial a}\right]_{\Delta z} = 0. \tag{6.20}$$

The variation in strain energy with contact radius is given by Johnson (1985: eq. (5.47)):

$$\left[\frac{\partial U_E}{\partial a}\right]_{\Delta z} = \frac{\pi^2 a^2}{E_c}p_0'^2, \tag{6.21}$$

where E_c is the composite elastic modulus (eq. (3.8)). The variation in surface energy loss with contact radius is

$$U_S = -W_{AB}\pi a^2$$
$$\frac{\partial U_S}{\partial a} = -2W_{AB}\pi a, \tag{6.22}$$

where W_{AB} is the work of adhesion between materials A and B and the minus sign comes from surface energy being lost as the contact area grows. Using eqs. (6.21) and (6.22) in eq. (6.20) and with some manipulation, we obtain the following expression for p_0':

$$p_0' = \left(\frac{2W_{AB}E_c}{\pi a}\right)^{1/2}. \tag{6.23}$$

The net force normal to the contacting interface (the externally applied loading force L_{ext}) is found by integrating the $p(r)$ over the contact area

$$L_{ext} = \int_0^a 2\pi r\, p(r) dr = \left(\frac{2}{3}p_0 + 2p_0'\right)\pi a^2. \tag{6.24}$$

Substituting for p_0 (eqs. (3.10) and (3.11)) and for p_0' (eq. (6.23)) and rearranging gives the relationship between the normal force L_{ext} and the contact radius a:

$$\left(L_{ext} - \frac{4E_c a^3}{3R}\right)^2 = 8\pi W_{AB}E_c a^3. \tag{6.25}$$

If no external loading force is applied ($L_{ext} = 0$), only the attractive forces derived from the work of adhesion pull the sphere toward the flat, and eq. (6.25) predicts that the stresses from this attractive interaction generate a finite contact radius

$$a_0 = \left(\frac{9\pi W_{AB}R^2}{2E_c}\right)^{1/3}. \tag{6.26}$$

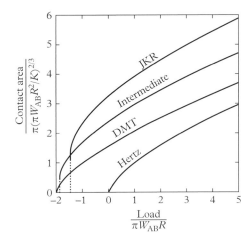

Figure 6.5 *Comparison of contact areas for the sphere-on-flat geometry predicted by the Johnson–Kendal–Roberts (JKR) theory, Derjaguin–Müller–Toporov (DMT) theory, and an intermediate theory. These approach the Hertz curve in the limit of $W_{AB} \to 0$. Adapted with permission from Carpick and Salmeron (1997). Copyright 1997 American Chemical Society.*

The contact radius for nonzero loading forces L_{ext} can be determined by manipulating eq. (6.25) to yield

$$a^3 = \frac{3R}{4E_c}\left(L_{ext} + 3\pi RW_{AB} + \sqrt{6\pi RW_{AB}L_{ext} + (3\pi RW_{AB})^2}\right), \qquad (6.27)$$

which reduces when $W_{AB} = 0$ to eq. (3.7) from the Hertz analysis.

Figure 6.5 plots how the JKR contact area varies with the external loading force L_{ext}. (Figure 6.5 also plots the contact areas from the Derjaguin–Müller–Toporov (DMT) theory and an intermediate theory, both of which will be discussed shortly.) The JKR contact area is finite when the $L_{ext} = 0$, and this area continues to decrease as the loading becomes more negative, until the contact becomes unstable and the surfaces separate; at this critical point, the release rate for stored elastic energy exceeds the creation rate for surface energy. At separation, the magnitude of the external loading force corresponds to an adhesive force at pull-off. The external loading force at separation is referred to as the pull-off force $F_{pull-off}$, and, with the sign convention that we are using here, it has a negative sign, since it acts in the opposite direction of the adhesive force L_{adh}. Within the JKR theory, it is given by

$$\begin{aligned} F_{pull-off} &= -L_{adh} \\ &= -\tfrac{3}{2}\pi W_{AB}R \end{aligned} \qquad (6.28)$$

For the case where the sphere and flat are the same material ($W_{AB} = 2\gamma_S$), eq. (6.28) becomes

$$F_{\text{pull-off}} = -3\pi\gamma_S R. \tag{6.29}$$

At separation, the contact radius predicted by JKR theory is still finite and given by

$$a_S = \frac{a_0}{4^{1/3}} = 0.63a_0. \tag{6.30}$$

A major shortcoming of JKR theory is that it assumes perfectly smooth surfaces. For materials with a very low elastic modulus, such as rubber or silicone elastomers, this is not much of a problem, as the force required to flatten the surface roughness is less than the attractive forces. (This is why rubber and elastomer surfaces often feel tacky to the touch.) Indeed, Johnson, Kendall, and Roberts verified their theory by pressing together rubber spheres with a low elastic modulus of 0.8 MPa (dramatically lower than a material such as glass with a modulus of 10^5 MPa) and measuring the resulting contact radius as a function of applied load. Their data agreed well with eq. (6.27) when contacting in air and immersed in water. From these experiments, they determined that $\gamma_S = 35$ mJ/m^2 and $\gamma_{SL} = 3.4$ mJ/m^2 for rubber in air and water, respectively. From these values, Young's equation (eq. (5.18)) predicts a contact angle of 64° for water on rubber, close to the experimental value of 66° of Johnson et al. (1971), providing the first experimental confirmation of Young's equation where all the variables are measured independently.

Many subsequent studies with low elasticity surfaces have further verified JKR theory (Roberts and Othman 1977, Lee 1978, Chaudhury and Whitesides 1991, 1992). One particular situation where JKR theory applies, that deserves special mention, is the contact geometry of the SFA (Israelachvili et al. 1980, Horn et al. 1987, Frantz et al. 1997); in this apparatus, the low elasticity is provided by a soft glue used to attach the thin mica sheets to a rigid cylindrical substrate.

6.2.2.2   *Derjaguin–Müller–Toporov (DMT) theory*

Another shortcoming of the JKR theory is that it only works well for materials with low elasticity, since only for these compliant materials does the elastic response mimic that expected for an infinitesimally short ranged attractive force. For more realistic attractive forces with an extended range, the induced stresses are spread out over a larger volume, leading to lower elastic strains than predicted by the JKR theory.

In the limit where the materials are nearly rigid, a different theory developed by Derjaguin et al. (1975) is used. In the DMT theory, the stress profile is assumed be Hertzian (described by eq. (3.11)) so the stress is always compressive within the contact zone, but the normal loading force L is the external loading force L_{ext} plus the total adhesive force L_{adh}, that is, $L = L_{\text{ext}} + L_{\text{adh}}$. The adhesive force can come from the work of adhesion, as given by eqs. (6.10) and (6.11), and/or from the meniscus force discussed later. Within the DMT model, the equation for the radius of contact a is same as for a Hertzian contact but a L_{adh} term is added to the external loading force:

$$a^3 = \frac{3R}{4E_c}(L_{ext} + L_{adh}).\qquad(6.31)$$

At separation (which occurs when $L_{ext} = -L_{adh}$), the DMT theory predicts that the contact radius goes to zero, in contrast to the JKR theory which predicts a finite contact radius given by eq. (6.30). Figure 6.5 compares the prediction of the DMT and JKR theories for the contact area with load for a sphere on a flat.

Like the JKR theory, the DMT theory also predicts a finite pull-off force, $F_{pull-off} = -L_{adh}$. Therefore, for the case of a sphere on a flat with no liquid meniscus present, the pull-off force is related to the work of adhesion by

$$F_{pull-off} = -2\pi W_{AB}R,\qquad(6.32)$$

which is a factor 4/3 higher than the JKR prediction (eq. (6.28)).

Although their predictions are different, the JKR and DMT theories are not in conflict. Rather, they represent limiting cases, where the JKR limit corresponds to one extreme that has strong, short-range adhesion coupled with highly compliant materials while DMT corresponds to the other extreme of low, long-range adhesion coupled with rigid materials. To determine which extreme is most applicable, one can use the Tabor parameter (Tabor 1977) defined by

$$\mu = \frac{R^{1/3} W_{AB}^{2/3}}{E_c^{2/3} D_0},\qquad(6.33)$$

where D_0 is the equilibrium separation between the two surfaces in contact. Typically, $D_0 \sim$ interatomic spacing a_0 (though as discussed in Section 7.3.4 these are not quite the same thing) and represents the typical distance that a *short-range* adhesion force would act over. The Tabor parameter was developed by Prof. David Tabor who noticed that the neck that forms around an elastic JRK contact (as illustrated in Figure 6.4) has a height on the order $(RW_{AB}^2/E^2)^{1/3}$ and that the JKR analysis is only valid when this height is greater than the range of the attractive surface force (Tabor 1977, Greenwood 1997). When large values of the Tabor parameter μ occur: the sphere radius is large enough, the materials are compliant enough, and the adhesive forces are strong and short-range enough for JKR theory to apply; while for small values of μ: the DMT theory should be used.

A practical challenge arises if one is determining W_{AB} from a measurement of $F_{pull-off}$ using either eq. (6.28) or eq. (6.32), since one needs to know the Tabor parameter to know which equation to use, but one cannot determine this without first knowing W_{AB} and D_0, which are usually uncertain. One approach is to check if the JKR or DMT regimes are appropriate by using extreme limiting values (Grierson et al. 2005). A method for determining D_0 and W_{AB} has recently been proposed where the additional needed information comes from the *snap-in* distance measured in an AFM experiment (Pastewka and Robbins 2014, Jacobs et al. 2015a, 2015b).

For those contact situations between the two extremes of JKR and DMT, Maugis (1992) has developed a comprehensive analysis to handle these intermediate cases, including approximate practical equations to determine the work of adhesion, contact area, and contact deformation for these intermediate cases (Carpick et al. 1999, Schwarz 2003).

In all these analyses of JKR, DMT, and intermediate cases, it is important to note that the materials are assumed to be homogeneous, isotropic, linear, and elastic; that all strains are assumed to be small (well below 0.1); and that the loads are small enough that the contact radius and elastic deformations are small compared to the radius of curvature. Also, the surfaces are assumed to smooth enough for intimate atom-to-atom contact to occur throughout the apparent contact area. We discuss how roughness reduces the adhesive forces predicted by JKR and DMT in Section 6.2.3.

6.2.2.3   *Adhesion deformation in nanoscale contacts*

For nanoscale asperity contacts, researchers have been able to verify that adhesive forces increase with contact area for a number of cases. These experiments involve sliding sharp AFM tips, with radii of a few tens of nanometers, over surfaces with different loads and measuring friction. The contact area was determined by assuming that friction was directly proportional to the contact area, and applying a suitable contact mechanics model to fit the friction vs. load measurements. Using this method, contact has been categorized as either JKR, DMT, or intermediate:

- *JKR behavior*—observed for platinum-coated tips sliding over a muscovite mica surface in ultrahigh vacuum (UHV) (Carpick et al. 1996a, 1996b) and for a silicon tip sliding over NaCl(001) in dry nitrogen (Meyer et al. 1996).

- *DMT behavior*—observed for a silicon nitride tips sliding in humid air over mica (Carpick et al. 1997) and over silicon wafers with SiO_2 layers (Lessel et al. 2013), and for tungsten carbide tips sliding over diamond in UHV (Enachescu et al. 1998).

- *Intermediate behavior*—observed for Si tips sliding on $NbSe_2$ in UHV (Lantz et al. 1997), and for amorphous carbon tips sliding on diamond in UHV (Gao et al. 2007).

These results show that, even for micro-asperity contacts only a few nanometers in size, adhesive forces can lead to contact areas greater than what would be expected from a Hertzian analysis.

Even though AFM tips typically have very small radii of curvature and are made of stiff materials, they still often experience JKR behavior when operated in vacuum conditions. This occurs in vacuum conditions because there one can clean off all the oxide and contamination layers from the tip and sample surfaces that would normally lower their surface energies outside of the vacuum environment. Once cleaned, these surfaces retain their intrinsically high values of the work adhesion W_{AB} for extended periods of time and, consequently, have high values of the Tabor parameter even though R is small and E_c is high.

6.2.3 Effect of roughness on adhesion in a dry environment

When we introduced the concept of work of adhesion between two solid surfaces back in Section 5.3, we defined it as the energy per unit area needed to separate to infinity two surfaces from *intimate contact*. By intimate contact, we mean that atom-to-atom bonding occurs everywhere across the interface between the two materials, which cannot occur if the (real area of contact) < (apparent area of contact). Consequently, to get the full adhesive force predicted by JKR or DMT in eqs. (6.28) and (6.32) the surfaces either have to be initially atomically smooth or have to be sufficiently elastically compliant to allow the surfaces to come into intimate contact throughout the apparent contact area.

While adhesive interactions are present between all surfaces, it is rare to observe that macroscopic surfaces are "sticky," a situation referred to as the "adhesion paradox" (Kendall 2007). Indeed, we know from everyday experience that, if we bring two nominally flat surfaces into contact, they most likely won't stick together even if the surfaces have fairly significant surface energies that should result in a fairly high intrinsic work of adhesion between the two surfaces. The paradox is resolved when one realizes that the residual roughness on most surfaces prevents intimate contact from occurring when the surfaces are brought into contact, as illustrated in Figure 6.6. Consequently, microscale roughness lowers the adhesion below what would be predicted from the JKR and DMT theories. Indeed, strong adhesion only occurs between contacting surfaces if at least one of the materials is elastically compliant and both are smooth enough so that a large enough real area of contact is formed when pressed into contact to generate a substantial adhesion force.

For elastically stiff materials, an rms roughness on the order a few nanometers is often enough to reduce adhesion forces to a negligible amount. An important technological example where roughness is deliberately used to reduce adhesion to a negligible level is in microelectromechanical systems (MEMS) structures that need to come into contact

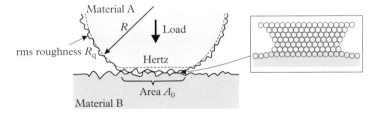

Figure 6.6 *A sphere-on-flat contact geometry with a high degree of surface roughness and materials sufficiently stiff to prevent JKR type of contact. On the macroscale, the sphere undergoes Hertzian elastic deformation (dashed line). On the microscale, JKR type of contact can potentially occur at the individual asperity contacts, as illustrated in the inset occurring at the atomic level for one of the contacts.*

to function, but must not become stuck (Houston et al. 1997, Ashurst et al. 2003, DelRio et al. 2005).

For elastically soft materials, however, a larger rms roughness is needed to minimize the adhesion. For example, Fuller and Tabor (1975) measured adhesion between millimeter-sized silicone rubber balls of different moduli, and nominally flat acrylic surfaces with different roughness levels. For a rubber with a Young's modulus of 2.4 MPa, the pull-off force dropped by approximately 90% when the roughness was increased from the smoothest baseline up to $R_a = 0.65$ μm approximately. However, for a rubber nearly 10 times less stiff (Young's modulus 0.22 MPa), a 90% drop in pull-off force was not obtained until reaching approximately $R_a = 1.4$ μm.

Various theoretical approaches have been advanced to analyze how surface roughness affects adhesion. For example, within the Greenwood–Williamson model (Section 3.4.1), each asperity–asperity contact can be treated as a small JKR (Fuller and Tabor 1975) or DMT (Maugis 1996) contact. Some of these methods will be discussed further in the Section 7.3.6 in relationship to how roughness impacts the van der Waals interaction between surfaces.

A more general way to analyze the effect of roughness on adhesion is to introduce the concept of an effective work of adhesion per unit area W_{eff} into eq. (6.19) for the total energy U_T of the interface (Persson 2002, Persson and Scaraggi 2014):

$$U_T = U_S + U_E = -W_{eff}A_0, \tag{6.34}$$

where A_0 is the nominal or apparent area of contact, U_E is the elastic strain energy that is stored in the elastic deformation of the asperity contacts, and U_S is the surface energy lost through contact due to the formation of adhesive bonds across the interface. The adhesion energy U_S corresponds to

$$U_S = -W_{AB}A_{contact}, \tag{6.35}$$

where $A_{contact}$ is total real area of contact, and W_{AB} is the *intrinsic* work of adhesion (introduced in Section 5.3) from the intimate atomic-to-atomic bonding that is occurring throughout the real area of contact.

Once the W_{eff} has been determined for a contacting interface, the JKR and DMT theories can be applied to predict the adhesive force by substituting the W_{eff} for W_{AB} in eqs. (6.28) and (6.32). For sphere-on-flat contact geometries composed of rough surfaces (such as illustrated in Figure 6.6) the adhesive force is then given by

$$\text{JKR}: \quad L_{adh} = \frac{3}{2}\pi W_{eff}R \tag{6.36}$$

$$\text{DMT}: \quad L_{adh} = 2\pi W_{eff}R \tag{6.37}$$

Physically eq. (6.34) indicates that the elastic strain energy U_E is counteracting the adhesive binding energy U_S, so that when the loading force is removed, the elastic energy stored in the deformation of asperities is released and works to break the adhesive bonds over the contact area. If the asperities are stiff enough and the adhesion weak enough, sufficient elastic energy exists to break the adhesion so that a negligible contact area occurs at zero load and the materials do not stick to each other ($W_{eff} \sim 0$).

For contact situations where adhesion occurs at zero load, the Persson contact mechanics model for rough surfaces that was discussed in Section 3.4.2 can be extended to include the effect of these adhesion forces. The details can be found in Persson (2002), Peressadko et al. (2005), and Persson and Scaraggi (2014).

6.2.3.1   *Example: effect of roughness on AFM adhesion*

Experimental evidence that roughness as a strong effect on adhesion at the nanoscale has been demonstrated by Jacobs et al. (2013), who investigated how roughness affects the pull-off force of AFM tips with different levels of surface roughness. The measurement of the pull-off force of an AFM tip from a flat single crystal diamond can be considered as the measurement of the force required to separate a sphere from a flat, but with the sphere having a nanoscale radius ($R = 17$–150 nm). The tip in this experiment consisted of a Si cone coated with a film of amorphous hydrogenated carbon (known as a-C:H and is a type of diamond-like carbon or DLC). From these experiments, Jacobs et al. (2013) were able to infer the effective work of adhesion W_{eff}. In addition to the experiments, Jacobs et al. (2013) also preform molecular dynamic simulations to extend the range of surface roughness to the sub-angstrom level.

As shown in Figure 6.7, the effective work of adhesion W_{eff} drops sharply as the tip's rms roughness R_q increases from 0.03 to 1.5 nm. Both experiments and simulations of W_{eff} follow a consistent trend and agrees well with a simple analytical model suggested by Rabinovich et al. (2000), where a rough tip is replaced by a smooth tip, and the flat surface is replaced by a flat surface with a single hemispherical-shaped asperity on it, whose radius depends only on the tip's rms roughness R_q; this hemisphere keeps the tip from approaching the rest of the flat and thus accounts for the tip's roughness. The resulting equation for the work of adhesion is:

$$W_{eff} = \frac{A_H}{12\pi D_0^2}\left[\left(1 + \frac{R_{tip}}{1.48R_q}\right)^{-1} + \left(1 + \frac{1.48R_q}{D_0}\right)^{-2}\right]. \tag{6.38}$$

where A_H (the Hamaker constant discussed in Chapter 7) represents the strength of the van der Waals forces between the tip and sample materials.

Since the primary effect of the roughness is to increase overall effective separation between the surfaces, which reduces the contributions from those adhesive interactions that fall off with distance, the model behind eq. (6.38) essentially replaces of the flat sample surface with a hemispherical so as to increase the separations away from the contact point in order to mimic the effect of the roughness. This effect is strong for the results in Figure 6.7: the effective work of adhesion drops from the intrinsic value of $W_{AB} = 140$ mJ/m^2 by a factor of 10 for an increase in rms roughness of just a few tenths of a nanometer.

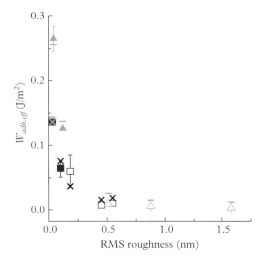

Figure 6.7 *The effective work of adhesion W_{eff} for two types of diamond-like carbon (DLC) AFM tips against a smooth diamond sample as determined from experiments and molecular dynamics simulations. W_{eff} decreases by more than an order of magnitude as tip roughness increases from being atomically smooth ($R_q = 0.3$ nm) to a nanometer scale roughness of $R_q = 1.6$ nm. Results from molecular dynamic simulations (filled data points) and from experimental testing (hollow data points) are shown for sharp tips made of ultrananocrystalline diamond (UNCD) (gray triangles) and of DLC (black squares). The "−" and the "x" points indicate the fits to eq. (6.38) using Hamaker constants, respectively, of $A_H = 5.5 \times 10^{-19}$ J for UNCD and $A_H = 2.7 \times 10^{-19}$ J for DLC. Reproduced with permission from Jacobs et al. (2013). Copyright 2013, Springer.*

Note that, for these experiments, both the a-C:H film and the diamond countersurface have rather low surface energies, as the surfaces are passivate by the adsorption of oxygen and hydrogen when exposed to air. Indeed, Jacobs et al. (2013) showed that fitting their data to eq. (6.38) produces a Hamaker constant A_H consistent with values measured for passivated diamond interfaces.

6.2.4   Criterion for sticky surfaces

Pastewka and Robbins (2014) have developed an analytical theory, backed up with extensive numerical simulations, to examine how the combination of roughness, adhesion, elasticity, and load impacts adhesion of large rough contacts. In this theory, two contacting rough surfaces are pulled into sticky contact by adhesion if the total attractive force exceeds the total repulsive force when the surfaces first make contact. An attractive pressure acts on the regions round each asperity contact zone (illustrated as the dark gray regions in Figure 6.8) over which the adhesive interaction is exerted across the gap (adhesive interaction is assumed to extend a distance Δr away from the surface.) Inside the asperity contact zones (medium gray in Figure 6.8), the normal pressure is repulsive with an average value p_{rep}. If we take A_{adh} to be the total area of all these regions where

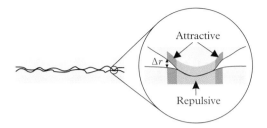

Figure 6.8 *Two rough surfaces in contact (left). Expanded view of one of the asperity contacts where a rigid asperity on the upper surfaces causes deformation on the bottom surface (right). Within the contact zone the pressure is repulsive. Just outside the contact zone, an attractive pressure is a exerted from adhesive interactions between the two surfaces where their separation distance is less than Δr.*

adhesion acts around the asperity contacts and if we take A_{rep} to be the total area where repulsive contact occurs, then the condition for the surfaces being pulled together by adhesion—total attractive force exceeding total repulsive force—can be restated as

$$p_{\mathrm{adh}} A_{\mathrm{adh}} > p_{\mathrm{rep}} A_{\mathrm{rep}}. \tag{6.39}$$

As discussed in Section 3.4.2, for surfaces with a self-affine type of roughness, the mean repulsive pressure acting over the apparent contact area is related to the loading force by eq. (3.59). This equation can be rewritten to yield the following expression for p_{rep}:

$$p_{\mathrm{rep}} = \frac{E_{\mathrm{c}} \nabla z_{\mathrm{rms}}}{\kappa} \tag{6.40}$$

where κ is on the order 2.

Pastewka and Robbins (2014) estimate the average adhesive pressure p_{adh} from adhesion to be

$$p_{\mathrm{adh}} = \frac{W_{\mathrm{AB}}}{\Delta r}. \tag{6.41}$$

For those situations where $A_{\mathrm{adh}} < A_{\mathrm{rep}}$, we can rewrite eq. (6.39) as

$$\frac{p_{\mathrm{adh}}}{p_{\mathrm{rep}}} = \frac{W_{\mathrm{AB}}/\Delta r}{E_{\mathrm{c}} \nabla z_{\mathrm{rms}}/\kappa} = \frac{\kappa l_{\mathrm{a}}}{\Delta r \nabla z_{\mathrm{rms}}} > 1, \tag{6.42}$$

where ∇z_{rms} is the rms slope of the surface roughness (eqs. (2.19) and (2.23)) and l_{a} is called the *adhesion length* and is defined by

$$l_{\mathrm{a}} = \frac{W_{\mathrm{AB}}}{E_{\mathrm{c}}}. \tag{6.43}$$

Usually, $\kappa \sim 2$ and the range the attractive force is on the order an atomic spacing: $a_0 \sim \Delta r$; so, using these values, eq. (6.42) reduces to

$$\frac{W_{AB}}{E_c a_0} = \frac{l_a}{a_0} > 0.5 \nabla z_{rms}. \tag{6.44}$$

Equation (6.42) corresponds to a necessary condition, at zero load, for how smooth surfaces need to be for them to be considered "sticky" (i.e., for a noticeable pull-off force to be required to separate the surfaces.) From this equation we see that the rms slope ∇z_{rms} is the dominant geometrical parameter in determining whether surfaces with self-affine roughness are sticky or not, and not the rms roughness R_q. We also see that stickiness is promoted by short ranges of the adhesive forces (smaller Δr) and larger adhesion lengths l_a (due to lower elastic moduli or higher work of adhesion).

While eq. (6.42) corresponds to a necessary condition for stickiness, this condition is not always sufficient, particularly if $A_{adh} > A_{rep}$. Using the analytical expression for area ratio A_{adh}/A_{rep} developed by Pastewka and Robbins (2014), one can derive the following more general criterion:

$$\frac{(\nabla z_{rms})^7}{(\nabla^2 z_{rms})^2} < 8 \frac{l_a^3}{\Delta r}, \tag{6.45}$$

where $\nabla^2 z_{rms}$ is the rms of the second derivative of the surface topography (eq. (2.13)). If eq. (6.45) is satisfied, then appreciable adhesion occurs at zero applied load, and a noticeable pull-off force is required to separate the surfaces. Again we have that short range adhesive forces (small Δr) and large adhesion lengths l_a (due to low elastic moduli or high work of adhesion) results in sticky surfaces. More importantly, eq. (6.45) provides (beyond eq. (6.42)) the further guidance for how surface roughness impacts adhesion in that small curvature radii (high $\nabla^2 z_{rms}$) also promote stickiness in addition to small slopes (low ∇z_{rms}), though the dependence on slope is much stronger than the dependence on curvature. The strong dependence on rms slope arises from two effects: (1) a surface with gentle slopes can make larger contact areas (imagine pressing a rubber block into a steeply- versus a gently-sloped set of conical peaks); (2) with gentler slopes, the larger perimeter with small gaps around the contacts means that the attractive forces are felt over a larger area.

6.2.4.1 *Examples of when surfaces are sticky*

As a practical example of how to use eq. (6.44), let us first consider the case of a diamond covalently bonding with another piece of diamond. For freshly cleaved diamond, $\gamma_{diamond} = 5$ J/m^2, so the intrinsic work of adhesion is $W_{diamond} = 2\gamma_{diamond} = 10$ J/m^2. Also, for diamond, $E_c = 1000$ GPa and $a_0 = 1.5$ Å. This leads to $l_a/a_0 = 0.06$ for when two clean diamond surfaces are brought into contact, so eq. (6.44) implies that adhesion with strong bonding across the interfaces can occur when $\nabla z_{rms} < 0.12$. This example illustrates how contacting surfaces with high surface energies can strongly adhere together even when the roughness is fairly significant (an $\nabla z_{rms} = 0.12$ corresponds to an average slope of 12%.)

However, surface energy values greater than 1 J/m^2 are generally not stable as, once a freshly prepared, clean surface is exposed to the environment, contaminants will adsorbed onto it and reduce the surface energy to typically <150 mJ/m^2. To illustrate the impact of this, we next consider the case of the passivated interface discussed in Section 6.2.3.1. This passivation of the diamond-like surfaces results in a much lower work of adhesion of $W_{AB} = 140$ mJ/m^2 (compared to that of covalently diamond with $W_{diamond} = 5$ J/m^2) and results in a value of $l_a/a_0 = 8.4 \times 10^{-4}$ for the passivated diamond-like surface. ∇z_{rms} then needs to be $<1.7 \times 10^{-3}$ for the surfaces to be considered sticky at zero load, a factor of 140 smoother than for covalently bonded diamond.

As another example, often surfaces are covered with hydrocarbon based contaminants, which lowers the surface energy to be comparable to that of pure hydrocarbons, or approximately 30 mJ/m^2. In general, moderately stiff surfaces that have been passivated to this degree need to be exceptionally smooth to achieve stickiness, as the surfaces slopes need to be <0.1%. One practical example where such complete adhesion does occur is during the wafer bonding of silicon wafers, which are first chemically mechanically polished to achieve sufficient smoothness.

If such smoothness cannot be achieved, an alternative is to reduce the effective elastic modulus E. This can be achieved with compliant elastomeric materials, or by using structures that are *effectively* compliant by using small and independent but dense structures like the fibers or spatulae found on the feet of insects and geckos, as described in Sections 6.7 and 7.3.7.

6.3　Wet environment

As most environments have some degree of humidity or wetness associated with them, surfaces tend not to be completely dry. As a consequence, the gaps between the contacting asperities sometimes become partially filled with liquid, affecting the overall adhesion force.

Some possible environmental sources for the liquid that forms menisci around the contact points or fills the gaps between them:

- *Thin films of lubricants and contaminants (Figure4.10)*—Even a sub-monolayer coverage of a contaminant can result in a meniscus a few nanometers in height if that contaminant has enough mobility to diffuse to the contact points.
- *Capillary condensation of water and organic vapors (Figures 5.7 and 5.18)*—As discussed in Section 5.2.2, for partial vapor pressures significantly below saturation, capillary condensation results in menisci a few nanometers in height. These nanoscale menisci can be difficult to detect, except by the significant adhesion force they generate.

6.3.1   Force for a sphere-on-flat in a wet environment

Let's assume that there is a sufficient amount of moisture or other liquid is present to form a meniscus around a sphere-on-flat contact, as illustrated in Figure 6.3(b). The extra loading force L_{men} that this liquid meniscus adds is the sum of two component forces: a surface tension force L_{ten} and a capillary pressure force L_{cap}.

The surface tension force L_{ten} in the axial direction comes from the vertical projection of the liquid surface tension pulling on the solid surfaces along the edges of the contact line of the meniscus. At the contact line of the meniscus on the flat, the vertical component of the surface tension is

$$L_{ten}/\text{unit perimeter length} = \gamma_L \sin \theta_2$$
$$L_{ten} = 2\pi\, r_2 \gamma_L \sin \theta_2. \tag{6.46}$$

In the sphere-on-flat geometry, L_{ten} tends to be a much smaller component of the total meniscus force than the capillary pressure component L_{cap}. Two effects lead to L_{ten} having a smaller value:

1. If the liquid wets the flat, $\sin \theta_2$ is small, and L_{ten} is then small from eq. (6.46).
2. If θ_2 is large, the radius r_1 is large, and r_2 needs to be small in order for the overall radius of curvature to stay constant at the Kelvin radius (eq. (5.11)).

Consequently, when a meniscus is small so it subtends a small angle φ (known as the "filling angle"), we can approximate the meniscus force L_{men} as being equal to the capillary pressure force $(L_{men} \approx L_{cap})$.

The capillary pressure force L_{cap} originates from the capillary pressure P_{cap} generated by the curvature of the meniscus surface acting over the area of the meniscus $A = \pi r_2{}^2$. Applying the Pythagorean theorem to the geometry in Figure 6.3(b) when φ is small, we have

$$(R - d)^2 + r_2^2 = R^2$$
$$R^2 - 2Rd + d^2 + r_2^2 = R^2$$
$$r_2^2 \simeq 2Rd. \tag{6.47}$$

leading to

$$A \approx 2\pi R d. \tag{6.48}$$

For the geometry shown in Figure 6.3(b), where $|r_1| \ll |r_2|$ and γ_L is the surface tension of the liquid, the capillary pressure in the meniscus is

$$P_{cap} = \gamma_L \left(\frac{1}{r_1} + \frac{1}{r_2} \right) \simeq \frac{\gamma_L}{r_1}. \tag{6.49}$$

To determine the capillary force acting along the axial direction, we multiply the capillary pressure P_{cap} times the area A of the meniscus parallel to the flat surface,

$$L_{cap} \approx A \times P_{cap} \simeq (2\pi Rd) \left(\frac{\gamma_L}{r_1} \right). \tag{6.50}$$

When φ is small,

$$d \simeq r_1 \left(\cos\theta_1 + \cos\theta_2 \right) \tag{6.51}$$

and eq. (6.50) becomes

$$L_{men} \simeq L_{cap} \simeq 2\pi R\gamma_L \left(\cos\theta_1 + \cos\theta_2 \right). \tag{6.52}$$

So, for a small meniscus, the meniscus force does not depend on the size of the meniscus. If the liquid wets both surfaces ($\theta_1 \simeq 0$ and $\theta_2 \simeq 0$), eq. (6.52) reduces to

$$L_{men} \simeq 4\pi R\gamma_L. \tag{6.53}$$

Conveniently, eq. (6.53) has the same form as eq. (6.11), so one only has to remember one formula for the sphere-on-flat geometry, and to remember to use the surface energy of the solid in a dry environment, and to use the surface tension of the liquid if the environment is wet enough for a meniscus to form.

6.3.1.1 Example: lubricant meniscus force on an AFM tip

Figure 6.9 shows the force acting on an AFM tip as three samples with thin liquid lubricant films are brought into contact with the tip and then withdrawn. Before each sample makes contact with the lubricant, the force is near to zero; when contact occurs, the lubricant wicks up around the tip, generating a sudden attractive meniscus force of $\sim 2 \times 10^{-8}$ N. Since this liquid perfluoropolyether polymer has a low surface tension of $\gamma_L = 24$ mN/m, it wets the sample and tip surfaces with a contact angle near zero, and we can use eq. (6.53) to estimate the mean macroscopic tip radius R from the measured jump in meniscus force:

$$\begin{aligned} R &\simeq \frac{L_{men}}{4\pi\gamma_L} \\ &\simeq \frac{2\times 10^{-8} \text{ N}}{4\pi (24 \text{ mN/m})} \\ &\simeq 66 \text{ nm} \end{aligned} \tag{6.54}$$

As the tip pushes through the thin liquid film, the meniscus force increases linearly with penetration distance u. This increase in force occurs as the meniscus area scales with u, which can be seen in a modified eq. (6.48):

$$A \approx \pi r_2^2$$
$$\approx 2\pi R d$$
$$\approx 2\pi R \left[r_{\text{eff}} \left(1 + \cos \theta \right) + u \right] \tag{6.55}$$

where r_{eff} is the effective capillary radius:

$$r_{\text{eff}} = \left(\frac{1}{r_1} + \frac{1}{r_2} \right)^{-1}. \tag{6.56}$$

When $\theta \sim 0°$, as it is for the liquid films in Figure 6.9, the dependence of meniscus force on penetration distance u is given by

$$F_{\text{men}} \simeq 4\pi R \gamma_{\text{L}} \left(1 + \frac{u}{2 r_{\text{eff}}} \right). \tag{6.57}$$

So, the effective capillary radius r_{eff}—and, consequently, the capillary pressure in the meniscus—can be determined by measuring the slope of the force vs. distance curve when the tip is in contact with the liquid (Mate et al. 1989, Mate and Novotny 1991). From the force vs. distance curves in Figure 6.9, we see that r_{eff} is larger for the thicker films. In a later chapter, we will discuss the reason for this, which is due to the balance of disjoining and capillary pressures.

Once the tip pushes through the liquid film, it experiences hard wall repulsive contact with the underlying solid surface. When the tip is withdrawn, an extra force, approximately 3×10^{-8} N in excess of the meniscus force, has to be applied to overcome the solid-solid adhesive force.

6.3.1.2   Solid–solid adhesion in the presence of a liquid meniscus

Equations (6.52), (6.53), and (6.57) give only the capillary pressure component to the adhesive force. To get the total adhesive force $L_{\text{adh-tot}}$ acting on the sphere in contact, one needs to add the solid–solid adhesive force to L_{men} by combining eqs. (6.10) and (6.52):

$$L_{\text{adh-tot}} = 2\pi R \gamma_{\text{L}} \left(\cos \theta_1 + \cos \theta_2 \right) + 2\pi R W_{\text{ALB}}, \tag{6.58}$$

where W_{ALB} is the work of adhesion between the two solids A and B in the presence of a liquid L. So, in Figure 6.9, where $R = 66$ nm, the 3×10^{-8} N extra force to overcome the solid-solid adhesion implies that $W_{\text{tip–liquid–sample}} \sim 72$ mN/m for this situation.

When the sphere and flat are the same material ($W_{\text{ALA}} = 2\gamma_{\text{SL}}$ and $\theta_1 = \theta_2 = \theta$), eq. (6.58) becomes

$$L_{\text{adh-tot}} = 4\pi R \left(\gamma_{\text{L}} \cos \theta + \gamma_{\text{SL}} \right) \tag{6.59}$$

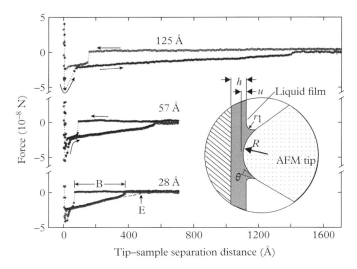

Figure 6.9 *Force acting on an AFM tip as a function of tip–sample separation distance for three samples with different thicknesses (28, 57, and 125 Å) of liquid lubricant films as they are brought into contact with the tip and then withdrawn. The liquid is a perfluoropolyether polymer Fomblin-Z deposited onto silicon wafers. Zero separation distance is defined to be the position where the force on the tip is zero when in contact with the sample. A negative force corresponds to an attractive force between the tip and sample. The inset illustrates the meniscus formation around the tip and the penetration distance u into the liquid film. In the inset, the film thickness h is greatly exaggerated relative to the tip radius R and meniscus radius r_1. Adapted with permission from Mate and Novotny (1991). Copyright 1991, American Institute of Physics.*

which becomes, using Young's equation (5.19),

$$L_{\text{adh-tot}} = 4\pi R \gamma_S \qquad (6.60)$$

indicating that, if the sphere makes good contact with a flat of the same material, the net adhesion force is the same with and without the liquid meniscus present.

Experimentally, however, the presence of liquid around a contact point is frequently found to alter the adhesive force even though eq. (6.60) suggests that it shouldn't. There are several reasons for this:

1. Equation (6.60) involves many assumptions, including that the sphere and flat are the same materials and have negligible elastic deformation in response to the adhesive forces; that the filling angle φ is small ($r_1, r_2 \ll R$), that $r_1 = r_2$, and that the asperity is a sphere (whereas an AFM tip or a surface asperity may have a spherical end, but it may transition to a cone or another shape at a height that is within the meniscus). A more general solution that relaxes these assumptions has been provided by Asay et al. (2010).

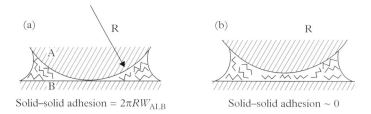

(a) R

(b) R

Solid–solid adhesion = $2\pi R W_{ALB}$ Solid–solid adhesion ~ 0

Figure 6.10 *Expanded view showing the molecules in a liquid meniscus around a sphere sitting on a flat (a) when the liquid molecules are displaced and solid–solid contact occurs and (b) when a layer of liquid molecules remains separating the two solid surfaces.*

2. The loading and adhesive forces may not be sufficient to squeeze all the liquid molecules out from the region where the two solids come into contact, as illustrated in Figure 6.10(b). In such a situation, the solid surfaces are separated by one or two layers of liquid molecules rather than making intimate contact. In this case, the work of adhesion is smaller than W_{ALB}, due to the separation of the surfaces by the molecular layer. In the next chapter, we discuss how the attractive force between solid surfaces varies with separation distance. In these situations, the adhesive force becomes more dominated by the meniscus force.

3. Most solid surfaces have a microscale roughness that greatly reduces the expected solid–solid adhesive force (Section 6.2.3). The liquid meniscus force is less likely, however, to be reduced by surface roughness, an effect perhaps best illustrated by the example of water adhering sand particles either to flat surfaces or to each other, such as when forming sandcastles.

6.4 Menisci in sand and colloidal material

If you look at sand underneath a microscope, you notice that the particles are not spherical, but instead have an irregular shape with a fair amount of microroughness. So, a sand particle sitting on a flat, rather being a sphere-on-flat geometry, is more like the situation shown in Figure 6.11, where solid–solid contact occurs only at the summits of the irregular protrusions.

Sand grains are typically 50–2000 μm in diameter, much larger than the Kelvin radius at low humidity. So, at low but finite humidity, capillary condensation causes minute menisci of water to form around the particle–flat contact points, as shown in Figure 6.11(a). Since these protrusions or asperities have radii of curvatures much smaller than the mean macroscopic radius R of the particle, the net adhesive force acting on the particle is much smaller than that predicted by $4\pi R\gamma_S$. Consequently, if we sprinkle dry sand onto a flat surface and turn the surface over, most of the sand falls off since the adhesive force on each particle is less than the force of gravity.

(a)   (b)

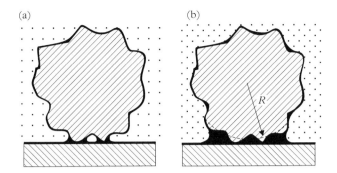

Figure 6.11 *A particle of sand sitting on a flat surface at different humidities. (a) At low humidity, a small amount of water forms menisci around the sharp, jagged protrusions or asperities that contact the flat surface. (b) At high humidity, more water condenses around the contact points, flooding the gaps between them.*

The situation, however, is quite different at high humidity, such as might be experienced at an ocean beach. At high humidity, the large Kelvin radius leads to the condensation of water menisci large enough to flood the gaps between the contacting asperities, as illustrated in Figure 6.11(b). For this flooded situation, the meniscus force is approximated by eq. (6.53), with R = the mean radius of the sand particle. So, if you spend a day at the beach, you will notice that sand sticks to most surfaces that it comes into contact with, but eventually falls off when moved to a drier environment like the inside of your home.

Let's estimate the meniscus force for a typical sand particle with a diameter of 0.4 mm sitting on a flat with a flooded contact region, and θ_1 and $\theta_2 = 0°$:

$$L_{\text{men}} = 4\pi R \gamma_{\text{L}}$$
$$= 4\pi \times (0.2 \text{ mm}) \times (73 \text{ mN/m})$$
$$= 0.18 \text{ mN}, \tag{6.61}$$

which is much larger than the weight of the sand particle:

$$\text{Weight} = \frac{4}{3}\pi g \rho R^3$$
$$= \frac{4}{3}\pi \times \left(9.8 \text{ m/s}^2\right) \times \left(2400 \text{ kg/m}^3\right) \times (0.0002 \text{ m})^3$$
$$= 0.0008 \text{mN}. \tag{6.62}$$

So, if the flat is turned upside down, the particle does not fall off.

We can also understand why at the beach the partially wet strip between the water and the sand is the easiest place to walk on. Since the water only partially covers the sand, the meniscus forces are strongest, making it relatively firm. Further away from the water, the sand is only slightly wet with only a few menisci forming between the particles, so the sand is loose, soft, and difficult to walk on. Right at the water's edge, the gaps between the sand particles are mostly flooded, so very few menisci are present to hold the sand together, again making it difficult to walk on. (This explanation is attributed to Einstein (Mermin 2005).)

Table 6.1 illustrates and describes what is occurring at the microscale as the amount of water is varied in sand or in other granular media. The schematic diagrams in third column illustrate how, as small amounts of water are added to sand, liquid bridges first form between the grains. Then, as the amount of liquid is increased, the gaps between the grains fill up with water, which slowly reduces the cohesion from the capillary pressure. Finally, this cohesion form capillary action is completely gone once all the grains are immersed in the liquid, forming a slurry. Intuitively, the maximum amount of capillary cohesion should occur for the granular media between the pendular and funicular states (second and third row of Table 6.1), when the number of meniscus bridges is at a maximum. Indeed, experiments show that the maximum in elastic modulus (and hence optimum strength) occurs when sand has a very low volume fraction of water (~1%) (Pakpour et al. 2012). Consequently, if you add a little water to sand, it becomes quite sticky and can be used to build beautiful sandcastles; but, if you keep adding water to the sand, eventually all the space between the particles fills up with water, menisci no longer exist to hold the sand together, and now all you have is unattractive silt.

Table 6.1 *Granular or colloidal media with various amounts of liquid. In the schematic diagrams, the solid circles represent the individual grains of sand while the gray represents the liquid. Reproduced with permission from Mitarai and Nori (2006). Copyright 2006, Taylor and Francis Ltd.*

Liquid content	State	Schematic diagram	Physical description
No	Dry		Cohesion between grains is negligible.
Small	Pendular		Liquid bridges are formed at the contact points of graing. Cohesive forces act through the liquid bridges.
Middle	Funicular		Liquid bridges around the contact points and liquid-filled pores coexist. Both give rise to cohesion between particles.
Almost saturated	Capillary		Almost all the pores are filled with the liquid, but the liquid surface forms menisci and the liquid pressure is lower than the air pressure. This suction results in a cohesive interaction between particles.
More	Slurry		The liquid pressure is equal to, or higher than, the air pressure. No cohesive interaction appears between particles.

Adding water to sand can also reduce the coefficient of friction for a sled being pulled over it. Experiments by Fall et al. (2014) indicate that the coefficient of friction of dry sand can be lowered by a factor of two by adding enough water to achieve the funicular regime illustrated in Table 6.1, which corresponds to a ~5% volume fraction of water. These experiments also indicate the reason why the friction is reduced: the added water makes the sand more rigid, which prevents sand from piling up in front of the sled as it is pulled over, thereby reducing the contribution from plowing friction.

That adding water reduces the friction of sand may help explain how ancient monuments were erected thousands of years ago in the Egyptian desert. One of the reasons why the construction of these monuments is so impressive is that it involved moving huge blocks of stones, colossal statues, and other massive objects across the desert using only human power. Figure 6.12 shows one of the few records from the ancient world of how this was done. This wall painting from the tomb of the monarch Djehutihotep depicts a colossal statue (~70 tons in weight) sitting on a sled being pulled across the desert by 172 men. Also depicted is a solitary man pouring some sort of liquid out in front of the sled. One of the hypothesis is that the man is pouring water onto the desert sand in order to reduce the coefficient of friction, which the experiments of Fall et al. (2014) indicate could achieve a friction coefficient value as low as $\mu \sim 0.3$. Another hypothesis is that the sled was slid over wooden rails that were lubricated by the water being poured, as wet wood-on-wood contacts can have similar a similar range of coefficients of friction as wet sand (Dowson 1978, Ayrinhac 2016).

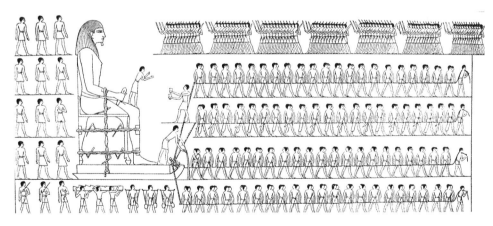

Figure 6.12 *Wall painting from the tomb of Djehutihotep from 1880* BC *depicting a colossal statue being dragged across the desert by 172 men (Newberry 1895). Standing on the front edge of the sled is a man pouring a liquid in front of the sled, presumably a lubricant that reduces the friction forces on the bottom of the sled.*

6.5 Meniscus force for different wetting regimes at contacting interfaces

The trend discussed in the previous section for water in sand can be generalized to other situations when liquid menisci form around the contact points. Figure 6.13 illustrates the different ways liquids can wet two contacting surfaces separated by surface roughness. For this situation, a different terminology is used—*toe dipping, pillbox, flooded* and *immersed*—to described the different regimes of fill levels of liquid than that in Table 6.1 for granular media. The meniscus force between the surfaces depends on which of the four regimes occurs at the contacting interface:

6.5.1 Toe dipping regime

The *toe dipping* regime occurs when the amount of liquid is sufficiently small that menisci form only around the summits of the asperities in contact with the opposing surface

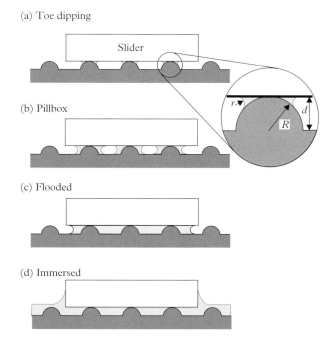

Figure 6.13 *Four wetting regimes for a liquid at the interface of a smooth slider surface contacting a rough surface (Mate and Homola 1997, Mate 1998). The surface roughness separates the surfaces by a distance d, and the liquid contact angles are assumed ~0°. Following the nomenclature introduced by (Matthewson and Mamin 1988), these wetting regimes are labeled toe dipping, pillbox, flooded, and immersed. Reproduced from Mate and Homola (1997) with permission from Springer Science and Business Media. Copyright 1997, Kluwer Academic Publishers.*

(Figure 6.13(a)). The summits are the "toes" that dip slightly into the liquid menisci. For contact angles ~0°, the toe dipping regime occurs when $r < d/2$, where r is the outer meniscus radius and d is the typical separation distance between the two surfaces induced by the surface roughness. If the ith summit has a mean radius of curvature R_i, then from eq. (6.53) the meniscus force L_{men-i} at this asperity contact is

$$L_{men-i} = 4\pi R_i \gamma_L \qquad (6.63)$$

and the total meniscus force acting between the two surfaces $= \sum L_{men-i}$. So, if all the asperities have the same height and radius of curvature, as shown in Figure 6.13, the total meniscus force is

$$L_{men-tot} = n4\pi R \gamma_L, \qquad (6.64)$$

where n is the number of asperities in contact.

6.5.1.1 *Example: toe dipping adhesion with an exponential distribution of summit heights*

The case illustrated in Figure 6.13 is unrealistic in that asperity summits on real surfaces do not all have the same height, but rather a range of heights described by some distribution, such as a Gaussian or self-affine distribution (see Chapter 2). In Section 3.4.1.3, we discussed the example of laser textured surfaces in disk drives, where laser pulses are used to make bumps on disk surfaces, with a well-controlled height. When the recording head slider sits on the laser textured surface, one might expect that this situation could be modeled by assuming that the slider makes contact with all the bumps as in Figure 6.13. However, the laser textured bumps are not all exactly the same height, but rather have a slight variation in height with a standard deviation σ^s of a few nanometers. Consequently, at low to moderate loading forces, the slider only contacts the highest bumps, a small fraction of the bumps underneath the slider surface.

In the present example, we determine the magnitude of the meniscus force for a recording head slider sitting on a laser textured disk surface with the same parameters as discussed in Section 3.4.1.3. (i.e., $L_{ext} = 20$ mN, $R = 200$ μm, and $\sigma^s = 1.5$ nm). In this situation, contact is only made with the highest bumps, and we can assume that these bumps have a height distribution approximated by an exponential. Using eq. (3.32) to estimate the number of bumps in contact:

$$n = \frac{L_{ext} + L_{men-tot}}{\pi^{1/2} E_c \sigma_s^{3/2} R^{1/2}},$$

where the total load L now is expressed as the sum of the externally applied load L_{ext} and a load $L_{men-tot}$ from the total meniscus force acting on those asperities in contact. Using eq. (6.64) for $L_{men-tot}$ and solving for n,

$$n = \frac{L_{\text{ext}}}{\pi^{1/2} E_c \sigma_s^{3/2} R^{1/2} - 4\pi R \gamma_L}. \tag{6.65}$$

Using $\gamma_L = 24$ mN/m for the perfluoropolyether liquid applied to disk surface as a thin lubricant film:

$$n = \frac{20 \text{ mN}}{\pi^{1/2} (108 \text{ GPa}) (1.5 \text{ nm})^{3/2} (200 \text{ } \mu\text{m})^{1/2} - 4\pi (24 \text{ mN/m}) (200 \text{ } \mu\text{m})} = 206, \tag{6.66}$$

considerably higher than the $n = 127$ determined in Section 3.4.1.3 for the situation with no meniscus force. The total meniscus force can then be determined using eq. (6.64):

$$L_{\text{men-tot}} = n4\pi R \gamma_L$$
$$= 206 \times (6.0 \times 10^{-5} \text{N}). \tag{6.67}$$
$$= 12 \text{ mN}$$

For this example, the total loading force $L = L_{\text{ext}} + L_{\text{men-tot}}$ is 60% higher than the externally applied loading force L_{ext}. So, we can see that the existence of toe dipping meniscus forces between contacting asperities substantially increases the total loading force acting at a contacting interface.

6.5.2   Pillbox and flooded regimes

As the amount of liquid between the surfaces increases, the liquid menisci start to flood the gaps between the asperity contacts (Figure 6.13(b)). When this occurs, the menisci around the contact points become very short, wide cylinders—a shape reminiscent of the old fashioned, cylindrically shaped containers in which pills used to be dispensed—leading to the name *pillbox* to describe this regime.

As the amount of liquid between the surfaces further increases, the pillbox menisci eventually run together, flooding the areas between the contact points: the *flooded* regime in Figure 6.13(c). As the pillbox and flooded regimes correspond to low and high fill levels of the gaps between the contacting surfaces, they represent the different extremes of a single wetting regime: the *pillbox–flooded* regime.

If the liquid has a contact angle $0°$, the radius of curvature of the edge of the menisci in the pillbox–flooded regime is $r = d/2$; this sets the capillary pressure $P_{\text{cap}} = -2\gamma_L/d$. Consequently, the total meniscus force is given by

$$L_{\text{pillbox-flooded}} = -A_{\text{men}} P_{\text{cap}}$$
$$= A_{\text{men}} \frac{2\gamma_L}{d} \tag{6.68}$$

where A_{men} is the total flooded area.

A drop of water sandwiched between two glass slides (Figure 5.5(a)) provides a simple example of an interface in the pillbox–flooded regime: The capillary pressure pulls to the two slides together, and the water floods the gap to form one large pillbox shaped meniscus. As discussed in Section 5.2.1, the slides end up with a finite separation distance determined either by the surface roughness or by the heights of the particles trapped between the slides. If this separation distance is $d = 1\ \mu m$, the capillary pressure within the pillbox meniscus is $-14.6\ N/cm^2$ or nearly 1.5 atmospheres, making it extremely difficult to separate the slides.

6.5.3   Immersed regime

When the liquid between contacting surfaces comes from capillary condensation of vapors, the pillbox–flooded regime is thermodynamically unstable since the equilibrium Kelvin radius is greater than the meniscus radius $d/2$. This imbalance drives more vapors to condense, growing the flooded area until the entire contact area becomes flooded, and a meniscus forms on the sides of the contacting bodies, as shown in Figure 6.13(d). This is the *immersed* regime. Since the gaps between contacting bodies tend to be very small, it can take an exceptionally long time (days, months, even years) for vapor molecules to diffuse into these gaps and condense. During this time, the meniscus force contribution to the adhesive force, as given by eq. (6.68), slowly increases; this is a major reason why static friction increases with rest time as described in Section 4.3. When the meniscus starts to grow around the sides of the slider, as shown in Figure 6.13(d), the capillary pressure and the meniscus force start to decrease. So, the transition from the flooded to immersed wetting regimes corresponds to a maximum in the meniscus force between the contacting surfaces. Eventually, the meniscus radius in the immersed regime equilibrates with the Kelvin radius, and the meniscus force reaches an equilibrium value.

For non-volatile liquids in the pillbox–flooded regime, the menisci grow because of the capillary pressure pulling in from the molecularly thin liquid films that coat the solid surfaces. This continues until the disjoining pressure associated with the liquid film equilibrates with the capillary pressure in the menisci. The disjoining pressure and the mechanism of equilibration are discussed further in Chapters 7 and 10.

6.6   Example: liquid adhesion of a microfabricated cantilever beam

In recent years, fabrication of micromechanical structures from thin films has become a commercially viable method for fabricating MEMS such as accelerometers for airbag deployment, ink jet print heads, and digital mirror displays for video projection. As the gaps that separate the free structures are typically only a few microns, adhesion of microstructures to nearby surfaces is an inherent problem in MEMS.

An adhesion problem can appear during the final stages of fabrication of MEMS devices, where an etchant is used to free the structure and then the etchant is rinsed away with deionized water. If the water is removed by simply drying through evaporation,

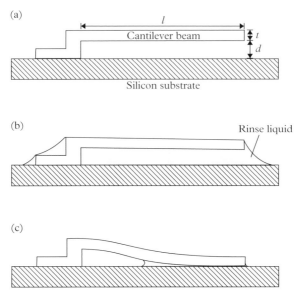

Figure 6.14 *Side view of a MEMS cantilever beam. (a) A free standing beam. (b) The beam during the evaporation of the rinse liquid. (c) The beam when the capillary force leads to adhesion of the beam to the substrate.*

the capillary force exerted by a drying meniscus can be sufficient to bend the flexible microstructures so that they contact neighboring surfaces. This effect is illustrated in Figure 6.14 for a simple microfabricated cantilever beam structure. During the early stages of drying, the immersed regime occurs (Figure 6.14(b)) and the capillary force is fairly low. As the liquid volume around the cantilever decreases, the meniscus radius decreases, increasing the capillary pressure and causing the cantilever to bend toward the substrate until it eventually touches (Figure 6.14(c)). This situation corresponds to the flooded regime where the capillary force reaches a maximum.

Several methods have been developed either to avoid meniscus formation or to alleviate the resulting adhesion or "stiction" (Maboudian and Howe 1997, Kim et al. 1998, Romig et al. 2003):

- *Supercritical drying*—Liquid CO_2 is used for the final rinse step, followed by pressurizing and heating above the supercritical point, and then venting at constant temperature so that no meniscus forms during drying.
- *Low surface tension solvents*—Rather than drying after the water rinse step, the water is rinsed away with a low surface tension solvent such as methanol to reduce the magnitude of the capillary force in the final drying phase.
- *Polymer standoffs*—The microstructures are fabricated with polymer standoffs that hold them apart during the drying process and that are removed with a final ashing step.

- *Vapor etching*—Highly reactive vapors, such as the vapor from hydrofluoric acid, can quickly and effectively etch sacrificial MEMS materials like silicon oxide, thus preventing the need for using a liquid etchant or rinsing step.
- *Surface roughening*—The microstructure and substrate surfaces are roughened to increase the separation of the surface means during contact so as to reduce the capillary pressure when in the pillbox-flooded wetting regime. Increased roughness also helps reduce the number of surface asperities in contact, and, correspondingly, the meniscus force for the toe dipping phase of drying.
- *Low energy surface monolayers*—The microstructure and substrate surfaces are coated with a monolayer film that provides for a high water contact angle to reduce the meniscus force.

The last two methods are also effective for reducing the likelihood that the microstructures will become stuck after fabrication and during use, a potential problem in high humidity conditions.

One might think that if one did a thorough job of drying the microstructure, for example, by heating it above the boiling point of water, that those structures stuck together by the meniscus forces would become unstuck once the meniscus forces are gone. Generally though, once stuck, the structures do not become unstuck since solid–solid adhesion at the contacting asperity summits, along with the van der Waals and electrostatic forces acting across the narrow gap separating the contacting surfaces, are sufficient to hold them together. These forces are discussed in the following chapter.

6.7 Example: surface forces in biological attachments

We have all noticed and perhaps marveled at how small insects and lizards are able to climb up vertical structures and even walk upside down on ceilings. So, it shouldn't come as a surprise that, for the past few centuries, scientists have been studying the mechanisms for how these creatures manage to achieve adhesion and traction on these surfaces.

While the microscale details are still under debate, the basic mechanisms for biological attachment are now established. For insects such as flies and beetles, adhesion is typically dominated by meniscus forces, with the liquid being secreted by glands near their attachment pads. Other animals such as spiders and geckos use dry adhesion, which can be purely van der Waals interactions (as discussed in Section 7.3.7 for the case of geckos) or a combination of van der Waals and electrostatic interactions (Izadi and Penlidis 2013).

For all these animals, adhesion takes place on specialized attachment pads on their legs or toes that can be either smooth pads or, more typically, covered by a fine pattern of hairy protuberances that help maximize the contact area. Figure 6.15 shows scanning electron microscopy (SEM) images of the terminal ends of these hairy pads for animals ranging from the small beetle to the much larger gecko. Having hairy structures that are highly flexible helps to bring many points into contact even on rough surfaces, enabling

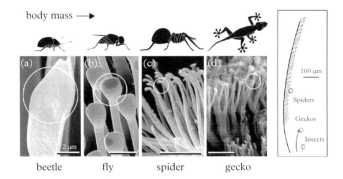

Figure 6.15 *SEM images of the hair structures, called setae, on the attachment pads of various animals. Note that as the animals become larger, the setae become finer. Adhesion takes place at the terminal ends called spatula (circled). The geometry of these spatula and their high density at the ends of very flexible setae allows for many of points of contacts with rough surfaces increasing amount of adhesion force possible. Reproduced with permission from Arzt et al. (2003). Copyright 2003, National Academy of Sciences, USA.*

the animals to maximize the interfacial adhesion and shear stresses. Splitting the contact into numerous subcontacts called spatulae, each at the end of a highly flexible hair called setae, alleviates the effect of internal elastic coupling intrinsic to bulk solids where pushing down on one region of a solid with a contact force also cause neighboring regions to move downward, making it less likely for these surrounding regions to make contact. Instead, having numerous contacts at the ends of setae allows them to deform independently, in similar manner to the multiple springs in a mattress helping its surface to conform to the shape of a body resting on it. These attachment pads with the hairy structure of flexible setae also help these small creatures control the amount of surface area brought into contact as the judicious manipulation of the loading force and pad orientation relative to the surface controls the number of spatula brought into contact.

The following analysis by Arzt et al. (2003) helps us to understand how having a hairy structure on the attachment pad helps to maximize the adhesion force. If the attachment pad is modeled as a single hemispherical protuberance, as shown in Figure 6.16(a), then the dry adhesive force would be given by

$$L_{\text{adh}-\text{single}} = \frac{3}{2}\pi RW$$

$$= \frac{3}{4}\pi DW, \tag{6.69}$$

where eq. (6.28) is used, since most biological attachments follow JKR contact mechanics due to the relatively low elastic compliance of biological material (Spolenak et al. 2005). So, if an ordinary house fly with a mass of a 12 mg were to hang by one of its legs from the ceiling by a smooth attachment pad with a radius $R = 100$ μm, the work of adhesion W from this single hemispherical protrusion would need to be

$$W = \frac{2L_{\text{adh-single}}}{3\pi R}$$

$$= \frac{2\,(12\text{ mg})\,(9.8\text{ m/s}^2)}{3\pi\,(100\text{ μm})}$$

$$= 250\text{ mJ/m}^2, \tag{6.70}$$

which is many times more than the 10–50 mJ/m^2 range of van der Waals interaction energies for biological materials.

If the attachment pad, rather than a single hemispherical contact, is composed of n contacts of diameter s covering an area D^2, as illustrated in Figure 6.16(b), then the density of contacts is $N_A = n/D^2 = 1/s$, and the adhesive force if all these setae make contact is

$$L_{\text{adh-n setae}} = n\frac{3}{4}\pi s W$$

$$= \sqrt{n}\frac{3}{4}\pi DW$$

$$= \sqrt{n}L_{\text{adh-single}}. \tag{6.71}$$

From this analysis we see that splitting up the contact into n subcontacts with radius $R/\sqrt{n}$ can increase the adhesion force by a factor of $\sqrt{n}$. So even in the dry adhesion case, a house fly can hang from one attachment pad provided that the number of setae in contact is on the order 10^2–10^3. Since the capillary meniscus force scales with radius R

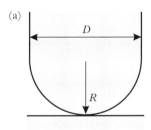

Figure 6.16 *(a) Model of a biological attachment pad as consisting of a single protrusion with a hemispherical shape with diameter D when contacting a flat surface. (b) Model of a biological attachment pad with area D^2 as consisting of setae with diameter s and ends with radius R$\prime$. Also shown are how the individual setae conform over a protrusion when contacting a surface. Adapted with permission from Arzt et al. (2003). Copyright 2003, National Academy of Sciences, USA.*

in the same manner as the JKR force, a similar $\sqrt{n}$ increase in meniscus adhesion force is achieved by having n setae within the apparent area of contact of the attachment pad.

....

6.8 PROBLEMS

1. Using the analysis in the paper by Pastewka and Robbins (2014), derive eq. (6.45).

2. In Chapter 3, Problem 2, we considered an array of asperities making contact with a flat surface. Consider exactly the same set of asperities, but now assume that the two surfaces have a work of adhesion W and the contact behavior falls into the DMT regime and that no external load is applied. As a reminder, the lower surface is flat and rigid. The upper surface is flat but with a set of $n \times n$ protruding asperities with spherical shaped apexes, all having the same height. The surfaces span an area $100b \times 100b$ where b is a length. The asperities have a radius of curvature of R/n and are spaced apart by a linear distance of $100b/n$. The asperities also have modulus E and Poisson's ratio ν. The asperities are spaced far enough apart that their stress and strain fields do not interact with each other. Calculate and plot on a log-linear plot the following quantities as a function of the number of asperities n^2 for $1 < n^2 < 10,000$. Normalize your calculations to the value of the quantity for $n^2 = 1$ of:

   (a) the number of asperities per unit area
   (b) the pull-off force per asperity and the total pull-off force
   (c) the contact area for each asperity
   (d) the total contact area
   (e) the ratio of the total contact area to the apparent contact area $(100b \times 100b)$ between the two surfaces
   (f) the elastic deformation d of each asperity
   (g) the maximum contact stress at the interface and acting perpendicular to the interface of an individual asperity contact
   (h) the absolute maximum shear stress (which is not necessarily at the interface!) at an individual asperity contact

   Based on these plots, comment on the effect that splitting up a single asperity into an array of asperities has on the adhesion force between two surfaces.

3. In Chapter 3, Problem 1, we considered the normal contact stiffness k_N. As a reminder, this is defined as $k_N = \partial L(\Delta z)/\partial z$ where $L(\Delta z)$ is the normal loading force that elastically deforms the surface a distance Δz along the z axis perpendicular to the surface. We showed that the normal contact stiffness for the Hertz model is $2aE_c$.

   (a) Show that the normal contact stiffness for the DMT model is also $2aE_c$.
   (b) Find the normal contact stiffness for the JKR model.

(c) Plot the normal contact stiffness for a 50 nm radius silicon AFM tip ($E_c = 200$ GPa) pressed into contact with a rigid substrate as the load varies from a maximum of 100 nN down to the pull-off force, given that the work of adhesion is 0.1 J/m^2 and assuming first that the DMT model is valid and then assuming that the JKR model is valid.

4. Consider a small silica colloidal sphere placed in good contact with a smooth silica glass slide in a regular lab environment. The slide is turned upside down. How small does the sphere need to be so that it does not detach due to gravity? Assume that the solid surface energies of both the silica glass and the colloidal sphere is 0.20 J/m^2. Explain any assumptions you make in your analysis.

··

6.9 REFERENCES

Arzt, E., S. Gorb and R. Spolenak (2003). "From micro to nano contacts in biological attachment devices." *Proceedings of the National Academy of Sciences* **100**(19): 10603–6.

Asay, D. B., M. P. De Boer and S. H. Kim (2010). "Equilibrium vapor adsorption and capillary force: exact Laplace–Young equation solution and circular approximation approaches." *Journal of Adhesion Science and Technology* **24**(15–16): 2363–82.

Ashurst, W. R., M. P. De Boer, C. Carraro and R. Maboudian (2003). "An investigation of sidewall adhesion in MEMS." *Applied Surface Science* **212**: 735–41.

Ayrinhac, S. (2016). "The transportation of the Djehutihotep statue revisited." *Tribology Online* **11**(3): 466–73.

Carpick, R. W. and M. Salmeron (1997). "Scratching the surface: fundamental investigations of tribology with atomic force microscopy." *Chemical Reviews* **97**(4): 1163–94.

Carpick, R. W., N. Agrait, D. F. Ogletree and M. Salmeron (1996a). "Measurement of interfacial shear (friction) with an ultrahigh vacuum atomic force microscope." *Journal of Vacuum Science & Technology B* **14**(2): 1289–95.

Carpick, R. W., N. Agraït, D. F. Ogletree and M. Salmeron (1996b). "Variation of the interfacial shear strength and adhesion of a nanometer-sized contact." *Langmuir* **12**(13): 3334–40.

Carpick, R. W., D. F. Ogletree and M. Salmeron (1997). "Lateral stiffness: a new nanomechanical measurement for the determination of shear strengths with friction force microscopy." *Applied Physics Letters* **70**(12): 1548–50.

Carpick, R. W., D. F. Ogletree and M. Salmeron (1999). "A general equation for fitting contact area and friction vs load measurements." *Journal of Colloid and Interface Science* **211**(2): 395–400.

Chaudhury, M. K. and G. M. Whitesides (1991). "Direct measurement of interfacial interactions between semispherical lenses and flat sheets of poly(dimethylsiloxane) and their chemical derivatives." *Langmuir* **7**(5): 1013–25.

Chaudhury, M. K. and G. M. Whitesides (1992). "Correlation between surface free-energy and surface constitution." *Science* **255**(5049): 1230–2.

DelRio, F. W., M. P. de Boer, J. A. Knapp, E. D. Reedy, P. J. Clews and M. L. Dunn (2005). "The role of van der Waals forces in adhesion of micromachined surfaces." *Nature Materials* **4**(8): 629–34.

Derjaguin, B. V. (1934). "Friction and adhesion. IV. The theory of adhesion of small particles." *Kolloid Zeits* **69**: 155–64.

Derjaguin, B. V., V. M. Muller and Y. P. Toporov (1975). "Effect of contact deformations on adhesion of particles." *Journal of Colloid and Interface Science* **53**(2): 314–26.

Dowson, D. (1978). *History of tribology*. London: Longman.

Enachescu, M., R. J. A. van den Oetelaar, R. W. Carpick, D. F. Ogletree, C. F. J. Flipse and M. Salmeron (1998). "Atomic force microscopy study of an ideally hard contact: the diamond(111)–tungsten carbide interface." *Physical Review Letters* **81**(9): 1877–80.

Fall, A., B. Weber, M. Pakpour, N. Lenoir, N. Shahidzadeh, J. Fiscina, C. Wagner and D. Bonn (2014). "Sliding friction on wet and dry sand." *Physical Review Letters* **112**: 175502.

Fogden, A. and L. R. White (1990). "Contact elasticity in the presence of capillary condensation. I. The nonadhesive Hertz problem." *Journal of Colloid and Interface Science* **138**(2): 414–30.

Frantz, P., A. Artsyukhovich, R. W. Carpick and M. Salmeron (1997). "Use of capacitance to measure surface forces. 2. Application to the study of contact mechanics." *Langmuir* **13**(22): 5957–61.

Fuller, K. N. G. and D. Tabor (1975). "The effect of surface roughness on the adhesion of elastic solids." *Proceedings of the Royal Society of London, Series A: Mathematical, Physical and Engineering Sciences* **345**(1642): 327–42.

Gao, G., R. J. Cannara, R. W. Carpick and J. A. Harrison (2007). "Atomic-scale friction on diamond: a comparison of different sliding directions on (001) and (111) surfaces using MD and AFM." *Langmuir* **23**(10): 5394–405.

Greenwood, J. (1997). "Adhesion of elastic spheres." *Proceedings of the Royal Society of London, Series A: Mathematical, Physical and Engineering Sciences* **453**(1961): 1277–97.

Grierson, D. S., E. E. Flater and R. W. Carpick (2005). "Accounting for the JKR-DMT transition in adhesion and friction measurements with atomic force microscopy." *Journal of Adhesion Science and Technology* **19**(3–5): 291–311.

Horn, R. G., J. N. Israelachvili and F. Pribac (1987). "Measurement of the deformation and adhesion of solids in contact." *Journal of Colloid and Interface Science* **115**(2): 480–92.

Houston, M. R., R. T. Howe and R. Maboudian (1997). "Effect of hydrogen termination on the work of adhesion between rough polycrystalline silicon surfaces." *Journal of Applied Physics* **81**(8): 3474–83.

Israelachvili, J. N. (2011). *Intermolecular and surface forces* (3rd ed.). Burlington, MA: Academic Press.

Israelachvili, J. N., E. Perez and R. K. Tandon (1980). "On the adhesion force between deformable solids." *Journal of Colloid and Interface Science* **78**(1): 260–1.

Izadi, H. and A. Penlidis (2013). "Polymeric bio-inspired dry adhesives: van der Waals or electrostatic interactions?" *Macromolecular Reaction Engineering* **7**(11): 588–608.

Jacobs, T. D., K. E. Ryan, P. L. Keating, D. S. Grierson, J. A. Lefever, K. T. Turner, J. A. Harrison and R. W. Carpick (2013). "The effect of atomic-scale roughness on the adhesion of nanoscale asperities: a combined simulation and experimental investigation." *Tribology Letters* **50**(1): 81–93.

Jacobs, T. D. B., J. A. Lefever and R. W. Carpick (2015a). "Measurement of the length and strength of adhesive interactions in a nanoscale silicon–diamond interface." *Advanced Materials Interfaces* **2**(9): 1400547.

Jacobs, T. D. B., J. A. Lefever and R. W. Carpick (2015b). "A technique for the experimental determination of the length and strength of adhesive interactions between effectively rigid materials." *Tribology Letters* **59**(1): 1.

Johnson, K. L. (1985). *Contact mechanics*. Cambridge: Cambridge University Press.

Johnson, K. L., K. Kendall and A. D. Roberts (1971). "Surface energy and contact of elastic solids." *Proceedings of the Royal Society of London, Series A: Mathematical and Physical Sciences* **324**(1558): 301–13.

Kendall, K. (2007). *Molecular adhesion and its applications: the sticky universe*. Berlin: Springer Science & Business Media.

Kim, C. J., J. Y. Kim and B. Sridharan (1998). "Comparative evaluation of drying techniques for surface micromachining." *Sensors and Actuators A: Physical* **64**(1): 17–26.

Lantz, M. A., S. J. O'Shea, M. E. Welland and K. L. Johnson (1997). "Atomic-force-microscope study of contact area and friction on $NbSe_2$." *Physical Review B* **55**(16): 10776.

Lee, A. E. (1978). "Role of elastic deformation in adhesion of solids." *Journal of Colloid and Interface Science* **64**(3): 577–9.

Lessel, M., P. Loskill, F. Hausen, N. N. Gosvami, R. Bennewitz and K. Jacobs (2013). "Impact of van der Waals interactions on single asperity friction." *Physical Review Letters* **111**(3): 035502.

Maboudian, R. and R. T. Howe (1997). "Critical review: adhesion in surface micromechanical structures." *Journal of Vacuum Science & Technology B* **15**(1): 1–20.

Mate, C. M. (1998). "Molecular tribology of disk drives." *Tribology Letters* **4**: 119–23.

Mate, C. M. and A. M. Homola (1997). "Molecular tribology of disk drives." In: *Micro/nanotribology and its applications*, B. Bhushan, Ed. Dordrecht: Kluwer Academic Publishers, pp.647–61.

Mate, C. M. and V. J. Novotny (1991). "Molecular-conformation and disjoining pressure of polymeric liquid-films." *Journal of Chemical Physics* **94**(12): 8420–7.

Mate, C. M., M. R. Lorenz and V. J. Novotny (1989). "Atomic force microscopy of polymeric liquid-films." *Journal of Chemical Physics* **90**(12): 7550–5.

Matthewson, M. J. and H. J. Mamin (1988). "Liquid-mediated adhesion of ultra-flat solid surfaces." *Proceedings of Materials Research Society Symposia* **119**: 87–92.

Maugis, D. (1992). "Adhesion of spheres: the JKR–DMT transition using a Dugdale model." *Journal of Colloid and Interface Science* **150**(1): 243–69.

Maugis, D. (1996). "On the contact and adhesion of rough surfaces." *Journal of Adhesion Science and Technology* **10**(2): 161–75.

Mermin, N. D. (2005). "My life with Einstein." *Physics Today* **58**(12): 10–11.

Meyer, E., R. Luthi, L. Howald, M. Bammerlin, M. Guggisberg and H. J. Guntherodt (1996). "Site-specific friction force spectroscopy." *Journal of Vacuum Science & Technology B* **14**(2): 1285–8.

Mitarai, N. and F. Nori (2006). "Wet granular materials." *Advances in Physics* **55**(1–2): 1–45.

Muller, V. M., B. V. Derjaguin and Y. P. Toporov (1983). "On two methods of calculation of the force of aticking of an elastic sphere to a rigid plane." *Colloids and Surfaces* **7**(3): 251–9.

Muller, V. M., V. S. Yushchenko and B. V. Derjaguin (1980). "On the influence of molecular forces on the deformation of an elastic sphere and its sticking to a rigid plane." *Journal of Colloid and Interface Science* **77**(1): 91–101.

Newberry, P. E. (1895). *El Bersheh: the tomb of Tehuti-Hetep*. London: Egypt Exploration Fund.

Pakpour, M., M. Habibi, P. Møller and D. Bonn (2012). "How to construct the perfect sandcastle." *Scientific Reports* **2**: 549.

Pastewka, L. and M. O. Robbins (2014). "Contact between rough surfaces and a criterion for macroscopic adhesion." *Proceedings of the National Academy of Sciences* **111**(9): 3298–303.

Peressadko, A. G., N. Hosoda and B. N. J. Persson (2005). "Influence of surface roughness on adhesion between elastic bodies." *Physical Review Letters* **95**(12): 124301.

Persson, B. N. J. (2002). "Adhesion between an elastic body and a randomly rough hard surface." *The European Physical Journal E* **8**(4): 385–401.

Persson, B. N. J. and M. Scaraggi (2014). "Theory of adhesion: role of surface roughness." *The Journal of Chemical Physics* **141**(12): 124701.

Pugno, N. M. (2007). "Towards a spiderman suit: large invisible cables and self-cleaning releasable superadhesive materials." *Journal of Physics: Condensed Matter* **19**(39): 395001.

Rabinovich, Y. I., J. J. Adler, A. Ata, R. K. Singh and B. M. Moudgil (2000). "Adhesion between nanoscale rough surfaces. I. Role of asperity geometry." *Journal of Colloid and Interface Science* **232**(1): 10–16.

Roberts, A. D. and A. B. Othman (1977). "Rubber adhesion and dwell time effect." *Wear* **42**(1): 119–33.

Romig, A. D., M. T. Dugger and P. J. McWhorter (2003). "Materials issues in microelectromechanical devices: science, engineering, manufacturability and reliability." *Acta Materialia* **51**(19): 5837–66.

Schwarz, U. D. (2003). "A generalized analytical model for the elastic deformation of an adhesive contact between a sphere and a flat surface." *Journal of Colloid and Interface Science* **261**: 99–106.

Spolenak, R., S. Gorb, H. J. Gao and E. Arzt (2005). "Effects of contact shape on the scaling of biological attachments." *Proceedings of the Royal Society of London, Series A: Mathematical, Physical and Engineering Sciences* **461**(2054): 305–19.

Tabor, D. (1977). "Surface forces and surface interactions." *Journal of Colloid and Interface Science* **58**(1): 2–13.

7

Physical Origins of Surface Forces

In the preceding two chapters, we discussed the concept of surface energy and how some forces between surfaces arise from surface energy and surface tension. Ultimately though, surface energy originates from forces between atoms; so, if we truly want to understand surface forces, we need to understand the various forces that can act between pairs of atoms and the ways these forces sum up to obtain the surface energy or the net force acting between surfaces. This approach also lets us examine a wider range of surface forces than just those related to surface energies, and allows us to examine how surface forces, especially adhesive forces, depend on separation distance and other physical parameters.

Figure 7.1 compares the relative strengths of the different bonding mechanisms typically encountered within molecules and between molecules and solid surfaces. In this chapter, we focus on the forces between surfaces originating from these atomic and molecular interaction potentials, as well as from contact electrification. We will consider only the net force acting in a direction perpendicular to the two solid surfaces, leaving viscous and tangential (frictional) forces for later chapters where we will also delve

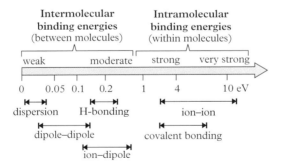

Figure 7.1 *Comparison of the magnitude of different bonding mechanisms that bind atoms together into molecules (intramolecular bonding) and that bind molecules together (intermolecular bonding). Note that the intermolecular binding energies in this figure are for very small molecules and that the intermolecular bonding energies scale either with size of the molecules (dispersion force) or the number of dipoles, H-bonding sites, etc., within the molecules.*

Tribology on the Small Scale: A Modern Textbook on Friction, Lubrication and Wear. Second edition. C. Mathew Mate and Robert W. Carpick. © Oxford University Press 2019. Published in 2019 by Oxford University Press.
DOI: 10.1093/oso/ 9780199609802.001.0001

into how these various forces acting between atoms contribute to friction, adhesion, and wear.

7.1 Normal force sign convention

Before beginning the discussion of forces acting in the normal direction to surfaces, we should first realize that the choice of sign for this force is somewhat arbitrary, with the choice usually depending on the context. For the forces between atoms and molecules, the typical convention is for an attractive force to have a negative sign and a repulsive force a positive sign, as shown in Figure 7.2. Since many forces between solids originate from the atomic and molecular potentials, it is natural to continue this sign convention when discussing those forces between surfaces originating from these potentials. (This is the convention adopted in this chapter, and results in an adhesive force F_{adh} having a negative sign.)

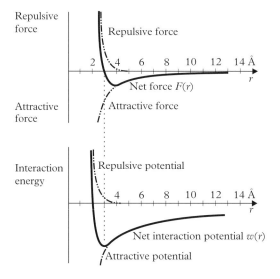

Figure 7.2 *The bottom plot shows how the potential energy w(r) between two molecules changes with separation distance r, while the top plot shows the force F(r) = −dw/dr on the molecules. The distances shown are typical values for atomic interactions.*

However, when discussing macroscopic tribological phenomena, an attractive normal force acts as a *loading force* L_{adh}, which by convention has a positive sign when pushing two surfaces towards each other, in contrast to the convention of using a negative sign for an adhesive force F_{adh}. So, when considering an adhesive force's contribution to the loading force in a macroscopic situation, one typically switches the sign of the adhesive force from negative to positive so that it matches the sign of the externally applied loading force (i.e., $L_{adh} = -F_{adh}$). This was the approach utilized in previous chapters, and in this chapter we will continue to use the nomenclature L_{adh} (with a positive sign) to represent an adhesive force contribution to the loading force, and we will use the letter F to represent forces arising from interaction potentials, with F being negative when attractive.

A similar convention issue occurs when energy is considered. When discussing the interaction between atoms and molecules, an attractive interaction results in a negative interaction energy, as shown in Figure 7.2. (This is consistent with the relationship $F = -dw/dr$, where F is the force as a function of the separation r generated by interaction potential energy w.) However, when the attractive interactions between atoms are summed to obtain the interaction energy between surfaces, the resulting interaction energy per unit area is often referred to as the *work of adhesion* W_{adh} (Section 5.3), which in this book has a positive sign.

7.2 Repulsive atomic potentials

In Section 3.1, we discussed how the forces between atoms lead to the mechanical properties of materials. Now, we discuss how these atomic and molecular forces combine to become the forces acting between solid surfaces. Figure 7.2 illustrates how the force between two atoms or molecules typically has two components: a short range repulsive force and a longer range attractive force. We start by considering the repulsive atomic potential (Israelachvili 2011: Section 7.2).

When two atoms are brought close enough together, their electron clouds overlap; as this overlap is energetically unfavorable, a repulsive force is generated to minimize the amount of overlap. This repulsive force not only arises from the electrostatic repulsion between like charges, but also results from the Pauli exclusion principle. Fundamentally quantum mechanical in nature, the Pauli exclusion principle states that two elections cannot occupy the same quantum state. The resulting atomic repulsion is called *exchange repulsion, hard core repulsion, steric repulsion*, or, for the case of ions, *Born repulsion*. The essential characteristic of an atomic repulsion potential is that it rises steeply with decreasing separation distance r. A number of expressions are frequently used for describing a repulsive potential:

- *Hard wall potential*

$$w(r) = \infty \text{ where } r \leq r_0$$
$$w(r) = 0 \text{ where } r > r_0 \tag{7.1}$$

and r_0 represents the hard wall radius

- *Inverse power-law potential*

$$w(r) = \frac{A}{r^n}, \tag{7.2}$$

where n is usually in the range 9–16. A common choice for the value of n is 12, which, when combined with the attractive potential for a van der Waals force, gives the well-known *Lennard-Jones potential* between atoms:

$$w_{L-J}(r) = \frac{A}{r^{12}} - \frac{B}{r^6}.$$ (7.3)

- *Exponential potential*

$$w(r) = ce^{-r/r_0},$$ (7.4)

where c and r_0 are adjustable parameters, with r_0 on the order 0.2 Å

These potentials are empirical relations with no physical theory underlying them. All three potentials, however, do a good job of describing the steeply rising character of the actual repulsive potentials. The inverse power and exponential potentials are more realistic potentials than the hard wall potential as they allow for some compressibility or softness of the potential, while the hard wall potential has the advantage of simplicity.

7.3   Van der Waals forces

7.3.1   Van der Waals forces between molecules

The van der Waals interaction describes the class of interactions between atoms and molecules, where the interaction energy $w(r)$ varies as $1/r^6$ and the force varies as $1/r^7$ with separation distance r. Three types of van der Waals forces can occur, which are illustrated in Figure 7.3:

1. *Orientation force (Keesom interaction)*—The electrostatic interaction between two polar molecules (such as two water molecules) in the liquid or gas phase, with both having freely rotating permanent dipoles, averaged over the thermally-excited probability distribution of rotation angles. Since the antiparallel alignment of the dipoles is more energetically favorable, an attractive force is generated when they adopt the antiparallel arrangement and a repulsive force when they adopt a parallel arrangement. Higher temperatures disorder this alignment, reducing the interaction force.

2. *Induction force (Debye interaction)*—The electrostatic interaction in the liquid or gas phase between a polar molecule with a freely rotating permanent dipole moment and a non-polar molecule, averaged over the thermally-excited probability distribution of rotation angles. As illustrated in Figure 7.3(d), the electric field from the permanent dipole moment induces a dipole moment in the non-polar molecule $\mu_{ind} = E\alpha$, where α is the polarizability of the non-polar molecule; the two dipoles then interact electrostatically to generate the induction force.

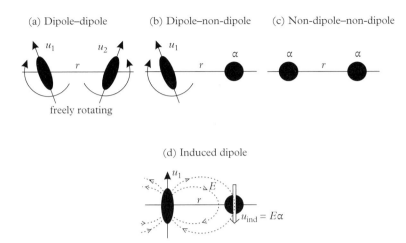

Figure 7.3 *Schematic of the three types of van der Waals interactions between molecules.*
(a) Orientation interaction between two freely rotating polar molecules with dipole moments u_1 and u_2.
(b) Induction interaction between a polar and a non-polar molecule with polarizability α. (c) Dispersion
interaction between two non-polar molecules. (d) Illustrates how the electric field E of a polar molecule
induces a dipole moment u_{ind} in a non-polar molecule.

3. *Dispersion force (London interaction)*—The induced dipole–dipole interaction between two non-polar molecules, which is quantum mechanical in origin. A simple explanation is that a non-polar atom, such as helium, will have an instantaneous dipole moment that naively can be associated with the instantaneous positions of the fluctuating electrons around its nucleus. The electric field from this instantaneous dipole polarizes any nearby atom, inducing in it a dipole moment. The electrostatic interaction between these two induced dipoles corresponds to an exchange of virtual photons, generating an instantaneous dispersion force. While the net dipole moment averages out to zero over time, the induced force is always attractive, and its time-average is finite.

Even though the dispersion force or London force is the weakest of the three van der Waals interactions, it is often considered the most important for the following reasons:

- The dispersion force present between all materials, while the orientation and induction forces require constituents with permanent dipole moments.
- The dispersion force does not decrease with temperature, unlike the orientation force.
- While the dispersion force may be the weakest of the van der Waals interactions, since most materials do not have a strong polar component, it is often the largest contributor to the overall van der Waals interaction. Only for the most polar of molecules, such as water, is the dispersion force not the largest contributor.

For two dissimilar polar molecules interacting across a vacuum, the van der Waals potential $w_{VDW}(r)$ can be expressed as the sum of these three interactions (Israelachvili 2011: eq. (6.17)):

$$
\begin{aligned}
w_{VDW}(r) &= -\frac{C_{VDW}}{r^6} \\
&= -\frac{C_{orient} + C_{ind} + C_{disp}}{r^6} \\
&= -\left[\frac{u_1^2 u_2^2}{3 k_B T} + \left(u_1^2 \alpha_{02} + u_2^2 \alpha_{01}\right) + \frac{3\alpha_{01}\alpha_{02} h \nu_1 \nu_2}{2(\nu_1 + \nu_2)}\right] \Big/ (4\pi\varepsilon_0)^2 r^6 \quad (7.5)
\end{aligned}
$$

where u_1 and u_2 are the permanent dipole moments of the two molecules, α_{01} and α_{02} are their polarizabilities, and $h\nu_1$ and $h\nu_2$ are their first ionization potentials (h and ν are, respectively the Planck constant and first ionization frequency). Equation (7.5) indicates that the van der Waals force F_{VDW} is always attractive across a vacuum (i.e., $F_{VDW} = -\partial w_{VDW}/\partial r < 0$). When a third medium (say a liquid solvent) fills the space between the two interacting molecules, the interaction of the molecular dipole moments within this medium must be taken into account when computing the van der Waals force. McLachlan (1963) developed a generalized theory for the van der Waals interaction that describes the effect of an intervening medium. From the McLachlan theory, an expression can be derived for the van der Waals interaction between dissimilar materials across an intervening third medium, in terms of the dielectric constants $\varepsilon(0)$ at zero frequency and refractive indices n determined at optical frequencies of the materials involved (Israelachvili 2011: Section 6.7). The van der Waals potential $w_{VDW}(r)$ is expressed as the sum of a "zero frequency" dipolar contribution $w(r)_{\nu=0}$ and "finite frequency" dispersion force contribution $w(r)_{\nu>0}$:

$$
w_{VDW}(r) = w(r)_{\nu=0} + w(r)_{\nu>0}. \quad (7.6)
$$

For two dissimilar spherical molecules or small particles labeled 1 and 2, with radii a_1 and a_2 with an intervening medium labeled 3, and assuming the electromagnetic adsorption frequency ν_e is the same for all three media, the zero and finite frequency contributions to the van der Waals interaction are

$$
w(r)_{\nu=0} = -\frac{3 k_B T a_1 a_2}{r^6}\left(\frac{\varepsilon_1(0) - \varepsilon_3(0)}{\varepsilon_1(0) + 2\varepsilon_3(0)}\right)\left(\frac{\varepsilon_2(0) - \varepsilon_3(0)}{\varepsilon_2(0) + 2\varepsilon_3(0)}\right) \quad (7.7)
$$

and

$$
w(r)_{\nu>0} = -\frac{\sqrt{3}\, h\nu_e a_1^3 a_2^3}{2r^6}\frac{\left(n_1^2 - n_3^2\right)\left(n_2^2 - n_3^2\right)}{\left(n_1^2 + 2n_3^2\right)^{1/2}\left(n_2^2 + 2n_3^2\right)^{1/2}\left[\left(n_1^2 + 2n_3^2\right)^{1/2} + \left(n_2^2 + 2n_3^2\right)^{1/2}\right]}, \quad (7.8)
$$

where it has been assumed that all three media have the same adsorption frequency ν_e.
The important points to notice about eqs. (7.7) and (7.8) are that:

- Since $h\nu_e >> k_B T$, the finite frequency dispersion force contribution $w(r)_{\nu > 0}$ is typically much greater than the zero frequency dipolar contribution $w(r)_{\nu = 0}$, especially when the intervening medium 3 is a gas or vacuum ($n_3 \sim 1$).
- The van der Waals force between two molecules is greatly reduced by having another material, such as a liquid, intervening between the two molecules. For the special case where all three media have the same refractive index ($n_1 = n_2 = n_3$), then $w(r)_{\nu > 0} = 0$ (van der Waals dispersive component is zero).
- For dissimilar molecules, $w(r)_{\nu > 0}$ is repulsive if n_3 is intermediate between n_2 and n_1. So, while the van der Waals force is always attractive when medium 3 is a gas or vacuum ($n_3 \sim 1$) and usually attractive when $n_3 > 1$, it can be repulsive (depending on the magnitude of $w(r)_{\nu = 0}$) in the special situation of n_1 and n_2 being unequal and n_3 being intermediate between these values.

7.3.1.1 *Retardation effects for dispersion forces*

In the above expressions for the van der Waals potential energies between molecules (eqs. (7.5), (7.7), and (7.8)), the interaction decays at the rate of $-1/r^6$ with separation distance. At sufficiently large distances, however, the dispersion potential component decays at a faster rate of $-1/r^7$. This comes about from the finite amount of time that it takes an electric field to travel between the molecules, which is determined by the speed of light in the intervening medium. At larger separations, the time that it takes for the electric field to travel from the first atom to the second atom and back again becomes comparable to the oscillation period of the fluctuating instantaneous dipole of the first atom. So, when the electric field from the induced dipole moment of the second atom returns to the first atom, the direction of the first atom's dipole moment has likely changed from where it originally started, weakening the interaction.

For two atoms in free space, this *retardation effect* begins at separations greater than 5 nm, and the potential approaches a $-1/r^7$ dependence at $r > 100$ nm. Since the dispersion force is already very weak at 5 nm separation, retardation effects are usually of little interest. When an intervening medium is present, however, the retardation manifests itself at smaller separations as the speed of light is slower. Only the dispersion force suffers from retardation; so, while the dispersion force may be the dominant van der Waals force at small separations, the induction and orientation contributions to the van der Waals force ultimately dominate at larger separation distances.

7.3.2 Van der Waals forces between macroscopic objects

In the 1930s, Hamaker showed that the van der Waals forces between molecules could be extended to macroscopic objects by performing a pairwise sum over the atoms in these objects (Hamaker 1937). This approach requires the assumption that the van der Waals interaction is *additive*, which is valid if no intervening material exists between the two objects (i.e., a gap of vacuum or gas). In the next couple of sections, we will use this approach of summing the molecular pair potentials, which is called the *microscopic approach* (following Israelachvili 2011: Chapters 11 and 13), to derive the van der Waals interaction potential and force for several practical geometries.

7.3.2.1   *Molecule–flat surface interaction*

If an atom or molecule is a distance D above a flat solid surface, as shown in Figure 7.4, it experiences an attractive van der Waals force between it and all the atoms in the solid. To evaluate the net interaction energy, we sum the individual pair potentials (Figure 7.4(a)):

$$w_{\text{net}}(r) = -\sum_i \frac{C_{\text{VDW}}}{r_i^6}. \tag{7.9}$$

Rather than summing up over all the individual atoms in the solid, it is simpler, of course, to convert this sum into an integral. For this integral, we take the element of integration to be a circular ring with radius x, as illustrated in Figure 7.4(b), where each part of the ring is a distance $r = \sqrt{z^2 + x^2}$ from the molecule, the cross-sectional area of the ring is $dxdz$, and the ring volume is $2\pi x dx dz$. If ρ is the number density of the atoms in the solid, the number of atoms in the ring is $2\pi\rho x dx dz$, and the net potential energy for a molecule a distance D above the surface is

$$
\begin{aligned}
w_{\text{molecule-flat}}(D) &= -2\pi\, C_{\text{VDW}}\,\rho \int_{z=D}^{z=\infty} dz \int_{x=0}^{x=\infty} \frac{x dx}{\left(\sqrt{z^2 + x^2}\right)^6} \\
&= \frac{\pi\, C_{\text{VDW}}\,\rho}{2} \int_D^\infty \frac{dz}{z^4} \\
&= \frac{\pi\, C_{\text{VDW}}\,\rho}{6D^3}
\end{aligned}
\tag{7.10}
$$

and the net van der Waals force acting on the molecule is

$$
\begin{aligned}
F_{\text{molecule-flat}} &= -\frac{\partial w_{\text{molecule-flat}}(D)}{\partial D} \\
&= -\frac{\pi\, C_{\text{VDW}}\,\rho}{2D^4}.
\end{aligned}
\tag{7.11}
$$

The force is directed along the z-axis, and the negative sign indicates that it is directed towards the surface, that is, attractive. (In doing this integral, we also made the approximation of an r^{-6} potential even at distances r where retardation effects may be significant; this approximation is valid when D is much less than the distance where the interaction transitions from non-retarded to retarded.)

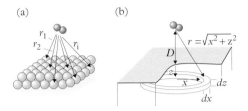

Figure 7.4 *(a) Interaction of a molecule above a surface represented as the sum of the individual pair potentials between that molecule and atoms in the solid. (b) This interaction of the molecule a distance D above the semi-infinite solid is now represented as an integral using a ring shaped integration element with radius x and with its center a distance z away from the molecule.*

7.3.2.2   Flat–flat interaction

Now that we have the interaction energy between an atom or molecule above a planar surface (eq. (7.10)), it is straightforward to derive the expression for the van der Waals energy between two planar surfaces. Rather than considering the case of two surfaces of infinite lateral extent (which would result in an infinite interaction energy), we consider the case illustrated in Figure 7.5(a) where the top surface has a unit area, so as to determine the van der Waals potential *per unit area*, while the bottom surface has a much larger area, effectively extending to infinity. For an atom in the top material (with a number density of atoms ρ_2) at a distance z away from the bottom surface (with a number density of atoms ρ_1), the interaction energy per atom is given by eq. (7.10). Multiplying this interaction energy by the number of atoms in a thin sheet of thickness dz gives the interaction energy for the sheet, and integrating from $z = D$ to infinity gives the van der Waals energy per unit area

$$W_{\text{flat-flat}}(D) = -\frac{\pi C_{\text{VDW}}\rho_1\rho_2}{6}\int_D^\infty \frac{dz}{z^3}$$

$$= -\frac{\pi C_{\text{VDW}}\rho_1\rho_2}{12D^2}. \qquad (7.12)$$

While eq. (7.12) was derived for one surface with unit area interacting with an infinitely larger one, it is also valid for the interaction between two unit area surfaces provided the separation distance D is much smaller than the lateral dimensions of the surfaces ($D << b$).

From eq. (7.12) the van der Waals force per unit area (or pressure) acting between two surfaces is readily determined:

$$F_{\text{flat-flat}}(D) = -\frac{\partial W_{\text{flat-flat}}(D)}{\partial D}$$

$$= -\frac{\pi C_{\text{VDW}}\rho_1\rho_2}{6D^3}. \qquad (7.13)$$

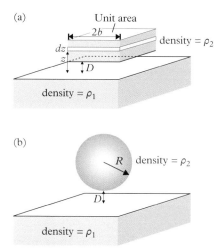

Figure 7.5 *(a) Geometry for calculating the interaction energy between a planar surface with unit area and a second planar surface of much larger area when D << b. (b) For sphere and planar surface with D << R.*

7.3.2.3 Sphere–flat interaction

For the sphere-above-flat geometry shown in Figure 7.5(b), the van der Waals potential again can be calculated by doing an integral for the molecule–surface and surface–surface potentials, as we did previously for eqs. (7.10) and (7.12). For situations where $D << R$, a simpler method is to use the Derjaguin approximation expressed in eq. (6.2) together with the van der Waals energy between two flat surfaces expressed in eq. (7.12) to determine the van der Waals force for a sphere above a flat surface:

$$F(D)_{\text{sphere-flat}} = 2\pi RW_{\text{flat-flat}}(D)$$
$$= -\frac{\pi^2 RC_{\text{VDW}}\rho_1\rho_2}{6D^2}. \qquad (7.14)$$

Integrating this force, we obtain the van der Waals interaction energy for the sphere-above-flat geometry:

For $D << R$,
$$W_{\text{sphere-flat}}(D) = \int_D^\infty F_{\text{sphere-flat}}\left(D'\right) dD'$$
$$= -\frac{\pi^2 RC_{\text{VDW}}\rho_1\rho_2}{6D} \qquad (7.15)$$

where the constant of integration is determined by setting $W_{\text{sphere-flat}}(\infty) = 0$.

For the other extreme of a sphere far away from the flat ($D >> R$), the van der Waals potential can be approximated by multiplying the potential per atom (eq. (7.10)) times the number of atoms in the sphere $(4\pi R^3 \rho_2/3)$:

$$\text{For } D >> R, \qquad W_{\text{sphere-flat}}(D) = -\left(\frac{4\pi R^3 \rho_2}{3}\right) \frac{\pi C_{\text{VDW}} \rho_1}{6D^3}. \qquad (7.16)$$

7.3.3   The Hamaker constant

Common to the above expressions for the van der Waals interaction between macroscopic objects is the combination of parameters $C_{\text{VDW}} \rho_1 \rho_2$, which together describe the strength of the van der Waals interaction between materials 1 and 2 across an empty gap. This combination is typically replaced with a new parameter A, which is called the *Hamaker constant* in recognition of H. C. Hamaker's contributions to the field. Within the microscopic approach described above, the relationship between the Hamaker constant A and the molecular level parameters is

$$A = \pi^2 C_{\text{VDW}} \rho_1 \rho_2. \qquad (7.17)$$

From eq. (7.17), we see that the Hamaker constant or the strength of the van der Waals interaction is proportional to the number densities and the polarizabilities of the atoms and molecules in the two interacting bodies. (Equation (7.5) shows that C_{VDW} is proportional to the polarizabilities.)

Figure 7.6 summarizes the expressions of the non-retarded van der Waals interaction potentials and forces, between macroscopic objects across an empty gap in terms of the Hamaker constant. While these expressions were derived in the previous sections using the microscopic approach, they still remain valid when the more rigorous quantum electrodynamics (QED) approach described in the next section is used.

7.3.3.1   *Determining Hamaker constants from Lifshitz's theory*

Rather than summing over point-to-point interaction potentials, a more fundamental (and consequently a more precise) approach to calculating the van der Waals interaction is to use QED. This was first done by Casimir (1948) who analyzed the retarded attractive dispersion force between two perfectly conducting plates, a force now referred to as the *Casimir force*. Lifshitz subsequently developed the general QED solution for the van der Waals interaction between two materials (which could be either conductors or non-conductors) separated by a third intervening material (Lifshitz 1956, Dzyaloshinskii et al. 1960, 1961).

In Lifshitz's theory, the materials are treated as a continuum, and the van der Waals force arises from fluctuations in the electromagnetic field between the two objects, modified by the intervening material due to the boundary conditions imposed by the

Geometry	van der Waals interaction energy	van der Waals interaction force
Two flat surfaces 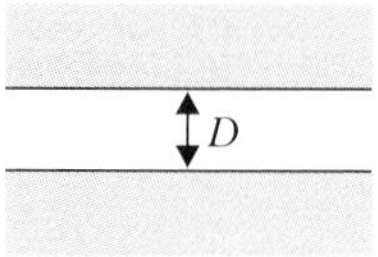 	$W = -\dfrac{A}{12\pi D^2}$ (per unit area)	$-\dfrac{\partial W}{\partial D} = -\dfrac{A}{6\pi D^3}$ (per unit area)
Sphere above flat $D \ll R$ 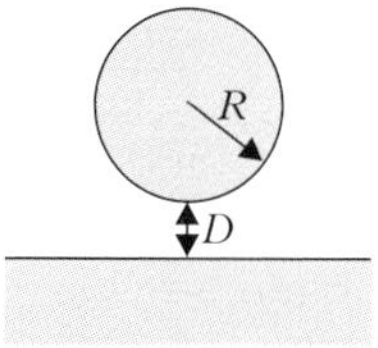 	$W = -\dfrac{AR}{6D}$	$\dfrac{\partial W}{\partial D} = -\dfrac{AR}{6D^2}$
Particle above flat $D \gg R$ 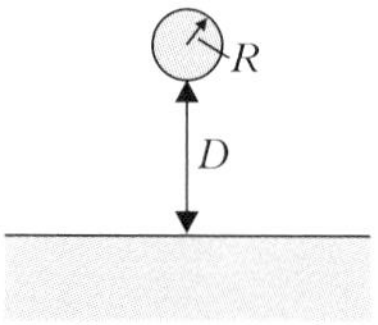 	$W = -\dfrac{2AR^3}{9D^3}$	$-\dfrac{\partial W}{\partial D} = -\dfrac{2AR^3}{3D^4}$

Figure 7.6 *Non-retarded van der Waals interaction energies and forces for selected geometries.*

geometry. The fluctuations or virtual photons take the form of standing waves that only occur at certain optical frequencies, with other frequencies from the continuum of virtual photons being excluded from the gap between the two materials. This exclusion results in a pressure that pushes the two materials together.

Using Lifshitz's theory, the van der Waals force between macroscopic objects can be calculated from the geometry of the interacting objects and from the frequency-dependent dielectric response function $\varepsilon(\omega)$ of the two interacting materials and any intervening materials. Typically though, the calculation is only done for a planar geometry, such as the one illustrated in Figure 7.7, and the result used to determine the Hamaker constant for the particular combinations of materials involved. The van der Waals force for other geometries is then computed by using the determined Hamaker constant in the appropriate expression (some of which are listed in Figure 7.6).

A major advantage of Lifshitz's theory over the earlier microscopic approach of Hamaker is that the van der Waals interaction can be calculated when an intervening material is present; Figure 7.7(a) illustrates such a case, where materials 1 and 3 interact

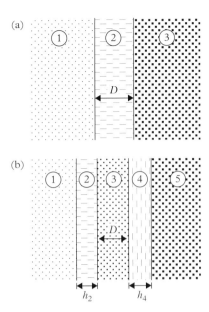

Figure 7.7 *(a) Schematic geometry for calculating the Hamaker constant between materials 1 and 3 separated by material 2. The Hamaker constant for this geometry is labeled A_{123}. (b) Geometry for a multilayer system where the half space 1 has an overlayer 2 with thickness h_2 and the half space 5 has an overlayer 4 with thickness h_4.*

across a gap filled with material 2. For these cases, we append subscripts to the Hamaker constant to identify the order of the materials across the gap; in the situation illustrated in Figure 7.7(a), for example, the Hamaker constant is labeled A_{123}.

The Lifshitz analysis is also readily extendable to cases where multiple layers of materials exist in the gap between two objects (Ninham 1970, Israelachvili 1972, White et al. 2005). Figure 7.7(b) illustrates such a case where a material 1 with an overlayer 2 interacts across a gap filled with medium 3 with another half space of material 5 with overlayer 4. For this situation, where the film thicknesses h_2 and h_4 are constant, the Lifshitz analysis can be used to calculate an effective Hamaker constant A_{12345}, but now A_{12345} is a function of the gap separation distance D (White et al. 2005). In the limit of very small or large separation distances, the Hamaker constant for the multilayer system in Figure 7.7(b) can be approximated by the Hamaker constant for the appropriate three-material system:

- for $D << h_2, h_4, A_{12345}(D) \rightarrow A_{234}$
- for $D >> h_2, h_4, A_{12345}(D) \rightarrow A_{135}$

though retardation effects often become important at these larger separation distances where $D >> h_2, h_4$.

As the accuracy of the Lifshitz approach depends on the accuracy of the dielectric spectra obtained from experiments and of their arithmetical representations, extensive work must first be done to obtain good representations of these optical spectra before the Lifshitz analysis can be applied. Fortunately, advances in vacuum ultraviolet spectroscopy and valence-electron-energy-loss spectroscopy have led to improved spectra over the wide range of optical frequencies needed for such an analysis (French et al. 1995, 1998, Palik and Ghosh 1998). With good mathematical representations of optical spectra now available, precise calculations of Hamaker constants can now be done (certainly more precise than those from the McLachlan theory mentioned previously). Several tabulations of Hamaker constants can be found in the literature (Hough and White 1980, Bergstrom 1997, French 2000).

Tables 7.1, 7.2, and 7.3 give values for the Hamaker constant for material combinations sometimes encountered in tribology. We can see that:

- The Hamaker constants across a vacuum or air gap are all within an order of magnitude of each other. For the same materials acting across a gap, these range at the low end from 3.8×10^{-20} J, for the low surface energy material PTFE (commercial name Teflon^{TM}) ($\gamma = 18$ mN/m), to 2.96×10^{-19} J at the high end for the high atomic density material diamond ($\rho = 1.76 \times 10^{23}$ cm^{-3}).

- The Hamaker constants for polymers (3.80–6.26×10^{-20} J) are at the lower end of this range, due to the lower density and polarizability of these materials. These

Table 7.1 *Non-retarded Hamaker constants for few inorganic materials and metals calculated from Lifshitz theory. Two identical materials separated by water or a vacuum (or air) gap, that is, A_{121}, where material 2 is vacuum or water. Hamaker constants for inorganics from Bergstrom (1997) and for metals from Klimchitskaya et al. (2000) and Eichenlaub et al. (2002).*

Material	Hamaker constant (10^{-20} J)	Hamaker constant (10^{-20} J)
	Across vacuum or air	Across water
α-Al$_2$O$_3$	15.2	3.67
C (diamond)	29.6	13.8
Mica	9.86	1.34
β-SiC	24.6	10.7
Si$_3$N$_4$ (amorphous)	16.7	4.85
SiO$_2$ (silica)	6.50	0.46
Au	44	—
Ag	39	—
Al	36	—
Cu	28	—

Table 7.2 *Non-retarded Hamaker constants for polymer materials. Two identical materials separated by water or a vacuum (or air) gap, that is, A_{121}, where material 2 is vacuum or water. Calculated from the Lifshitz theory (Hough and White 1980, French et al. 2007).*

Polymer	Hamaker constant (10^{-20} J)	
	Across vacuum or air	Across water
Poly(methyl methacrylate) (PMMA)	5.84	0.147
ET-MAA	4.77	0.216
Polyester (PEST)	6.09	0.405
Polycarbonate (PCARB)	5.08	0.350
Polystyrene (PSTY)	7.09	0.771
Polytetrafluoroethylene (PTFE)	3.80	0.333
Polyvinyl chloride (PVC)	7.80	1.30

Table 7.3 *Non-retarded Hamaker constants (10^{-20} J) for inorganic materials interacting against four different types of materials across vacuum/water. Calculated from Lifshitz theory (Bergstrom 1997).*

Material	Silica	Silicon nitride	Alumina	Mica
C (diamond)	13.7/1.71	22.0/7.94	21.1/7.05	17.0/4.03
Mica	8.01/0.69	12.8/2.45	12.2/2.15	9.86/1.34
6H-SiC	12.6/1.52	20.3/7.22	19.2/6.05	15.5/3.54
β-Si$_3$N$_4$	10.8/1.17	17.3/5.13	16.5/4.43	13.3/2.61
SiO$_2$ (quartz)	7.59/0.63	12.1/2.07	11.6/1.83	9.35/1.16

lower Hamaker constants help to account for polymers having weaker interactions and lower surface energies than other materials.

- Materials with high atomic density and polarizability are at the upper end of the range (1–3 $\times$ 10^{-19} J). Metals are examples of highly polarizable materials. Interacting across a vacuum, metals typically have a Hamaker constant of 3–5 $\times$ 10^{-19} J (Israelachvili 2011: Section 13.7).

- The Hamaker constants are greatly reduced when water fills the gap between the two materials. This is a fairly general phenomenon: introducing a liquid into the gap between materials typically reduces the van der Waals interaction by up to an order of magnitude below the vacuum/air value. This should not be surprising, as it was pointed out during the discussion of the McLachlan theory (eqs. (7.7) and (7.8)).

Tables 7.1 and 7.2 show the Hamaker constants for two identical materials acting across a gap. From these Hamaker constants (which have the form A_{121} or A_{1v1}, where the subscript v represents a vacuum or vapor gap), we can approximate the Hamaker constant for pairs of dissimilar materials by using the combining relations (Israelachvili 2011: Section 13.12):

$$A_{121} = A_{212}, \tag{7.18}$$

$$A_{123} \simeq (A_{121}A_{323})^{1/2}, \tag{7.19}$$

$$A_{1v3} \simeq (A_{1v1}A_{3v3})^{1/2}, \tag{7.20}$$

$$A_{121} \simeq A_{1v1} + A_{2v2} - 2A_{1v2}. \tag{7.21}$$

7.3.3.2   *Example: van der Waals force on a polystyrene sphere above a PTFE flat*

In Section 6.2.1.2, we calculated the adhesive force, arising from surface energy, for a polystyrene sphere sitting on a PTFE flat. Since the interactions between the molecules within and between polystyrene and PTFE are mainly van der Waals, we can also calculate this adhesive force directly from the van der Waals force, if we know the appropriate value to use for the separation distance D. (Simply using $D = 0$ would result in the unphysical value of $F_{VDW} = -\infty$.) Our approach here will be to calculate how the van der Waals adhesive force varies with separation distance D, and then determine at what separation distance the adhesive force is the same as that determined in Section 6.2.1.2 using surface energies.

First, we estimate the Hamaker constant using eq. (7.20):

$$
\begin{aligned}
A_{PTFE-v-PSTY} &= (A_{PTFE-v-PTFE}A_{PSTY-v-PSTY})^{1/2} \\
&= \left[\left(3.8 \times 10^{-20} \, J\right) \times \left(7.09 \times 10^{-20} \, J\right)\right]^{1/2} \\
&= 5.2 \times 10^{-20} \, J
\end{aligned}
\tag{7.22}
$$

The van der Waals force is then calculated using the equation for a sphere above a flat (Figure 7.6, assuming $D << R$):

$$
\begin{aligned}
F_{VDW} &= -\frac{A_{PTFE-v-PSTY}R}{6D^2} \\
&= -\frac{\left(5.2 \times 10^{-20} \, J\right) \times (0.5 \, \mu m)}{6D^2} \\
F_{VDW} \, (\mu N) &= -\frac{0.43}{D\left(\overset{\circ}{A}\right)^2}.
\end{aligned}
\tag{7.23}
$$

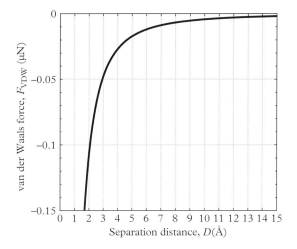

Figure 7.8 *Attractive van der Waals force F_{VDW} for a 1 μm diameter polystyrene sphere at a distance D above a PTFE flat.*

Figure 7.8 plots the van der Waals force between this 1 μm polystyrene sphere and a PFTE flat as function of separation distance D.

In Section 6.2.1.2, we found that $L_{adh} = 0.15$ μN (implying that $F_{VDW} = -0.15$ μN, when the sphere contacts the flat). From Figure 7.8 and eq. (7.23), we see that this value of adhesive force occurs at $D_0 = 1.69$ Å. As discussed in the next section, this is close to the value of $D_0 = 1.65$ Å generally considered the van der Waals separation distance where the atoms on two opposing surfaces are in intimate contact.

As discussed in Section 6.2.3, surface roughness can substantially reduce the magnitude of the adhesive force. One simple approach for estimating how surface roughness reduces the van der Waals adhesive force is to set the van der Waals separation distance D equal to the separation of surface means when the two rough surfaces are in contact plus D_0. For example, if we assume that roughness separates the surface means by 8.31 Å (so $D = 10$ Å) for this 1 μm diameter particle sitting on a flat, we see from Figure 7.8 that F_{VDW} decreases to -3 nN, or a magnitude fifty times smaller than what is expected for perfectly smooth surfaces contacting with $D = D_0$.

7.3.4   Surface energies arising from van der Waals interactions

In prior sections, we summed the van der Waals interaction energies between the atoms in one material with all the atoms in the other material to obtain the *object–object* interaction energy. Summing instead over the interactions between *all* atoms in both materials determines, this time, the *total* van der Waals interaction energy. Consequently, if we consider two identical materials with parallel surfaces separated by a distance D, then the total van der Waals interaction energy per unit area can be expressed as

$$W_{\text{total-per-unit-area}}(D) = (\text{bulk cohesive energy from van der Waals})$$

$$+ 2\gamma_{\text{VDW}} - \frac{A}{12\pi D^2}, \tag{7.24}$$

where the first term is component of the bulk cohesive energy arising from van der Waals interactions, the second term is the surface energy of the two surfaces arising from van der Waals energy, and the last term is the flat–flat van der Waals interaction energy. If D_0 is the separation distance where the atoms on the two surfaces come into intimate contact such that $W_{\text{total-per-unit-area}}(D_0) = (\text{bulk cohesive energy})$, this implies that

$$\gamma_{\text{VDW}} = \frac{A}{24\pi D_0^2}. \tag{7.25}$$

At first, one might be tempted to equate D_0 with the interatomic distance a_0; but this approach ends up underestimating the surface energy γ, as eq. (7.25) is essentially derived from a continuum theory where the surfaces are treated as ideally smooth and ignores the change in coordination of atoms when a surface is produced; consequently, eq. (7.25) is not really valid for separation distances on the order of an atomic diameter. It can be shown, however, that $D_0 = a_0/2.5$ provides a good estimate for the value of D_0. If we use $a_0 = 4$ Å as a typical atomic spacing, then $D_0 = 1.65$ Å, and eq. (7.25) becomes

$$\gamma_{\text{VDW}} = \frac{A}{24\pi\left(1.65\ \text{Å}\right)^2}. \tag{7.26}$$

Equation (7.26) can be rewritten to estimate the Hamaker constant if we know the component of the material's surface energy arising from van der Waals interactions:

$$A\,(\text{J}) \approx 2.1 \times 10^{-21}\gamma_{\text{VDW}}\left(\text{mJ/m}^2\right). \tag{7.27}$$

For materials where van der Waals interactions are the dominant contributor to the bulk cohesive energy, which is the case for many polymers, then we can equate the total surface energy with the component from van der Waals ($\gamma \approx \gamma_{\text{VDW}}$) and eq. (7.26) provides a reasonable estimate of the total surface energy. For materials with significant contributions from other types of bonding (hydrogen bonding, covalent bonding, metallic bonding, etc.), however, the above analysis is more difficult to apply. One approach is to use the Fowkes method discussed in Section 5.5.2.2 to determine the dispersive component of the surface energy γ_d and use that in place of γ_{VDW} in eq. (7.27) to determine the Hamaker constant A.

7.3.5   Van der Waals adhesive pressure

If we imagine an arbitrary plane dividing a van der solid, the force acting across this plane is an adhesive pressure that holds the solid together and is balanced by the atomic level repulsive forces. This adhesive pressure for a van der Waals solid is given by

$$p_{adh} = \frac{A}{6\pi D_0^3}.$$
(7.28)

So, picking a lower end value of the Hamaker constant of $A = 5 \times 10^{-20}$ J, we find, in materials weakly bound by van der Waals interactions with $D_0 = 1.65$ Å, that the adhesive pressure is

$$p_{adh} = \frac{5 \times 10^{-20} \text{ J}}{6\pi \left(1.65 \times 10^{-10}\text{m}\right)^{-3}}$$
$$= 590 \text{ MPa}$$
$$= 5823 \text{ atm.}$$
(7.29)

So, the adhesive pressure holding these weakly bound materials together is impressively large!

Even greater adhesive pressures occur for materials where stronger types of bonding (covalent, metallic, ionic, etc.) exists in additions to the van der Waals pressure. Therefore, one might expect for a similarly large adhesive pressure to occur over the apparent contact area when two objects are brought into contact. However, it is only for materials like elastomers, which have elastic moduli low enough for the van der Waals adhesive forces to deform the surfaces to the extent that intimate contact occurs over the whole area, that such high pressures can develop over the whole contact area. (These are also those situations where the JKR theory of Section 6.2.2.1 works best for estimating the adhesive force from surface energies.) As discussed in Section 6.2.3, for other situations the stiffness of the surface roughness prevents intimate contact from occurring except at asperity summits, so the high adhesive pressures are localized to these small contact zones (where they can still dramatically influence the resulting tribology).

As discussed in the previous chapter, the gaps between the solid–solid contacts can become filled with soft materials or liquids from the condensation of vapors or from preexisting lubricant or contamination layers; for this situation, the adhesive pressure exists over the entire contact area, but is the average of the adhesive pressures at solid–solid contacts and the soft material between the contacts.

The next section discusses in detail how surface roughness influences the average van der Waals adhesive pressure in the absence of intervening material between the solid–solid contacts.

7.3.6 Van der Waals interaction between contacting rough surfaces

So far, the expressions for the van der Waals interaction between surfaces (such as those in Figure 7.6) have assumed that the surface roughness is much smaller than the separation distance D. Often, though, we need to know the van der Waals contribution to the adhesive force acting between contacting rough surfaces. Intuitively, we can see that increasing roughness lowers the van der Waals adhesive force as it results in a larger fraction of the opposing surfaces being farther away from each other. The challenge is to come up with a suitable analytical or numerical method for estimating the van der Waals interaction between these rough surfaces. Since the nature of roughness can vary dramatically from one situation to the next, a single modeling approach is unlikely to be applicable for all cases. In Sections 6.2.3 and 6.2.4 we discussed the newer methods of Persson (2002) and Pastewka and Robbins (2014) for analyzing the adhesion between randomly rough surfaces. In what follows, we discuss three older but simpler approaches for handling the van der Waals forces between rough surfaces, which fortunately apply to a wide range of rough contacting interfaces including those without random roughness:

1. *Separation of surface means approach*

The simplest approach is to model the surfaces as being relatively smooth except for a few exceptionally high asperities that set the average separation distance D_{ave} when contact occurs. Figure 7.9(a) illustrates this situation where the rms surface roughness $\sigma_q \ll D_{ave}$ and the van der Waals force across the non-contacting area dominates over the force acting at the contacting asperities. The van der Waals interaction energy per unit area can be approximated by the expression for the van der Waals energy between two parallel surfaces:

$$W_{\text{adh-eff}} = \frac{A}{12\pi D_{ave}^2},\tag{7.30}$$

where we assume that $D_{ave} < 5$ nm so we can use the non-retarded expression. In eq. (7.30), we equated van der Waals interaction energy with the effective work of adhesion $W_{\text{adh-eff}}$, which, as discussed in Section 6.2.3, represents a work of adhesion value that has been lowered by surface roughness. For this situation, the average adhesive pressure is given by

$$p_{\text{adh}} = -\frac{\partial W_{\text{adh-eff}}}{\partial D}$$

$$= \frac{A}{6\pi D_{ave}^3}.\tag{7.31}$$

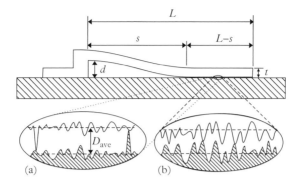

Figure 7.9 *A microfabricated cantilever beam that has become stuck to the substrate due to the van der Waals adhesive force. The expanded views show two possible types of surface roughness. (a) Where contact only occurs at a few exceptionally high asperities and the surface roughness in between these asperities is much less than the separation of surface means, D_{ave}. (b) Where the heights of surface asperities follow a more typical distribution, such as a Gaussian distribution.*

2. Greenwood–Williamson approach

The next approach applies to rough surfaces with spherical or elliptically shaped asperities summits whose heights follow approximately a Gaussian distribution. For this situation illustrated in Figure 7.9(b), we can use the Greenwood–Williamson (G–W) method (described in Section 3.4.1) to model the two rough surfaces in terms of an equivalent rough surface contacting a flat surface. For the equivalent rough surface, the asperities are all assumed to have the same radius of curvature R_{ave}, which is the average radius of curvature of asperities on the top and bottom surfaces, and the distribution of asperity heights is assumed to be Gaussian with a standard deviation given by $\sigma_s = \left(\sigma_{top}^2 + \sigma_{bottom}^2\right)^{1/2}$.

The next step is to determine the mean separation D_{ave} where the repulsive forces from the elastic or plastic deformation of the contacting asperities balances the adhesive pressure from van der Waals forces and meniscus forces, plus any externally applied load. This is accomplished by determining the repulsive contact and adhesive pressures as a function of average separation distance D_{ave} through numerical integration over the distribution of asperity heights. The van der Waals force acting on an asperity with a height z from the surface mean is given by using the sphere-above-flat expression in Figure 7.6 with $D_0 = 1.65$ Å used for those asperities in contact and $(D_{ave} - z + D_0)$ used for those not in contact. Typically, D_{ave} is found to be in the range 3–5σ, depending on the strength of the adhesive pressures, the externally applied load, and the contact stiffness.

Once the D_{ave} has been determined, the total van der Waals interaction energy per unit area (the effective work of adhesion $W_{adh\text{-}eff}$) can be determined by integrating the individual sphere-above-flat energies over the asperity height distribution. This approach can be extended to account for a multitude of other possible effects at the contacting asperities such as elastic and plastic deformations and the capillary effects of lubricant menisci (Polycarpou and Etsion 1998).

3. *Comprehensive approach*

The last approach is a more comprehensive, numerical method that can be applied to most types of surface roughness (DelRio et al. 2005). First, a representative sample of the surface topographies of the opposing surfaces is collected using atomic force microscope (AFM) images with sufficiently high resolution to capture the details of individual asperities. Next, pairs of these images are placed facing each other in computer software with their surfaces means separated by varying D_{ave} distances. For each D_{ave}, the local separation distance $d_{local-i}$ for each pixel i is computed by subtracting from D_{ave} the sum of the z-heights measured relative to the respective surface means:

$$d_{local-i} = D_{ave} - (z_{substrate-i} + z_{cantilever-i}). \tag{7.32}$$

(For those cases where contact occurs $(z_{substrate-i} + z_{cantilever-i}) > D_{ave}$, $d_{local-i}$ is set equal to zero.) Then, the effective work of adhesion per unit area is determined as a function of D_{ave} by treating each pixel location as a parallel surface geometry and averaging over the image area:

$$W_{adh\text{-}eff}(D_{ave}) = \frac{1}{N_{pixels}} \left(\sum_i \frac{A}{12\pi (d_{local-i} + D_0)^2} \right) \tag{7.33}$$

where N_{pixels} is the number of pixels in an image and D_0 is the effective separation distance at contact. Differentiating eq. (7.33) with respect to D_{ave} determines the average van der Waals adhesive pressure as a function of D_{ave}. The equilibrium separation distance D_{ave} is determined by the balance between the adhesive pressure and the repulsive contact pressure, which can be determined, for example, by the Greenwood–Williamson (Section 3.4.1) or the Persson (Section 3.4.2) methods.

7.3.6.1   *Example: stuck microcantilevers*

Now let's apply the previous analysis to the practical example of a microfabricated cantilever beam of length L that was initially free standing, as shown in Figure 6.14(a), but which, for some reason, has become stuck to the substrate surface, as shown in Figure 7.9. This stuck cantilever is similar to the one discussed in Section 6.6, but now we consider the case where no liquid is present to generate a meniscus force and only van der Waals forces act across the contacting interface. Figure 7.10 shows the results of DelRio et al. (2005), who experimentally realized this van der Waals stuck cantilever geometry by first coating the surfaces with a hydrophobic monolayer to prevent the capillary condensation of water vapor.

As well as being of technical importance to the design of microelectromechanical system (MEMS) devices, the stuck cantilever geometry also has scientific value in that

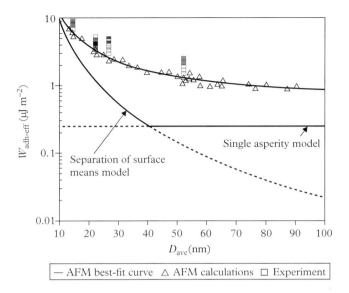

Figure 7.10 *Effective work of adhesion* $W_{adh\text{-}eff}$ *for silicon oxide cantilever beams stuck on silicon oxide substrates, both coated with a hydrophobic monolayer to prevent water absorption. The squares are experimental values for cantilevers with four different roughness values (and thus, four different values of D_{ave}) where the stored elastic energy is determined by finite element method simulations of the unstuck portion of the beam. The triangles are values calculated according to the comprehensive approach using $5 \times 5 \, \mu m^2$ AFM images of the cantilever and substrate surfaces and accounting for retardation effects, with the line being the best fit through the calculated points. Also shown are the predictions for two smooth, parallel surfaces (the separation of surface means model, which was discussed in Section 7.3.6), and for a single sphere-on-flat contact (single asperity model). Reprinted by permission from DelRio et al. (2005); copyright 2005, Macmillan Publishers Ltd: Nature Materials.*

the effective work of adhesion can be experimentally measured from the lengths of the unstuck and stuck portions, s and $L - s$, respectively, making it possible to compare theory with experiment (Mastrangelo and Hsu 1993, Houston et al. 1997, De Boer and Michalske 1999, DelRio et al. 2005). An equilibrium adhered length $L - s$ occurs when the sum of the elastic energy U_E stored in the unattached portion and the adhesion energy U_S stored in the adhered portion is minimized, that is, when

$$\frac{d}{ds} (U_E + U_S) = 0. \tag{7.34}$$

For the case shown in Figure 7.9 where the attached portion lies flat on the substrate, the elastic energy stored in the bent cantilever can be expressed as (De Boer and Michalske 1999):

$$U_E = \frac{Ewt^3 d^2}{2s^3}, \tag{7.35}$$

where w is the width of the cantilever beam and E its elastic modulus. The adhesion energy stored in the attached portion is simply the effective work of adhesion per unit area $W_{\text{ad-eff}}$ times the apparent contact area:

$$U_S = -W_{\text{adh-eff}} w (L - s).\tag{7.36}$$

Using eqs. (7.35) and (7.36) in eq. (7.34) and solving for $W_{\text{adh-eff}}$ as a function of the length of the unattached portion and the beam parameters yields

$$W_{\text{adh-eff}} = \frac{3}{2}\left(\frac{Et^3 d^2}{s^4}\right).\tag{7.37}$$

Thus, the work of adhesion can be determined by measuring s, and using known parameters (E, t, d).

Figure 7.10 shows that the calculated results (using the comprehensive approach) agree well with experimentally determined values of the $W_{\text{ad-eff}}$ from using eq. (7.37) for silicon oxide cantilever beams stuck on silicon oxide substrates with varying amounts of surface roughness (DelRio et al. 2005). DelRio et al. also compare the results from the comprehensive approach with those of the separation of surface means approach and a single-asperity model.

An important thing to notice in Figure 7.10 is that the values for the van der Waals interaction energy obtained from the separation of surface means approach are much lower than those from the comprehensive approach and experimental values. It is always the case that the separation of surface means approach underestimates the van der Waals interaction energies and forces. This is because those parts of the rough surface with surface heights above the surface mean contribute much more to the interaction energy than what is lost from those portions with surface heights below the surface mean, due to the D^{-2} dependence of the van der Waals interaction energy.

7.3.7   Example: gecko adhesion

Geckos are lizards that have the remarkable ability of being able to run up smooth walls and across ceilings with relative ease. Scientists and many casual observers have long puzzled over how geckos, weighing up to 50 g, are able successfully to scale such a wide variety of surfaces. As mentioned in Section 6.7, many insects also have the ability to cling to surfaces, by secreting liquids from glands in their feet so as to generate adhesion through capillary forces. Since these glands are absent in geckos, another mechanism must be at work. Capillary effects have also been considered for gecko adhesion due to an observed dependence of adhesion on humidity (Huber et al. 2005), but recent models suggest this effect is due to humidity induced softening of the beta keratin on the bottom of the gecko's toes (Chen and Gao 2010).

Autumn and coworkers determined that geckos exploit the weak, but always present van der Waals forces, to adhere to surfaces (Autumn et al. 2000, 2002, Autumn and Peattie 2002). The key to the effectiveness lies in the nanostructuring of the gecko's

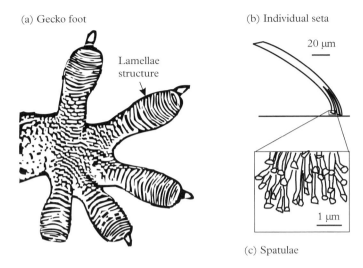

(a) Gecko foot

Lamellae
structure

(b) Individual seta

20 μm

1 μm

(c) Spatulae

Figure 7.11 *(a) Schematic of a gecko foot illustrating the lamellar structure of setae hairs on the toes (Autumn and Peattie 2002). (b) An individual seta hair as it comes close to contacting a flat surface. (c) The ends of the spatulae that must become oriented towards the flat surface if strong van der Waals adhesion is to occur.*

toe pads. With these extraordinary hierarchical structures on the pads of their toes terminating in nanoscale features, geckos are able to engage and disengage the van der Waals adhesive force. As illustrated in Figure 7.11, these toe pads consist of ridges of lamellae that are covered with arrays of hair-like beta-keratin bristles, called setae. For a typical tokay gecko, an individual seta has about a 100 μm length and a 5 μm diameter. Each seta branches into hundreds of tiny endings, called spatulae, that touch the surface to engage the van der Waals forces. The individual spatulae ends, where adhesion occurs, have a flattened shape, about 200 nm across and need to be properly oriented to achieve maximum adhesion.

Measurements with whole tokay gecko indicate that the two front feet can generate a shear force of 20.1 N (Irschick et al. 1996), which corresponds to an average force of 6.2 μN per seta. Measurements by Autumn et al. (2000) on individual seta indicate that the key to the achieving the maximum adhesion and shear forces is properly orienting the seta and its spatulae ends relative the contacting surface. First, the seta stalk needs to be oriented within a few degrees of the angles that align its end parallel with the contacting surface. Next, to maximize the contact area, the seta is first pushed toward the surface then dragged a short distance across it so as to bring the flat ends of the spatulae uniformly flush against the surface. With this procedure, a maximum shear force can be achieved for an individual seta of 200 μN, 32 times larger than the average obtained by measuring with the whole animal. Geckos are thought to use a similar procedure when they plant their toes on a surface to obtain the best traction and adhesion, then they disengage the forces rotating their toes so as to peel the setae ends off the contacting surface. The maximum adhesive force per seta is also found to be 200 μN. If all 6.5 million setae of a 50 g tokay gecko achieved this maximum value, the total adhesion force would be 1.3 kN, enough to support the weight of two humans.

The van der Waals force acting on an individual spatula can be estimated using the equation for two planar surfaces:

$$F_{\text{adh-spatula}} = -A_{\text{spatula}} \frac{A}{6\pi D^3}, \tag{7.38}$$

where $A_{\text{spatula}} = 2 \times 10^{-14}$ m^2 is the typical area of an end of a spatula. If we assume that the Hamaker constant $A = 5 \times 10^{-20}$ J and that the value of D corresponds to the value seen for atomically-smooth interfaces ($D = D_0 = 1.65$ Å, as discussed in Section 7.3.4), then

$$F_{\text{adh-spatula}} = -\left(2 \times 10^{-14}\text{m}^2\right) \frac{5 \times 10^{-20} \text{ J}}{6\pi (0.165 \text{ nm})^3}$$

$$= -12 \text{ } \mu\text{N}. \tag{7.39}$$

So, a single seta with a hundred contacting spatulae could potentially achieve an adhesion force as high as 1.2 mN. Since this an order of magnitude higher than the measured value for an individual seta, it is likely that a small amount of roughness between the contacting surfaces leads to an average spacing $D_{\text{ave}} > D_0$. For example, if we assume 100 spatula per seta make contact, a 200μN adhesion force for an individual seta can be achieved when $D_{\text{ave}} = 3.0$ Å, suggesting that spatula ends come fairly close to achieving the most intimate possible contact.

7.3.8   Van der Waals contribution to the disjoining pressure of a liquid film

So far, we have been considering only the van der Waals interaction between two solid bodies separated either by a vacuum, a gas, or a liquid. The situation described in Figure 7.7 is fairly general, however, and materials 1, 2, and 3 can be any combination of gas, liquid, or solid. Consequently, the Hamaker constant can be calculated for many other situations besides the interaction between two solid bodies. One combination frequently encountered in tribology is the situation illustrated in Figure 7.12, where a liquid lubricant film with thickness h covers a solid surface, with a vapor (e.g., air) above the liquid. Since the vapor–liquid interface and the liquid–solid interface are two parallel flat surfaces, the same expression derived in Section 7.3.2 for two solid objects can be used for this situation. Therefore, the van der Waals interaction between the lubricant film and the solid in excess of the cohesive and interfacial energies is

$$\pi(h) = \frac{A_{\text{SLV}}}{6\pi h^3}, \tag{7.40}$$

where $\pi(h)$ is called the *disjoining pressure* of the liquid film, and A_{SLV} is the Hamaker constant for the solid–liquid–vapor geometry. The term "disjoin," coming from the

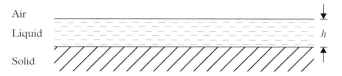

Figure 7.12 *Liquid film with thickness h on a solid surface. Following the convention of Figure 7.7, the Hamaker constant for this solid–liquid–vapor geometry is labeled A_{SLV}.*

combination of "dis" and "join," means to pull apart or separate; the disjoining pressure can be considered as a pressure acting to separate the solid–liquid and liquid–vapor interfaces.

In eq. (7.40), the sign convention is chosen so that the Hamaker constant A_{SLV} is positive when an attractive van der Waals interaction exists between the liquid film and the solid surface, so that an attractive interaction leads to a positive value for the disjoining pressure $\pi(h)$. Such an attractive interaction works to *thicken* the liquid film. This convention differs from that used for two solids separated by a gap, where an attractive interaction between the solids, which also provides for a positive Hamaker constant, leads to a negative force that acts to thin the gap separating the solids.

The liquid to thicken a film can come from either the condensing from vapor or from liquid drawn from a nearby source such a from droplet on the surface or liquid residing in a nearby pore or crevice in the surface. The mechanism driving this thickening is easy to understand: since the liquid and the surface are attracted to each other through dispersion forces, then a thicker liquid leads to a greater total attractive energy. As the liquid film thickens, the net energy gain lessens since the additional liquid being added is farther away from the solid. This movement of liquid lubricant around surfaces due to the action of disjoining pressure is discussed in more detailed in Chapter 10.

Figure 7.13 shows an example of measurements of how the disjoining pressure varies with film thickness for a liquid lubricant deposited on a silicon wafer (these perfluoropolyether lubricants are frequently used to lubricate disk drives and MEMS devices) (Fukuzawa et al. 2004a, 2004b). Since this particular perfluoropolyether lubricant does not have any functional groups, the interaction with the substrate is dominated by van der Waals interactions. Since the silicon wafer typically has a 1 nm thick silicon oxide surface layer (as illustrated in the inset of Figure 7.13), a somewhat different equation than eq. (7.40) needs to be used to account for the van der Waals contribution to the disjoining pressure for this three-interface system (air–film–SiO_x–Si):

$$\pi(h) = \frac{A_{SiOx\text{-}LV}}{6\pi h^3} + \frac{A_{Si\text{-}LV} - A_{SiOx\text{-}LV}}{6\pi (h + d_{SiOx})^3}. \tag{7.41}$$

The solid line in Figure 7.13 is obtained by fitting eq. (7.41) to the measured values of disjoining pressure. This fitting yields the values of $A_{SiOx\text{-}LV} = 9.1 \times 10^{-21}$ J and $(A_{Si\text{-}LV} - A_{SiOx\text{-}LV}) = 7.0 \times 10^{-20}$ J for the two Hamaker constants and $d_{SiOx} = 1.0$ nm for the silicon oxide thickness. From Figure 7.13, we see that, for a film thickness

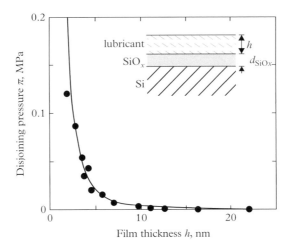

Figure 7.13 *The symbols are measured values of disjoining pressure for different film thicknesses of a perfluoropolyether lubricant deposited on a silicon wafer with a thin SiO_x layer, as illustrated in the inset. The solid line shows the fit to the data by assuming a van der Waals interaction between lubricant and the silicon wafer as described by eq. (7.41). The perfluoropolyether lubricant is a linear chain polymer with the chemical structure $CF_3-(CF_2O)_m-(CFO)_n-CF_3$ and is a liquid at room temperature. These measurements were done by measuring the meniscal radius of the liquid in a narrow groove on a silicon wafer when different thicknesses of this liquid lubricant were applied. From the meniscal radius, the capillary pressure of the liquid in the groove is determined, which equals the disjoining pressure of the liquid film on the silicon wafer around the groove as discussed in Chapter 10. The film thickness is determined by ellipsometry. Adapted from Fukuzawa et al. (2004a), with the permission of AIP Publishing.*

greater than 10 nm, the disjoining pressure is fairly small (<0.03 atm). For a thickness less than 2 nm (the relevant range for these perfluoropolyether lubricants when used in disk drives and MEMS devices), the disjoining pressure is quite large: eq. (7.41) indicates that $\pi = 2$ atm at $h = 2$ nm, rising to 10 atm for $h = 1$ nm.

One shortcoming of eqs. (7.40) and (7.41) is that they imply that $\pi(h) \to \infty$ as $h \to 0$. This unphysical situation of infinite disjoining pressure can be remedied by invoking the concept of the effective closest approach distance D_0 between the liquid film and solid substrate, as was done for calculating the van der Waals contribution to surface energy in Section 7.3.4. In this case eq. (7.40) is rewritten

$$\pi(h) = \frac{A_{SLV}}{6\pi(h+D_0)^3} \tag{7.42}$$

where D_0 is on the order of the sum of the van der Waals radii of the substrate and liquid atoms. Since D_0 is on the order of a few angstroms, the difference between eq. (7.40) and eq. (7.42) only becomes significant for film thicknesses h less than a few nanometers. For example, much better agreement has been found between experiment and theory by using eq. (7.42) and $D_0 = 3.17$ Å for the spreading of nanometer thick perfluoropolyether liquid films (Marchon and Karis 2006).

7.4   Liquid-mediated forces between solids

The presence of a liquid between two solids can greatly change the nature of the forces between the opposing surfaces. In Chapter 6, we discussed the forces arising from the surface tension of the liquid between the surfaces. In this section, we discuss some of the forces acting directly across the gap due to the presence of a liquid.

Previously in this chapter, we discussed how the presence of a liquid typically reduces the size of the van der Waals force by up to an order of magnitude, decreasing its significance. We also showed that the van der Waals force remains attractive, if the value of the liquid refractive index is not intermediate between those of the solid refractive indexes (a fairly rare event). Consequently, one might suspect that all dissolved particles would tend to stick together due to the van der Waals attraction and precipitate out as a solid mass; this does not generally happen as various repulsive and oscillatory forces occur in liquids to separate surfaces. This is fortunate: as our bodies are 75% water, we would all be subjected to a rather unpleasant fate if all our suspended solids precipitated out.

7.4.1   Solvation forces

When a flat solid surface is introduced into a liquid, the liquid molecules adjacent to the surface rearrange to pack well against it, as this helps lower the overall free energy. As illustrated in Figure 7.14, this packing leads to several features in the liquid density profile near the surface:

- The layer of molecules in contact with the surface has a density ρ_s that is higher than the bulk liquid density ρ_{bulk}.
- The liquid structure transitions from being ordered next to the solid to being disordered in the bulk liquid, with each subsequent layer being less ordered than those closer to the surface
- The liquid density profile has peaks spaced by a molecular diameter a, and the amplitude of these density oscillations decays exponentially away from the surface.

This ordering of liquid molecules next to a surface does not require any liquid–liquid or liquid–wall interaction, but instead is determined primarily by the geometry of the molecules and how they pack around the constraining boundary. Since a similar structural ordering also occurs for solvent molecules around a particle or solute molecule in a solution, forces arising from any disruption of this ordering are referred to as *solvation forces*.

When a second solid surface is brought near this first surface so as to sandwich the liquid between the two (Figure 7.15), the ordering at the individual surfaces is disrupted,

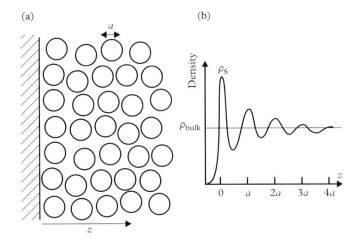

Figure 7.14 *(a) Illustration of how packing of spherical liquid molecules next to a solid surface leads to layering of molecules near the surface. (b) Density profile of the liquid near the surface.*

and the molecules rearrange to find the most energetically favorable packing geometry. Figure 7.15(a) illustrates how the molecular packing for spherical molecules between two parallel surfaces varies with the separation distance D. Since liquid molecules interact via some potential, perhaps resembling something like the Lennard-Jones potential (eq. (7.3)), it takes energy to disrupt the original ordering of the liquid. This disruption, caused by the second surface, generates an oscillatory solvation pressure $p(D)$ with attractive maxima when D is a multiple of the molecular diameter a. To a good approximation, the solvation pressure can be described by an exponentially decaying cosine function (Tarazona and Vicente 1985):

$$p(D) \approx -k_B T \rho_s \cos\left(2\pi D/a\right) e^{-D/a}, \tag{7.43}$$

which is shown in Figure 7.15(b). Equation (7.43) is just the expression for the oscillating solvation force contribution, which is superimposed on the other forces acting between solids such as the van der Waals and the electrostatic double-layer forces. These contributions are not necessarily additive; for example, within Lifshitz's theory, the van der Waals force depends on the density of the intervening liquid, and, in turn, this density is influenced by the solvation force.

One might think it necessary to have two parallel surfaces to observe the oscillatory solvation force, as any surface curvature would lead to a range of separation distances and an averaging out of the oscillatory component of the solvation force. Somewhat surprisingly, however, surface curvature does not eliminate the solvation forces. This can be understood as follows: if the ordering of liquid molecules between two parallel surfaces results in an interaction energy $W(D)$ that oscillates with separation distance D, then, by the Derjaguin approximation, the force $F(D)$ between two spheres, a sphere-on-

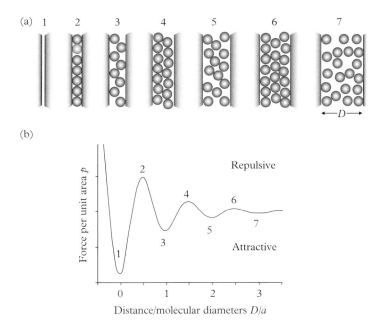

Figure 7.15 *(a) Ordering of spherically shaped liquid molecules between two parallel surfaces of finite extent; the molecules are free to exchange with those in the bulk liquid. The molecules are able to form well defined layers when the separation distance D is a multiple of the molecular diameter a. (b) Solvation pressure as a function of separation distance. Reproduced from Butt et al. (2005) with permission from Elsevier, copyright 2005.*

flat, or cross cylinders will also be oscillatory. For example, for a sphere-on-flat geometry, $F(D)_{\text{sphere/flat}} = 2\pi RW(D)$ (eq. (6.2)); so, the solvation force simply scales with the radius of curvature of the surface.

Oscillatory solvation forces were first observed experimentally in the surface force apparatus (SFA) by Israelachvili and Pashley (1983), who squeezed aqueous solutions between two atomically smooth sheets of mica and found that the pressure oscillated with a mean periodicity of 0.25 ± 0.03 nm at separations ≤ 1.5 nm, roughly the diameter of water molecules. Since then, the SFA has been used extensively to observe oscillatory solvation forces for many types of solvents and complex fluids (Israelachvili 2011: Section 15.7). The types of molecules that are most likely to order and lead to pronounced oscillatory forces are:

- Inert molecules that are roughly spherical and fairly rigid like CCl_4, benzene, toluene, and cyclohexane. The periodicity of the oscillatory forces is equal to the mean molecular diameter of the liquid molecules.
- Linear chain molecules like *n*-alkanes and short chain polydimethylsiloxanes. For these molecules, the periodicity of the oscillatory forces is equal to the chain

diameter, indicating that these types of molecules order in layers with their chains oriented parallel to the surfaces.

On the other hand, the formation of molecular layers becomes less pronounced and even suppressed as the molecular structure of the liquid becomes more flexible, polydisperse, or asymmetric (Israelachvili 2011: Section 15.7).

For oscillatory solvation forces to be observed, it is not only necessary for the liquid molecules to have the right structure, it is also important for the solid surfaces to have the right structure, namely that they need to be fairly smooth, as increasing the random roughness of the solid surfaces averages out the oscillatory forces. Typically, a roughness on the order of the molecular diameter of the liquid—generally less than a nanometer—is all that it takes to make the oscillatory force disappear (Gee and Israelachvili 1990, Gao et al. 2000, Granick et al. 2003, Samoilov et al. 2004, Yang et al. 2011).

It is important to remember, however, that roughness does not suppress the ordering of molecules into layers next to the solid surfaces, it just averages out the oscillatory nature of the solvation forces that might come about from this layer formation. Ordering of liquid molecules into layers may still persist between and around the individual asperity contacts. For example, Figure 7.16 shows results from molecular dynamic simulations of how molecules of hexadecane form not only ordered layers between two gold surfaces that are parallel to each other, but also form ordered layers around two micro-asperities on the top and bottom surfaces. The different frames also show how this ordering changes as the two micro-asperities move towards each other.

7.4.1.1   *Example: experimental observation of solvation forces by AFM*

In addition to the SFA, another method for experimentally measuring the solvation force is to use a sharp AFM tip contacting a flat surface in the presence of the liquid. Since a typical AFM tip has a radius of curvature of only a few tens of nanometers, this gets around the roughness issue somewhat, as the opposing surfaces need only be smooth over their effective contact area often only a few nanometers across. The observations of oscillatory solvation forces at these nanoscale AFM contacts indicates that such oscillatory forces should also occur in general when individual asperity contacts are separated by liquids.

The group of Prof. O'Shea at the National University of Singapore has done much of the AFM experimental work on oscillatory solvation forces, which they review in O'Shea et al. (2010). Figure 7.17 shows an example from this group, where the solvation forces acting on a tip are measured as the separation distance D is reduced between it and the surface of a graphite basal plane while immersed in liquid squalane (Gosvami et al. 2008). As the surfaces are pushed together, the squalane drains out without any measurable resistance until D is reduced to 3 nm. At this separation, it takes a measurable repulsive force to squeeze out the first of five squalane layers. Once the force is increased to the threshold needed to squeeze out each layer, the tip pops through that layer to the tip-sample separation distance corresponding to the top of the next layer, with each layer being progressively harder to squeeze out, until a force of $F_{1 \to 0} = 10$ nN is needed to squeeze out the last remaining $n = 1$ squalane layer. If in Figure 7.17 we assume that

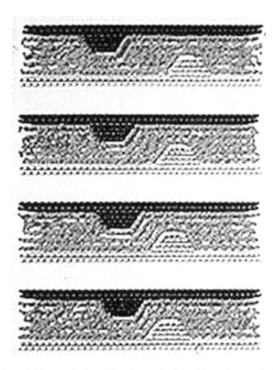

Figure 7.16 *Results from MD simulations showing ordering of hexadecane (a linear chain n-alkane) ordering between two sliding solid surfaces of gold and around the small asperities on the top and bottom surfaces. The bottom gold surface moves from right to left as one goes from the top frame to the bottom frame. This motion brings the asperities on the opposing surfaces closer together, during which the number of ordered hexadecane layers sandwiched between their facing side walls changes from five layers in the second frame from the top, to four layers in the third frame, to three layers in the fourth frame. Reprinted from Gao et al. (1995) with permission from the American Association for the Advancement of Science. Copyright 1995, AAAS.*

the tip radius of curvature is $R_{\text{tip}} = 20$ nm, then $F_{1\to0}/2\pi R_{\text{tip}} = 80$ mN/m, which by the Derjaguin approximation is the work per unit area needed to squeeze out this last squalane layer from two parallel surfaces.

7.4.2   Forces in an aqueous medium

The high dielectric constant and high degree of hydrogen bonding of water lead to some distinct forces when surfaces are separated by liquid water, that is, *electrostatic double-layer force*, *hydration repulsion*, and *hydrophobic attraction*. These forces have been extensively studied due to their importance to biological systems, colloidal suspensions, and other aqueous-based systems. In the next few sections, we provide a short overview of these forces specific to an aqueous medium.

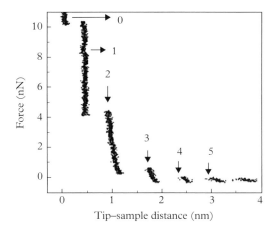

Figure 7.17 *Normal force versus separation distance for squalane being squeezed between an AFM tip and the basal plane surface of highly oriented pyrolytic graphite (HOPG). Jumps due to the oscillatory solvation force are clearly seen and marked by n = 1–5 corresponding to the number of layers of squalane molecules between the tip and graphite surface prior to the jump. A jump occurs when the force is high enough to reduce the number of layers in the gap by one. Reprinted with permission from Gosvami et al. (2008). Copyright 2008 by the American Physical Society.*

7.4.2.1 *Electrostatic double-layer force*

When immersed in water, a solid surface develops a surface charge by one of two mechanisms:

1. Ions dissolved in the water absorb onto the surface.
2. Chemical groups on the surface ionize or dissociate into ions. For example:

    (a) Organic surfaces frequently have carboxylic groups that dissociate in the presence of water ($-COOH \rightarrow -COO^- + H^+$).

    (b) Glass surfaces hydroxylate in the presence of water, and these surfaces become negatively charged when the resulting silanol groups dissociate ($SiOH \rightarrow SiO^- + H^+$).

Once these charges are present on a surface, they are balanced (so that the net charge is zero) by counterions dissolved in the region next to the surface, as illustrated in Figure 7.18. A few of these counterions bind to the surface to form what is called the *Stern* or *Helmholtz* layer. The remaining counterions form a diffuse *electric double layer* in the liquid close to the charged surface. In this electric double layer, two competing forces act on the counterions: the attractive electrostatic force pulling the ions toward the oppositely charged surface and the entropic force or thermal agitation that pushes these ions apart from each other and away from the surface, increasing entropy.

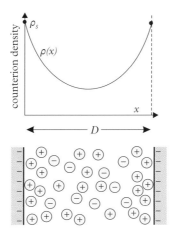

Figure 7.18 *Two negatively charged surfaces separated by a distance D in water (bottom). The counterion density profile (top).*

When two surfaces with similar charges are brought together in a liquid, as shown in Figure 7.18, their electric double layers overlap, increasing the concentration of counterions near the surfaces. Since this increase in concentration reduces entropy, it generates an entropic repulsive interaction that is greater than attractive electrostatic interaction. The resulting repulsive force is referred to as the *electric* or *electrostaticdouble-layer force*, even though the repulsion arises from entropic confinement. The contact value theorem expresses the repulsive pressure due to the electrostatic double-layer force in terms of the increased counterion density at the solid surfaces (for a derivation of this theorem see Israelachvili (2011: Section 14.7)):

$$p(D) = k_B T \left[\rho_s(D) - \rho_s(\infty) \right]. \tag{7.44}$$

At large distances, the electrostatic double-layer force decays exponentially due to screening of the electric field by the ion concentration; the decay length equals the Debye length:

$$\lambda_D = \sqrt{\frac{\varepsilon_r \varepsilon_0 k_B T}{2 c e^2}}, \tag{7.45}$$

where $\varepsilon_0 \varepsilon_r$ is the dielectric permeability of the liquid and c is the concentration of a monovalent salt in mol/L. If higher valency ions are present, $2c$ is replaced with $\sum c_i Z_i^2$ where c_i is the concentration of the ith ion species, Z_i its valence, and the sum is over all the types of ions present.

According to eq. (7.45), the higher the dielectric constant, the longer the Debye length. So, while the electrostatic force can occur in all liquids, this force only extends for significant distances in high dielectric constant liquids. Water has one of the highest dielectric constants of all liquids ($\varepsilon_r = 78.4$ at 25°C), resulting in a very long-range electrostatic double-layer force at low ion concentration. For water, eq. (7.45) reduces to $\lambda_D = 3.04\overset{\circ}{\text{A}}/\sqrt{c}$ and, at pH $= 7$ ($c = 10^{-7}$ M), $\lambda_D = 960$ nm $\sim 1\ \mu$m, a rather large decay distance. Adding salt to water increases the concentration of the dissolved ions, resulting in the electrostatic interaction being more effectively screened (i.e., a shorter Debye length). For example, the ion concentration (predominately from dissolved NaCl or KCl) in most fluids in animal bodies (including our own bodily fluids) is about 0.2 M, resulting in a Debye length $\lambda_D = 0.7$ nm for these aqueous fluids: much smaller than for pure water, but still large enough for electrostatic double-layer repulsion to play a major role in our bodily functions.

For colloids in aqueous medium, the electrostatic double-layer repulsion keeps particles in suspension. Adding salt, however, will cause many colloidal systems to coagulate. This behavior was first quantitatively explained by Derjaguin and Landau (1941) and Verwey and Overbeek (1948) in what is now called the *DLVO theory* of colloidal stability. In this theory, the interaction between two particles is assumed to consist of two parts: a van der Waals attraction and an electrostatic double-layer repulsion. At low salt concentrations, the double-layer repulsion keeps the particles separated. Increasing the salt concentration leads to increased screening of the double-layer repulsion and, at a critical concentration, the height of the repulsive barrier is reduced to the point that the van der Waals attraction causes the particles to coagulate.

7.4.2.2   *Hydration repulsion and hydrophobic attraction*

When water or aqueous salt solutions are in contact with surfaces, the highly polar character of water leads to two types of strong forces that are not well understood, even after many decades of intense study. These forces are either repulsive (the *hydration force*) or attractive (the *hydrophobic force*), and can be oscillatory or strongly monotonic (or a combination of these).

These forces are often thought to arise from the strong network of hydrogen bonding that exists between bulk water molecules. Anytime a foreign molecule or surface is introduced into water, the water molecules in the immediate vicinity reorient themselves to obtain the most favorable interaction in their new situation, even if the reorientation interferes with the hydrogen bonding between water molecules. When two surfaces are brought within a few molecular diameters of each other in water, the hydrogen bonding network becomes further disrupted; this reorientation is illustrated in the right side of Figure 7.19. This disruption of the hydrogen bonding network leads to either repulsive or attractive forces between the surfaces depending on whether the surfaces are hydrophilic or hydrophobic, respectively.

The precise nature of *hydration repulsion* between two hydrophilic surfaces is still unclear, but is thought to be due to water molecules adjacent to the surface forming strong hydrogen bonds with hydrophilic surface groups such as hydroxyl (–OH) groups and with hydrated surface ions. The repulsion then corresponds to the energy needed to disrupt the hydrogen bonding network sufficiently to squeeze the

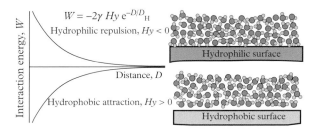

Figure 7.19 *The right side illustrates how water molecules reorient themselves near hydrophilic (upper) and hydrophobic (lower) surfaces to achieve the most favorable network for hydrogen bonding. For the hydrophilic surface, the hydrogen atoms on the water molecules near the surface are preferentially oriented towards the surface so as to maximize bonding to that surface. For the hydrophobic surface, not only do the hydrogen atoms on the water adjacent to the surface oriented away from the surface, but also a small gap opens up between this water layer and the hydrophobic surface. The left side shows generic potentials for hydrophilic repulsion and hydrophobic attraction as proposed by Donaldson et al. (2015). Reprinted with permission from Donaldson et al. (2015). Copyright 2015 American Chemical Society.*

hydrated water out from between the surfaces. Empirically, the interaction energy W for hydration repulsion is found to decay exponentially with the surfaces' separation distance D:

$$W = -2\gamma H_y e^{-D/D_H}, \qquad (7.46)$$

where γ is the hydrophilic-water interfacial energy, D_H ranges from 0.3 to 1 nm, and H_y is called the Hydra parameter, which ranges from -0.2 to 0 for hydrophilic surfaces (Donaldson et al. 2015). This short range makes hydration repulsion readily distinguishable from other types of interactions with longer range (electrostatic, van der Waals, and electrostatic double layer for salt concentrations below 0.1 M). For molecularly smooth surfaces, the hydration repulsion may also exhibit an oscillatory component due to the solvation force of layered water molecules. Since the hydration force extends away from the surface to distances more than two molecular layers of water, more factors must be contributing to the hydration force than just the hydrogen bonding of the water layer adjacent to the hydrophilic surfaces. One theory is that the adjacent, hydrated water molecules are more ordered than in bulk liquid and that this higher degree of ordering extends away from the surface for several water layers (Marcelja and Radic 1976). Further information about hydration forces can be found in the reviews by Cevc (1991), Leikin et al. (1993), Israelachvili and Wennerstrom (1996), Donaldson et al. (2015), and Israelachvili (2011: Section 15.8).

By definition, hydrophobic surfaces are those that repel water (water contact angles $\geq 90°$); so, we should not be surprised that, when two hydrophobic surfaces immersed in water are brought together, water is spontaneously repelled from between the surfaces when the gap is sufficiently small. Associated with this exodus of the water is an attractive force acting between the two hydrophobic surfaces, called the *hydrophobic attraction*.

Even though researchers have been studying the hydrophobic effect since the 1930s, how this interaction originates at the molecular level (as with the hydration force) is still largely unclear (Meyer et al. 2006, Hammer et al. 2010). Figure 7.20 illustrates some of the molecular mechanisms that have been proposed to explain hydrophobic forces; it is possible that more than one of these mechanisms may be operative in any particular situation. These mechanisms for the hydrophobic force can be divided into three categories (Israelachvili 2011: pp. 377–8):

- *Water structure*—Since water molecules have difficulty forming hydrogen bonds with a hydrophobic surface, they reorient themselves away from the surface to increase hydrogen bonding with nearby water molecules. This leads to a narrow depletion layer forming next to the hydrophobic surface, as illustrated in Figure 7.20(a). Evidence that water molecules form this depletion layer has been found in X-ray reflective measurements where a few angstroms wide decrease in water density is observed at a water–hydrophobic surface interface (Jensen et al. 2003, Poynor et al. 2006). Often this restructuring results in an attractive interaction that decays exponentially with separation distance as expressed in eq. (7.46), with $0.3 < D_H < 1$ nm and $0 < H_y < 1$. (Note that in this situation H_y is positive resulting in a negative interaction energy and an attractive interaction, while hydrophilic surfaces have a negative value for H_y resulting in a positive interaction energy and a repulsive interaction.)

- *Electrostatic models*—Figures 7.20(c) and (d) schematically illustrate some ways an electrostatic attraction can arise in the presence of water. In Figure 7.20(d), the presence of water has caused a deposited Langmuir–Blodgett film or self-assembled monolayer to rearrange to form patchy bilayer islands. These islands can then move under the influence of the electrostatic field fluctuations from the opposing surface, resulting in the attractive hydrophobic force.

- *Vapor bridges*—Dissolved air forms nanometer sized bubbles or "nanobubbles" on the hydrophobic surfaces; these bubbles coalesce to bridge the gap when the surfaces are brought together, as illustrated in Figure 7.20(e). Since the water contact angle is greater than 90°, the bridging bubbles form a meniscus that exerts an attractive force between the two surfaces. This nanobubble mechanism can lead to a very long-range hydrophobic attraction (tens of nanometers in the AFM experiments by Tyrrell and Attard (2001)).

If hydrophilic and hydrophobic surfaces are brought together in water, neither hydration repulsion nor hydrophobic attraction is observed. These "Janus" type interfaces, however, have other interesting tribological properties. For example, water films sheared in an SFA between a hydrophilic and a hydrophobic surface are observed to have much lower friction than water sheared between two hydrophilic or two hydrophobic surfaces (Zhang et al. 2002).

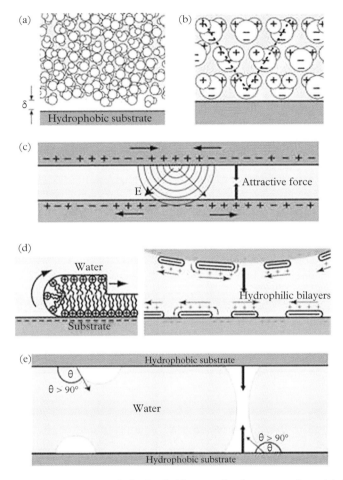

Figure 7.20 *Possible mechanisms for the hydrophobic attraction between surfaces. (a) A depletion layer, with thickness δ of one or two water molecules, occurs next to a hydrophobic surface. (b) The presence of a hydrophobic solute (or ion) that affects the local orientation of the surrounding water molecules, which can propagate many molecular layers into the bulk water. (c) Local charge fluctuations on one surface induce charge fluctuations on the opposing surface causing a long-range attractive electrostatic interaction. (d) A possible source for these charge fluctuations is the formation and movement of bilayers on the surfaces. (e) Dissolved air in the water can nucleate out as nanobubbles on the surfaces; when these nanobubbles bridge across the opposing surfaces, they generate an attractive Laplace pressure. Reproduced with permission from Meyer et al. (2006). Copyright 2006, National Academy of Sciences, USA.*

7.5   Contact electrification

The ancient Greeks were the first to document the phenomenon of *contact electrification* or *triboelectricity*. One of the Seven Sages of Greece, Thales of Miletus (624-547 BC), described experiments of rubbing of amber with cat's fur, which then could be used to pick up small objects. We now know that this phenomenon is due to the transfer of electric charges during contact, leaving the surfaces with net electrical charges that electrostatically attract other objects. While some make the distinction of *triboelectricity* being from rubbing and *contact electrification* being from simple contact, here we treat the two situations as being the same contact electrification phenomenon.

Contact electrification is easily observed in everyday life. For example, when the humidity is low, walking across a carpeted floor generates a net charge on your body, that discharges with a slight shock when you touch a conductor like a metal doorknob. Likewise, your plastic comb will start to attract hair after a single pass through dry hair. If the hair is wet, water conducts away the surface charges, neutralizing the effect.

Since contact electrification frequently occurs during contact of dissimilar materials, its potential impact needs to be taken into account when designing tribological systems. For example, an adhesive electrostatic force can contribute to a loading force on rubbing surfaces leading to higher friction and wear and perhaps premature failure. Also, the electrostatic charge built up by contact electrification can suddenly discharge (an event called *electrostatic discharge* or *ESD*) damaging surrounding materials and electrical circuits.

Contact electrification is not always detrimental, however, as evidenced by the several important technologies developed to exploit the phenomenon:

- Photocopiers and laser printers use contact electrification to charge toner particles, which are then manipulated with electrostatics.
- Some mining processes use contact electrification to separate the ore.
- In electrostatic spray painting, contact electrification gives the paint droplets a net charge before they are directed towards a grounded conducting surface to be painted.
- A van de Graaff generator uses contact electrification between a roller and an insulating belt to generate a high electric field that is then used to generate positively charged ions and to deposit them onto the belt. These charges are later removed from the belt to create a constant current source capable of achieving extremely high voltages—as high as 20 million of volts.

One interesting new device that has been proposed with prototypes developed is a microsized triboelectric power generator (Wang 2013), that converts small amounts of mechanical energy into electricity that can power small electrical devices or sensors. This is done by repeatedly making contact between dissimilar materials to separate charges that are then bled off through the circuit that is to be powered. One potential use of such a "nanogenerator" is that the motion of a human body during everyday life (e.g., by having several of these triboelectric power generators embedded in the soles of one's shoes) could potentially be tapped to power one's portable electronics or a biomedical device. Such triboelectric nanogenerators could potentially eliminate the need to periodically recharge or change batteries for these devices.

7.5.1 Conductor–conductor mechanism of contact electrification

The charge density σ on a conductor surface such as a metal is related to the electric field E at the surface by $E = \sigma/\varepsilon_0\varepsilon_r$ where $\varepsilon_0\varepsilon_r$ is the dielectric permeability of the material outside the conductor. On a conductor, the electrostatic potential is constant over the surface, implying that the electric field is oriented perpendicularly to a conductor surface. To calculate the electrostatic force between two conductors, we first note that the stored electrostatic energy equals $CV^2/2$, where V is the electrostatic potential difference between the two conductors and C their capacitance. The gradient of the electrostatic energy with respect to the separation distance D gives the electrostatic force:

$$F_{el} = \frac{V^2}{2}\frac{\partial C}{\partial D}. \tag{7.47}$$

With the sign convention used in this chapter, the F_{el} is negative as the electrostatic force between two conductors is always attractive.

Figure 7.21 shows the simplest geometry for calculating the electrostatic force acting between two parallel plate conductors. If $D = \sqrt{A}$ where A is the area of the plates, the electric field between the plates is $E = V/D$, so the charge density at the conductor surfaces is $\sigma = \pm\varepsilon_0\varepsilon_r V/D$. Since the capacitance of two parallel conducting plates is

$$C_{plates} = \frac{\varepsilon_0\varepsilon_r A}{D}, \tag{7.48}$$

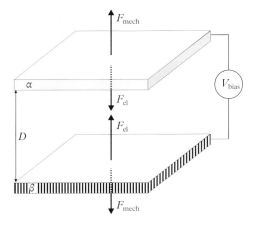

Figure 7.21 *Two parallel plate conductors α and β with bias voltage V_{bias} applied between them. The electrostatic potential from the contact potential and the bias voltage generates electrostatic forces F_{el} that act to decrease the separation d. The plates are held in equilibrium by mechanical forces F_{mech}.*

the electrostatic force F_{el} pulling the plates towards each other is given by

$$F_{\text{el-plates}} = -\frac{\varepsilon_o \varepsilon_r A}{2D^2} V^2. \tag{7.49}$$

The electrostatic potential difference V is the sum of the external applied bias voltage V_{bias} and the internal *contact potential* V_c:

$$V = V_{\text{bias}} + V_{\text{c}}. \tag{7.50}$$

The contact potential V_c is the potential difference between plates arising from contact electrification of the conducting surfaces.

Figure 7.22 illustrates how two metals with different work functions develop a contact potential between them, in terms of electron energy levels (Harper 1967, Lowell and Rose-Innes 1980). If initially, as in Figure 7.22(a), the metals are not electrically connected and have no potential difference between them, the electron energy is constant in the gap between the two metals. When an electrical connection is made between the two metals as shown in Figure 7.22(b), electrons flow to bring the system into thermodynamic equilibrium, which occurs when the metals' Fermi levels equilibrate to the same energy. This exchange of electrons leads to the surface with the higher work function having a net negative charge and the surface with the lower work function having a net positive charge. The opposite surface charge densities then generate an electrostatic potential difference V_c between the two metals proportional to the difference in work functions:

$$\sigma = \varepsilon_0 \varepsilon_r E = \varepsilon_0 \varepsilon_r \left(V_c / D \right), \tag{7.51}$$

$$V_{\text{c}} = \left(\varphi_A - \varphi_B \right) / e. \tag{7.52}$$

Work functions are difficult to predict in practical situations as they depend not only on the type of metal, but also on the presence of oxides, lubricants, and other surface contaminants.

The act of bringing two conductors into contact establishes an electrical connection, where electrons suddenly discharge across the interface at the point of contact. During this discharge, the contact potential and electrostatic attractive force are created. This new electrostatic force now needs to be overcome, along with the non-electrostatic adhesive forces, during separation of the surfaces, resulting in another component of adhesion hysteresis (previously discussed in Section 5.6).

Applying an external bias voltage equal and opposite to the contact potential ($V_{\text{bias}} = -V_c$), as shown in Figure 7.22(c), can be used to cancel the contact potential and the resulting surface charge density. When $V_{\text{bias}} \neq -V_c$, an attractive electrostatic force exists between the conducting surfaces. MEMS and other microactuator devices often use electrostatic force to actuate the motion of small mechanical transducers, so the magnitude of the contact potential needs to be taken into account when designing the bias voltage circuitry.

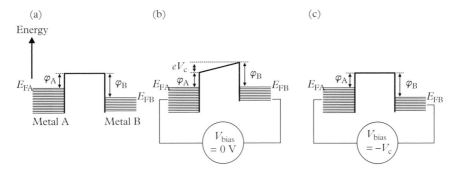

Figure 7.22 *Energy of an electron inside and outside two metals with different work functions φ_A and φ_B. (a) When no electrical connection or potential difference exists between the metals. (b) When electrical connection is made between the two metals, the Fermi level energies E_F become equal generating a potential energy difference $eV_c = (\varphi_B - \varphi_A)$ between the vacuum level energies of the two metals. One way that this electrical connection can be made is by bringing the metals into contact or close enough (a few nanometers) for electrons to tunnel through the energy barrier. (c) Applying a bias voltage $V_{bias} = -V_c$ cancels the potential difference and the electrostatic force between the two metals.*

7.5.2 Metal– and insulator–insulator mechanisms of contact electrification

For metal– and insulator–insulator contacts, the amount of charge present on the insulator surfaces tends to increase with each contact and with the amount of rubbing time before eventually reaching a maximum charge density. The continual accumulation of surface charges with repeated contacts or rubbing is thought to originate from the following mechanism. During each contact, the contact electrification is localized to those nanoscale areas where actual solid–solid contact occurs, and, for insulators, the surface charge is slow to leak to neighboring areas. Further surface charges develop when contacts or rubbing bring more areas into solid–solid contact.

While the contact electrification of conductors is reasonably well understood, the situation is less clear when one or both of the materials are insulators. Three of the proposed mechanisms are illustrated in Figure 7.23:

(a) The *electron transfer model* where electrons are transferred to the electronegative material leaving a positively charged holes in the other material.

(b) The *material transfer model* where slightly charged chunks of material detach during contact and adhere to the opposing surface. The bits of materials pulled out can range in size from as small as individual ions to as large as micro-sized particles.

(c) The *ion transfer model* where charge transfer occurs from the exchange of ions across a water bridge.

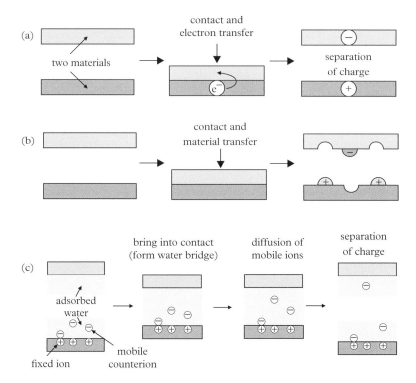

Figure 7.23 *Three possible mechanisms for transferring charge during contact electrification of insulating materials. (a) Electron transfer model, where electrons are transferred to the electronegative material leaving a positively charged hole in the other material. (b) Material transfer model, where bits of materials with a slight charge are pulled out by adhering to the opposing surface when the surfaces are detached from contact. The bits of materials can range in size from as small as individual ions to as large as micro-sized particles. (c) Ion transfer model, where charge transfer results from the redistribution of ions across a water bridge between two solid surfaces. (a) and (c) adapted with permission from McCarty and Whitesides (2008), copyright 2008, John Wiley and Sons.*

For electron transfer model, it may at first not seem obvious how electrons can be exchanged between insulating surfaces in the absence of ion and atom exchanges. To explain this mechanism of charge transfer, a *surface state theory* has been developed along lines similar to the metal–metal contact electrification theory (Davies 1967, Inculet and Wituschek 1967, Lowell and Rose-Innes 1980, Castle 1997). In the surface state theory illustrated in Figure 7.24, an insulator is assumed to have states localized at its surface and with energy levels within the band gap of the insulator. These states can be occupied by either electrons or ions, and the energy difference between the highest occupied state and the vacuum level corresponds to the insulator's effective work function. During contact, charges transfer from the surface with the lower effective work function Φ_A to the surface

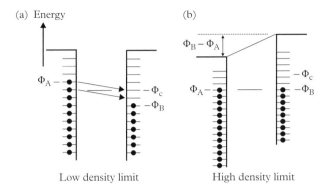

Figure 7.24 *Energy states for electrons or ions during insulator–insulator contact charging. A dash represents a surface state, and a dot represents a filled surface state. Reproduced from Castle (1997), copyright (1997) with permission from Elsevier.*

with the higher effective work function Φ_B, that is, towards the direction of equalizing the energy levels of the highest occupied states of the two materials.

If N is the density of surface states per unit energy per unit area, the maximum possible charge density σ_L that can occur from contact electrification is, in the limit of a low density of surface states,

$$\sigma_L = -eN\left(\Phi_A - \Phi_B\right). \tag{7.53}$$

In the low density limit, N is so small that the electrostatic potential difference stays near zero and the energy levels of the surface states barely change after the charge transfer (Figure 7.24(a)).

At the high density limit of surface states, the surface charge density generates enough electrostatic potential energy difference to shift the energy levels of the two materials to the point where the effective Fermi levels equalize as shown in Figure 7.24(b), resulting in an electrostatic potential difference $(\Phi_A - \Phi_B)/e$. In this high density limit, the maximum possible charge density σ_H that can occur from contact electrification is determined by the electric field between the surfaces:

$$\sigma_H = -\varepsilon_0 \frac{(\Phi_A - \Phi_B)}{eD}, \tag{7.54}$$

where D is the effective separation distance between the two insulators when in contact and ε_0 the permittivity of free space. Equation (7.54) predicts surface charge densities higher than ever observed in experiments, so it represents an upper limit of possible charge transfer. For both the low density (eq. (7.53)) and high density (eq. (7.54)) limits, the surface state theory predicts that the amount of charge exchange depends linearly on the effective work function difference.

7.5.2.1 *Electrostatic discharge and triboluminescence*

Typical maximum surface charge densities observed experimentally on insulators are on the order of 10^{-5}–10^{-4} C/m^2. The electrostatic voltages generated by such a high surface charge density can be sufficient to break down air and even some materials. Since the electric fields are strongest when the separation distances are small, the discharges are most prevalent just after separation when the electric field is still high. Several types of discharge processes can occur during separation:

1. Tunneling can occur through the narrow energy barrier when the surfaces are first separated and the gap is still less than a few nanometers.
2. At larger distances, field emission of electrons can occur if the electric field strength becomes comparable to typical field emission strengths of 0.6–2.3 $\times$ 10^9V/m (Horn and Smith 1992).
3. The combination of high electric fields and emitted electrons and charged particles during a tribo-event can lead to the breakdown of the gas separating the surfaces as described by Paschen's law. In this discharge process, an avalanche event occurs where ionized gas molecules emit electrons that ionize even more gas molecules; this process continues until these newly created charges neutralize the surface charges that are responsible for the high electric fields.

This last mechanism often leads to the very interesting phenomenon of *triboluminescence*, where light is emitted during fracture and rubbing of dielectric materials (Walton 1977, Jha and Chandra 2014). Triboluminescence is particularly strong in piezoelectric materials where the stresses required for fracture lead to polarization of the material before fracture and surface charges after fracture. (Since sugar is a piezoelectric material, triboluminescence can be readily observed by watching someone in a dark room chew on a brittle hard candy with their mouth open.) The high surface charge density then causes a discharge in the gas in the gap between the fractured surfaces, emitting a significant amount of UV and visible light.

Triboluminescence is also frequently observed with non-piezoelectric materials, though it is generally not as intense as for piezoelectric materials. For these materials, other mechanisms in addition to the gas discharge can occur (Chakravarty and Phillipson 2004). For example, rubbing two materials together frequently causes electrons and ions to be emitted and excite a light-emitting plasma in the surrounding gas (Nakayama and Hashimoto 1995).

A particularly interesting case of triboluminescence of non-piezoelectric materials occurs when a roll of adhesive tape is unrolled, where it is possible to observe not only UV and visible light, but also nanosecond flashes of X-rays (Camara et al. 2008). As the tape unrolls, the acrylic adhesive becomes positively charged while the polyethylene backing material becomes negatively charged. As the charged surfaces separate, the electric field is strong enough to eject electrons from the polyethylene surface and accelerate them towards the adhesive surface. At low air pressures, the electrons can be accelerated to energies high enough to generate Bremsstrahlung X-rays when the electrons impact the adhesive. This phenomenon has been used to make an X-ray source capable of X-ray imaging by Prof. Putterman's research group at the University of California, Los Angeles.

7.5.3   Example: contact electrification induced force for a PVC sphere contacting a PTFE flat

In this example, we will estimate the magnitude of the electrostatic adhesive force generated by contact electrification when a 1 μm diameter sphere made of polyvinyl chloride (PVC) contacts a polytetrafluoroethylene (PTFE) flat, with a loading force arising from surface energy.

As polystyrene and PVC have similar surface energies ($\gamma_{PVC} \approx \gamma_{polystyrene} = 33$ mN/m, which occurs since they have similar Hamaker constants, as indicated in Table 7.2), we can use the value of $L_{adh} = 0.15$ N calculated in Section 6.2.1.2. Assuming Hertzian contact, the contact area from this force is given by eq. (3.9):

$$A = \pi a^2$$
$$= \pi \left(\frac{3RL}{4E_c} \right)^{2/3}$$
$$= \pi \left[\frac{3\,(0.5\ \mu\text{m})\,(0.15\ \mu\text{N})}{4\,(0.5\ \text{GPa})} \right]^{2/3}$$
$$= 7.3 \times 10^3\ \text{nm}^2 \tag{7.55}$$

where we have used $E_c \approx E_{PTFE} = 0.5$ GPa, since, for rigid PVC, $E_{PVC} \sim 2$ GPa $>> E_{PTFE}$.

When PVC contacts PTFE, the typical surface charge density is $\sigma = 2 \times 10^{-5}$ C/m^2 with the PVC becoming positively charged and the PTFE negatively charged (Baytekin et al. 2012). As illustrated in Figure 7.25, this charge is distributed over the contact area A. For small separations ($D << R$), we can treat the electrostatic force as coming from a parallel plate capacitor with surface charge density σ and area A:

$$F_{elec} = -\frac{A\sigma^2}{2\varepsilon_0\varepsilon_r}$$
$$= -\frac{\left(7.3 \times 10^{-15}\text{m}^2\right)\left(2 \times 10^{-5}\ \text{C/m}^2\right)^2}{2\left(8.9 \times 10^{-12}\ \text{F/m}\right)}$$
$$= -1.6 \times 10^{-13}\ \text{N}. \tag{7.56}$$

So, for small spherical particles, the electrostatic force from contact electrification is an insignificant contribution to the adhesive force compared to the contribution from van der Waals interactions. If the radius of curvature becomes large enough, for example, the

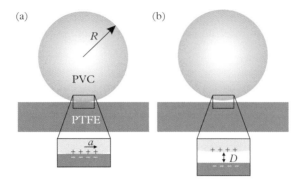

Figure 7.25 *(a) A sphere made of polyvinyl chloride (PVC) coming into contact with a polytetrafluoroethylene (PTFE or Teflon^{TM}) flat making a contact area with radius a. Due to contact electrification, a positive surface charge forms on the PVC surface and a negative surface charge on the PFPE surface over the area of contact. (b) When the two surfaces are separated, the surface charge density σ remains over the areas that were previously in contact.*

1–2 cm effective radius typically encountered in an SFA, then the electrostatic force can become comparable or larger than the adhesive force generated by the intrinsic work of adhesion, as $F_{elec}/F_{VDW} \propto R^{4/3}$. Another way to make the electrostatic force more comparable to the initial van der Waals adhesion force is to roll the particle around on the flat so that a charge density builds on most of the particle's surface area.

Another thing to notice about the electrostatic force from contact electrification is that, in the parallel plate geometry, the force does not depend on separation distance (eq. (7.56)). Consequently, the forces from contact electrification can be very long range, in contrast to the very short range nature of van der Waals interactions. So, while van der Waals interactions are greatly reduced by surface roughness, electrostatic interactions between insulators are not, which is often why they are more noticeable for macroscopic objects in contact.

7.6   PROBLEMS

1. Sometimes an AFM tip is modeled as a parabola of revolution of radius R. Assuming that the axis of the tip is centered on the z-axis with the end of the tip at the origin, the equation describing the outer surface of the tip is given by:

$$z = \frac{1}{2R}\left(x^2 + y^2\right) \text{ (Cartesian coordinates)}$$

$$z = \frac{r^2}{2R} \text{ (polar coordinates)} . \tag{7.57}$$

(a) Find an expression for the van der Waals energy between such a tip and a semi-infinite flat solid that is a distance D away from the tip (i.e., where the surface of the solid is located at $z = -D$). Assume that the tip is close enough to the surface so that $D \ll R$.

(b) In one or two sentences, compare the expression you obtain to the equation given in Figure 7.6 for a sphere of radius R above a surface. Explain how the similarities and differences in these equations make physical sense.

2. Equation (7.25) shows the relationship between surface energy and the Hamaker constant and how the surface energy depends on the separation distance D_0 for the two identical surfaces being considered. However, D_0 turns out to significantly smaller than the atomic spacing a_0 of the material with $D_0 \sim a_0/2.5$. Derive this relationship. The following steps can be followed for this purpose:

(a) Consider a close-packed arrangement of atoms in an fcc crystal with a one-atom basis. Consider the coordination number of an atom in the bulk. Imagine that the crystal is then split, creating two (111) surfaces that are now far apart. Consider the coordination number of an atom at one of the (111) surfaces and find the change in coordination number Δn before and after splitting the crystal. Thus, creating the two surfaces has reduced the number of atom-atom interactions by Δn, for each atom pair across the (111) plane.

(b) Write down the van der Waals energy w for two interacting atoms in terms of C_{VDW} and the atomic separation a_0. The surface energy per atom for the two new (111) surfaces will thus be given by this van der Waals energy multiplied by the change in coordination number for each atom Δn (i.e., for each pair of atoms at the two newly produced surfaces, Δn atom-atom interactions were eliminated.)

(c) Use the relationship between C_{VDW}, the Hamaker constant A, and the number density of a close-packed solid to show that $w = A/2\pi^2$.

(d) We will now have an expression for the van der Waals energy per surface atom required to produce the two surfaces, which will be given by $\Delta n w$. If we divide this by the *area* per atom on the (111) surface, we will have an expression for the energy per unit area required to produce the two surfaces. From this show that the surface energy $\gamma_{solid} = A/24\pi [a_0/\alpha]^2$ where $\alpha \approx 2.57$. Find the exact expression for α.

3. Consider a bowling ball made of polyurethane, which has

- Young's modulus $E_{polyurethane} = 30$ MPa,
- Poisson's ratio $\nu_{polyurethane} = 0.4$,
- mass density $= 1.20$ g/cm^3,
- surface energy $\gamma_{polyurethane} = 30$ mJ/m^2, and
- Hamaker constant $A_{polyurethane-vapor-polyurethane} = 7.5 \times 10^{-20}$ J.

Also consider a ceiling that is effectively a rigid material with a ceiling-vapor-ceiling Hamaker constant of 6.0×10^{-20} J and that the ceiling material and the polyurethane interact purely through van der Waals forces. Find how small the bowling ball must be for it to stick to the ceiling under the following conditions:

(a) In dry conditions (relative humidity ~0%). You will need to estimate the work of adhesion; an example in Section 6.2.1.2 provides a reasonable way to do this. Be sure to justify the particular contact mechanics model you use.

(b) In humid conditions (relative humidity ~90%). The contact angle of water on polyurethane is 30°; on the ceiling, 45°. Assume that eq. (6.58) is valid. Once again you will need to estimate the work of adhesion, this time in the presence of water. Assume the solid–water–solid Hamaker constant is seven times lower than the solid–vapor–solid Hamaker constant.

7.7 REFERENCES

Autumn, K. and A. M. Peattie (2002). "Mechanisms of adhesion in geckos." *Integrative and Comparative Biology* **42**: 1081–90.

Autumn, K., Y. A. Liang, S. T. Hsieh, W. Zesch, W. P. Chan, T. W. Kenny, R. Fearing and R. J. Full (2000). "Adhesive force of a single gecko foot-hair." *Nature* **405**(6787): 681–5.

Autumn, K., M. Sitti, Y. C. A. Liang, A. M. Peattie, W. R. Hansen, S. Sponberg, T. W. Kenny, R. Fearing, J. N. Israelachvili and R. J. Full (2002). "Evidence for van der Waals adhesion in gecko setae." *Proceedings of the National Academy of Sciences of the United States of America* **99**(19): 12252–6.

Baytekin, B., H. T. Baytekin and B. A. Grzybowski (2012). "What really drives chemical reactions on contact charged surfaces?" *Journal of the American Chemical Society* **134**(17): 7223–6.

Bergstrom, L. (1997). "Hamaker constants of inorganic materials." *Advances in Colloid and Interface Science* **70**: 125–69.

Butt, H. J., B. Cappella and M. Kappl (2005). "Force measurements with the atomic force microscope: technique, interpretation and applications." *Surface Science Reports* **59**(1–6): 1–152.

Camara, C. G., J. V. Escobar, J. R. Hird and S. J. Putterman (2008). "Correlation between nanosecond X-ray flashes and stick-slip friction in peeling tape." *Nature* **455**(7216): 1089–92.

Casimir, H. B. G. (1948). "On the attraction of two perfectly conducting plates." *Proc Koninklijke Nederlandse Akademie Van Wetenschappen* **51**: 793–5.

Castle, G. S. P. (1997). "Contact charging between insulators." *Journal of Electrostatics* **40–41**: 13–20.

Cevc, G. (1991). "Hydration force and the interfacial structure of the polar surface." *Journal of the Chemical Society, Faraday Transactions* **87**(17): 2733–9.

Chakravarty, A. and T. E. Phillipson (2004). "Triboluminescence and the potential of fracture surfaces." *Journal of Physics D: Applied Physics* **37**(15): 2175–80.

Chen, B. and H. Gao (2010). "Humidity induced softening leads to apparent capillary effect in gecko adhesion." *MRS Proceedings* **1274**: 81–6.

Davies, D. K. (1967). "Examination of electrical properties of insulators by surface charge measurement." *Journal of Scientific Instruments* **44**(7): 521–4.

De Boer, M. P. and T. A. Michalske (1999). "Accurate method for determining adhesion of cantilever beams." *Journal of Applied Physics* **86**(2): 817–27.

DelRio, F. W., M. P. de Boer, J. A. Knapp, E. D. Reedy, P. J. Clews and M. L. Dunn (2005). "The role of van der Waals forces in adhesion of micromachined surfaces." *Nature Materials* **4**(8): 629–34.

Derjaguin, B. V. and L. Landau (1941). "The theory of stability of highly charged lyophobic sols and coalescence of highly charged particles in electrolyte solutions." *Acta Physicochimica URSS* **14**: 633.

Donaldson, S. H., A. Røyne, K. Kristiansen, M. V. Rapp, S. Das, M. A. Gebbie, D. W. Lee, P. Stock, M. Valtiner and J. Israelachvili (2015). "Developing a general interaction potential for hydrophobic and hydrophilic interactions." *Langmuir* **31**(7): 2051–64.

Dzyaloshinskii, I. E., E. M. Lifshitz and L. P. Pitaevskii (1960). "Van der Waals forces in liquid films." *Soviet Physics JETP-USSR* **10**(1): 161–70.

Dzyaloshinskii, I. E., E. M. Lifshitz and L. P. Pitaevskii (1961). "The general theory of van der Waals forces." *Advances in Physics* **10**(38): 165–209.

Eichenlaub, S., C. Chan and S. P. Beaudoin (2002). "Hamaker constants in integrated circuit metalization." *Journal of Colloid and Interface Science* **248**(2): 389–97.

French, R. H. (2000). "Origins and applications of London dispersion forces and Hamaker constants in ceramics." *Journal of the American Ceramic Society* **83**(9): 2117–46.

French, R. H., R. M. Cannon, L. K. Denoyer and Y. M. Chiang (1995). "Full spectral calculation of nonretarded Hamaker constants for ceramic systems from interband transition strengths." *Solid State Ionics* **75**: 13–33.

French, R. H., H. Mullejans, D. J. Jones, G. Duscher, R. M. Cannon and M. Ruhle (1998). "Dispersion forces and Hamaker constants for intergranular films in silicon nitride from spatially resolved-valence electron energy loss spectrum imaging." *Acta Materialia* **46**(7): 2271–87.

French, R. H., K. I. Winey, M. K. Yang and W. Qiu (2007). "Optical properties and van der Waals–London dispersion interactions of polystyrene determined by vacuum ultraviolet spectroscopy and spectroscopic ellipsometry." *Australian Journal of Chemistry* **60**(4): 251–63.

Fukuzawa, K., J. Kawamura, T. Deguchi, H. Zhang and Y. Mitsuya (2004a). "Disjoining pressure measurements using a microfabricated groove for a molecularly thin polymer liquid film on a solid surface." *The Journal of Chemical Physics* **121**(9): 4358–63.

Fukuzawa, K., J. Kawamura, T. Deguchi, H. Zhang and Y. Mitsuya (2004b). "Measurement of disjoining pressure of a molecularly thin lubricant film by using a microfabricated groove." *IEEE Transactions on Magnetics* **40**(4): 3183–5.

Gao, J., W. D. Luedtke and U. Landman (1995). "Nano-elastohydrodynamics: structure, dynamics, and flow in nonuniform lubricated junctions." *Science* **270**(5236): 605–8.

Gao, J. P., W. D. Luedtke and U. Landman (2000). "Structures, solvation forces and shear of molecular films in a rough nano-confinement." *Tribology Letters* **9**(1–2): 3–13.

Gee, M. L. and J. N. Israelachvili (1990). "Interactions of surfactant monolayers across hydrocarbon liquids." *Journal of the Chemical Society, Faraday Transactions* **86**(24): 4049–58.

Gosvami, N. N., S. K. Sinha and S. J. O'Shea (2008). "Squeeze-out of branched alkanes on graphite." *Physical Review Letters* **100**(7): 076101.

Granick, S., Y. X. Zhu and H. Lee (2003). "Slippery questions about complex fluids flowing past solids." *Nature Materials* **2**(4): 221–7.

Hamaker, H. C. (1937). "The London–van der Waals attraction between spherical particles." *Physica* **4**(10): 1058–72.

Hammer, M. U., T. H. Anderson, A. Chaimovich, M. S. Shell and J. Israelachvili (2010). "The search for the hydrophobic force law." *Faraday Discussions* **146**: 299–308.

Harper, W. R. (1967). *Contact and frictional electrification*. Oxford: Clarendon Press.

Horn, R. G. and D. T. Smith (1992). "Contact electrification and adhesion between dissimilar materials." *Science* **256**(5055): 362–4.

Hough, D. B. and L. R. White (1980). "The calculation of Hamaker constants from Lifshitz theory with applications to wetting phenomena." *Advances in Colloid and Interface Science* **14**(1): 3–41.

Houston, M. R., R. T. Howe and R. Maboudian (1997). "Effect of hydrogen termination on the work of adhesion between rough polycrystalline silicon surfaces." *Journal of Applied Physics* **81**(8): 3474–83.

Huber, G., H. Mantz, R. Spolenak, K. Mecke, K. Jacobs, S. N. Gorb and E. Arzt (2005). "Evidence for capillarity contributions to gecko adhesion from single spatula nanomechanical measurements." *Proceedings of the National Academy of Sciences of the United States of America* **102**(45): 16293–6.

Inculet, I. I. and E. P. Wituschek (1967). "Electrification by friction in a 3×10^{-7} Torr vacuum." *Static Electrification, Institute of Physics Conference Series* **4**: 37–43.

Irschick, D. J., C. C. Austin, K. Petren, R. N. Fisher, J. B. Losos and O. Ellers (1996). "A comparative analysis of clinging ability among pad-bearing lizards." *Biological Journal of the Linnean Society* **59**(1): 21–35.

Israelachvili, J. and H. Wennerstrom (1996). "Role of hydration and water structure in biological and colloidal interactions." *Nature* **379**(6562): 219–25.

Israelachvili, J. N. (1972). "Calculation of van der Waals dispersion forces between macroscopic bodies." *Proceedings of the Royal Society of London, Series A: Mathematical and Physical Sciences* **331**(1584): 39–55.

Israelachvili, J. N. (2011). *Intermolecular and surface forces* (3rd ed.). Burlington, MA: Academic Press.

Israelachvili, J. N. and R. M. Pashley (1983). "Molecular layering of water at surfaces and origin of repulsive hydration forces." *Nature* **306**(5940): 249–50.

Jensen, T. R., M. O. Jensen, N. Reitzel, K. Balashev, G. H. Peters, K. Kjaer and T. Bjornholm (2003). "Water in contact with extended hydrophobic surfaces: Direct evidence of weak dewetting." *Physical Review Letters* **90**(8): 086101.

Jha, P. and B. P. Chandra (2014). "Survey of the literature on mechanoluminescence from 1605 to 2013." *Luminescence* **29**(8): 977–93.

Klimchitskaya, G., U. Mohideen and V. Mostepanenko (2000). "Casimir and van der Waals forces between two plates or a sphere (lens) above a plate made of real metals." *Physical Review A* **61**(6): 062107.

Leikin, S., V. A. Parsegian, D. C. Rau and R. P. Rand (1993). "Hydration forces." *Annual Review of Physical Chemistry* **44**: 369–95.

Lifshitz, E. M. (1956). "The theory of molecular attractive forces between solids." *Soviet Physics JETP-USSR* **2**(1): 73–83.

Lowell, J. and A. C. Rose-Innes (1980). "Contact electrification." *Advances in Physics* **29**(6): 947–1023.

Marcelja, S. and N. Radic (1976). "Repulsion of interfaces due to boundary water." *Chemical Physics Letters* **42**(1): 129–30.

Marchon, B. and T. E. Karis (2006). "Poiseuille flow at a nanometer scale." *Europhysics Letters* **74**(2): 294–8.

Mastrangelo, C. H. and C. H. Hsu (1993). "Mechanical stability and adhesion of microstructures under capillary forces. II. Experiments." *Journal of Microelectromechanical Systems* **2**: 44–55.

McCarty, L. S. and G. M. Whitesides (2008). "Electrostatic charging due to separation of ions at interfaces: Contact electrification of ionic electrets." *Angewandte Chemie International Edition* **47**(12): 2188–207.

McLachlan, A. D. (1963). "3-body dispersion forces." *Molecular Physics* **6**(4): 423–7.

Meyer, E. E., K. J. Rosenberg and J. Israelachvili (2006). "Recent progress in understanding hydrophobic interactions." *Proceedings of the National Academy of Sciences* **103**(43): 15739–46.

Nakayama, K. and H. Hashimoto (1995). "Effect of surrounding gas-pressure on triboemission of charged-particles and photons from wearing ceramic surfaces." *Tribology Transactions* **38**(1): 35–42.

Ninham, B. W. (1970). "Van der Waals forces across triple-layer films." *Journal of Chemical Physics* **52**(9): 4578–87.

O'Shea, S. J., N. N. Gosvami, L. T. W. Lim and W. Hofbauer (2010). "Liquid atomic force microscopy: solvation forces, molecular order, and squeeze-out." *Japanese Journal of Applied Physics* **49**(8S3): 08LA01.

Palik, E. D. and G. Ghosh (1998). *Handbook of optical constants of solids*. San Diego, CA: Academic Press.

Pastewka, L. and M. O. Robbins (2014). "Contact between rough surfaces and a criterion for macroscopic adhesion." *Proceedings of the National Academy of Sciences* **111**(9): 3298–303.

Persson, B. N. J. (2002). "Adhesion between an elastic body and a randomly rough hard surface." *The European Physical Journal E* **8**(4): 385–401.

Polycarpou, A. A. and I. Etsion (1998). "Static friction of contacting real surfaces in the presence of sub-boundary lubrication." *Journal of Tribology: Transactions of the ASME* **120**(2): 296–303.

Poynor, A., L. Hong, I. K. Robinson, S. Granick, Z. Zhang and P. A. Fenter (2006). "How water meets a hydrophobic surface." *Physical Review Letters* **97**(26): 266101.

Samoilov, V. N., I. M. Sivebaek and B. N. J. Persson (2004). "The effect of surface roughness on the adhesion of solid surfaces for systems with and without liquid lubricant." *Journal of Chemical Physics* **121**(19): 9639–47.

Tarazona, P. and L. Vicente (1985). "A model for density oscillations in liquids between solid walls." *Molecular Physics* **56**(3): 557–72.

Tyrrell, J. W. G. and P. Attard (2001). "Images of nanobubbles on hydrophobic surfaces and their interactions." *Physical Review Letters* **87**(17): 176104.

Verwey, E. J. W., J. T. G. Overbeek and K. Van Nes (1948). *Theory of the stability of lyophobic colloids: the interaction of sol particles having an electric double layer*. Elsevier Publishing Company.

Walton, A. J. (1977). "Triboluminescence." *Advances in Physics* **26**(6): 887–948.

Wang, Z. L. (2013). "Triboelectric nanogenerators as new energy technology for self-powered systems and as active mechanical and chemical sensors." *ACS Nano* **7**(11): 9533–57.

White, L. R., R. R. Dagastine, P. M. Jones and Y. T. Hsia (2005). "Van der Waals force calculation between laminated media, pertinent to the magnetic storage head–disk interface." *Journal of Applied Physics* **97**(10): 104503.

Yang, K., Y. Lin, X. Lu and A. V. Neimark (2011). "Solvation forces between molecularly rough surfaces." *Journal of Colloid and Interface Science* **362**(2): 382–8.

Zhang, X. Y., Y. X. Zhu and S. Granick (2002). "Hydrophobicity at a Janus interface." *Science* **295**(5555): 663–6.

8

Measuring Surface Forces

The basic concept for measuring the force F acting between two solid surfaces in the direction perpendicular or parallel to these surfaces is to convert the magnitude of the force into a measurable displacement. Figure 8.1 illustrates this approach for a force acting in the direction perpendicular or normal to the surface. First, one of the solids is attached to a flexible spring element with a known spring constant K_{spring}. Then, the two surfaces are brought incrementally closer together by moving either the base of the other solid or the base of the spring a distance ΔD_0. Measuring the spring element deflection ΔD_{spring} and applying Hooke's law determines the change in force ΔF due to the displacement ΔD_0:

$$\Delta F (\Delta D_0) = K_{spring} \Delta D_{spring}. \tag{8.1}$$

Rather than working with changes in force ΔF, it is more convenient to work with the total force F measured relative to some reference. By convention, the force F is set equal to zero at large separations $(D = +\infty)$.

Once the force F has been determined as a function of sample position, this can be converted into a plot of force versus separation distance (F vs. D) by using the following relationships to determine the separation distance D:

$$\begin{aligned} \Delta D &= \Delta D_0 - \Delta D_{spring} \\ D &= D_0 - D_{spring} \end{aligned}. \tag{8.2}$$

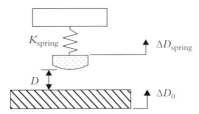

Figure 8.1 *Idealized setup for measuring the force acting between two surfaces in the direction perpendicular to the surfaces.*

Tribology on the Small Scale: A Modern Textbook on Friction, Lubrication and Wear. Second edition. C. Mathew Mate and Robert W. Carpick. © Oxford University Press 2019. Published in 2019 by Oxford University Press.
DOI: 10.1093/oso/ 9780199609802.001.0001

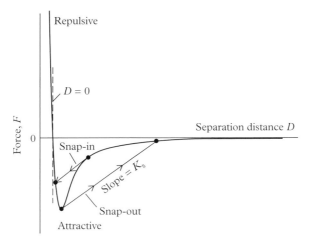

Figure 8.2 *A typical force-versus-separation distance (F vs. D) curve where the force is attractive at large separation distances and repulsive at small distances. As the two surfaces approach each other, the force gradient may exceed the spring constant (dF/dD > K$_{spring}$), and the upper surface illustrated in Figure 8.1 snaps into contact. Similarly, when the surfaces are pulled apart, the surfaces will snap apart where dF/dD > K$_{spring}$. (See also Figure 5.17.)*

Figure 8.2 shows the typical shape of a F vs. D curve during approach and retraction. For the separation distance, the convention is to choose $D = 0$ when $F = 0$ while the two solids are in repulsive hard wall contact, as is done in Figure 8.2. The rationale for this choice is that the measurement zero for separation distance then lies within an atomic diameter of the theoretical zero for the separation distance of the surfaces' outermost layer of atoms.

A similar method is used for measuring the frictional force acting tangential to the contacting surfaces, as illustrated in Figure 8.3. A flexible spring element is attached to the bottom surfaces and displaces laterally when a tangential force during contact with the upper surface. Once this displacement $\Delta D_{spring-lateral}$ is measured, Hooke's law is used to convert the displacement to a force:

$$\Delta F_{lateral} = K_{spring-lateral} \Delta D_{spring-lateral}. \tag{8.3}$$

One of the uses of surface force measurements is to confirm the atomic origins of the adhesive, repulsive, and friction forces between surfaces. Since many of these atomic scale forces act over just few angstroms, one needs to measure the distances—ΔD, ΔD_0, ΔD_{spring}, and $\Delta D_{spring-lateral}$—with a precision of an angstrom or better to determine these forces accurately. Fortunately, current techniques for measuring such displacement can routinely measure these sub-angstrom displacements.

Figure 8.2 also illustrates a frequent complication with force measurements in the direction normal to the surface. The attractive force causes the two surfaces to snap

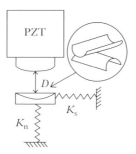

Figure 8.3 *Idealized setup of a surface force apparatus (SFA) that is configured to measure simultaneously the forces acting perpendicularly and tangentially on the bottom surface as the PZT moves the upper surface toward contact. The PZT is also capable of moving the upper surface laterally so as to induce a frictional force when in contact with the lower surface. The inset shows schematically how the two mica sheets are arranged in a crossed cylinder geometry. From Seror et al. (2015); copyright 2015, Springer Nature; licensed under a Creative Commons Attribution 4.0 International License.*

in or out of contact during either the approach or the retraction. This also leads to adhesion hysteresis, where the maximum attractive force during retraction is greater than the maximum experienced during approach (this is a type of mechanical adhesion hysteresis, as discussed in Section 5.6.1).

Snap-in/snap-out occurs when the spring element is too soft. As two surfaces approach, the attractive force increases as does the *gradient* of the attractive force. If this gradient exceeds K_{spring}, the rate of increase in attractive force during an approach is greater at this point than the rate of increase of the counteracting force of the spring element, and a sudden snap into contact occurs. Since the surfaces snap through those separation distances where high gradients occur, the force cannot be measured over these separations. A similar snap-out occurs when the surfaces are retracted. The snap-in/snap-out problem can be avoided by using a suitably stiff spring element, but at the price of losing force sensitivity. Alternately, dynamic methods (discussed in Sections 8.2.3 and 8.3), or special force-feedback instrumentation can be used to prevent these instabilities (Houston and Michalske 1992).

Next, we discuss two of the techniques, the surface force apparatus (SFA) and atomic force microscope (AFM), that have been developed for measuring normal and lateral forces between surfaces separated by atomic scale dimensions.

8.1 Surface force apparatus (SFA)

Since the mid-1970s, the SFA has been a major workhorse for measuring molecular scale forces.

Figures 8.3 and 8.4 illustrate how an SFA works (Israelachvili and Adams 1978, Parker et al. 1989, Israelachvili and McGuiggan 1990, Israelachvili et al. 2010). At the heart of an SFA are two curved pieces of mica (radius of curvature ~1 cm) that are held in a crossed cylinder geometry so as to form a circular or elliptical contact area when their surfaces touch. One mica sheet is attached with epoxy to a cylindrical lens mounted on a cantilever spring element, which deflects a distance $\Delta D_{cantilever}$ when a normal force acts between the two mica sheets. The other mica sheet is epoxied to another cylindrical lens mounted on a PZT tube that provides a few microns of motion in both the vertical (ΔD_0) and lateral directions.

The separation distance D between the two mica sheets is measured by multibeam interferometry. With this technique, slightly transparent silver films are deposited onto the backsides of the mica sheets ($\sim 2 \, \mu m$ thick); an intense source of white light is focused onto the bottom lens; and a small fraction of the incident light enters the optical cavity formed by the two silvered surfaces. Those wavelengths of light that constructively interfere within this cavity emerge from the optical cavity and enter the spectrograph. The spectrograph disperses these fringes of equal chromatic order (FECO) onto the CCD camera generating an output shown schematically in Figure 8.4. From the location of these fringes, the following can be determined: the separation distance between the silvered mica surfaces, the refractive indexes of the material between them, and the cross-sectional shape of the contact zone (Israelachvili 1973, Horn and Smith 1991). Sometimes optical coatings are used in place of the silver coatings, particularly, if other optical probes are to be used to further study the confined liquids (Mukhopadhyay et al. 2003). In addition to the FECO interferometry technique, two other techniques— capacitance (Tonck et al. 1988, Frantz et al. 1996) and optical fiber interferometry (Frantz et al. 1997)—have also been used for measuring the changes in the separation distance between SFA mica surfaces.

In an SFA, zero separation of the two facing mica surfaces ($D = 0$) is determined by bringing the surfaces into contact in the absence of liquid or adsorbed layers, so that the atoms on the opposing mica surfaces are in intimate contact. The distance between the silvered surfaces under this intimate contact condition is measured first using the FECO technique, and then this distance subtracted from future measurements when liquid or polymer molecules are present between the surfaces to provide the absolute separation distance. Using eq. (8.2), the SFA cantilever deflection $\Delta D_{cantilever}$ is determined from the separation ΔD measured by the FECO technique and the distance ΔD_0 that the top surface is moved (i.e., $\Delta D_{cantilever} = \Delta D_0 - \Delta D$); then, the normal force is determined from Hooke's law (eq. (8.1)). For most SFAs, a typical sensitivity for ΔD is 1 Å; so, with a relatively soft spring constant of $K_{spring} = 100$ N/m, the force sensitivity is 10 nN.

The SFA can also measure lateral friction forces by incorporating a lateral spring and lateral displacement sensing, as illustrated in Figures 8.3 and 8.4. Many examples of frictional phenomena have been studied using this method, including for dry surfaces in contact (Homola et al. 1989, Vigil et al. 1994), surfaces coated by thin films like self-assembled monolayers (Ruths et al. 2003, Gao et al. 2004), and for liquids under a range of confining pressures (Klein and Kumacheva 1995).

The mica sheets are a key feature of the SFA technique. These sheets are formed from single crystals of mica that can easily be cleaved along a certain crystal plane to form a surface atomically flat over centimeter sized areas. For such surfaces, effectively no roughness is present, except for the periodic arrangement of mica surface atoms (roughness < 1 Å). Since the surface roughness is negligible, this vastly simplifies the

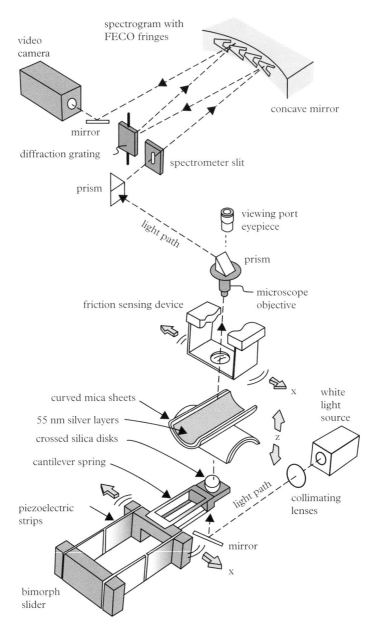

Figure 8.4 *Exploded diagram of the internal components of a surface force apparatus (SFA) capable of measuring normal and lateral forces acting between to crossed cylindrical surfaces of mica. The lower surface of mica is connected to a cantilever spring that measures the normal force (the double cantilever arrangement is used to ensure that this surface does not tilt when the cantilevers bend.) This cantilever is connected to piezoelectric strips that move the lower surface laterally for friction experiments. The sheets of mica are attached to two cylindrical lenses (labeled as silica disks in the figure) arranged in a crossed cylinder geometry. The sides of the mica sheets facing each other are cleaved to provide atomically smooth surfaces. The upper surface is connected to friction sensing strain gauges. The changes in separation between top and bottom surfaces are measured by optical interference from the semi-reflective silver films on the backsides of the mica sheets of white light that enters through a bottom viewport and exits through a microscope objective. (The dashed line shows the light path.) Those wavelengths of the white light that form standing waves between the surfaces are transmitted through to the spectrograph. The spectrograph measures the wavelengths transmitted, from which the distance between silvered surfaces can be determined as a function of lateral position. Reproduced from Israelachvili et al. (2010) with permission from IOP Publishing, copyright 2010.*

interpretation of SFA results. The lack of surface roughness also means that, at $D = 0$, the atoms on opposing surfaces are uniformly in contact with a spacing approximately that of the distance between the bulk atomic layers.

The SFA technique has been extended beyond mica surfaces by coating the cleaved mica surfaces with other materials such as molecular monolayers, oxides (Horn et al. 1989, Hirz et al. 1992, Ducker et al. 1994), and metals (Parker and Christenson 1988, Levins and Vanderlick 1995). These materials are usually deposited onto the mica in such a way that the resulting surfaces are nearly atomically smooth. With such smooth surfaces, it has become fairly straightforward to observe molecular level force effects that would otherwise be smeared out by roughness, such as the effects of the solvation forces discussed in Section 7.4.1.

Another feature of the SFA is illustrated in Figure 8.5(a). When the mica sheets squeeze a liquid between them, they elastically deform so that the facing surfaces become

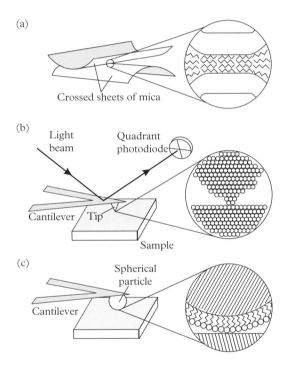

Figure 8.5 *Comparison of the contact regions for three different techniques for probing molecular scale forces: (a) Surface force apparatus (SFA) with liquid molecules forming layers between the surfaces. The repulsive forces from the molecular layers sandwiched between the mica surfaces elastically deform the mica sheets and the glue attaching them to the cylindrical lenses to form two parallel surfaces in the region of high contact pressure. In the figure, the scale normal to the surface is greatly exaggerated: the parallel region typically extends tens of microns across, while the gap between the surfaces is only a few nanometers. (b) Atomic force microscope (AFM) with an atomically sharp tip and no meniscus around the contact. The quadrant photodiode records cantilever deflections due to both normal forces and lateral forces simultaneously. Ultra-high vacuum conditions are typically required to achieve a contact without a meniscus of contaminants. (c) AFM with a spherical particle tip coated with a monolayer of molecules. This type of AFM, where the tip is coated with a well-defined molecular entity, is referred to as chemical force microscopy.*

parallel to each other. This flexing of the thin mica sheets is possible because of the easy compression of the soft epoxy used to attach the mica to the cylindrical lenses. Most liquids generate a repulsive force sufficient to deform the mica when confined to a few nanometers of thickness between the mica sheets. This allows these liquids to be studied in a confined geometry between two atomically smooth, parallel surfaces, a feature that has been extensively used to study lubricating properties of molecular films.

For modest applied loads, the contact region is typically tens of microns across. So, while the SFA has great force sensitivity to molecular forces at nanometer separations, measured with angstrom precision, these results are averaged over large nominal contact areas. If these forces need to be measured with a high lateral resolution, typically some type of an atomic force microscope is used, as described in the next section.

8.2 Atomic force microscope (AFM)

Since its introduction by Binnig et al. (1986), atomic force microscopy has evolved into a versatile technique for measuring the surface forces at small separation distances, in addition to measuring surface topographies (Section 2.2.2). Figure 8.5(b) shows a typical modern AFM where a normal force causes the cantilever to bend normal to the surface and a lateral force causes a torsional twist of the cantilever; these two types of deflections are typically measured by reflecting a light beam off the back of the cantilever and onto a position-sensitive photodiode.

8.2.1 AFM cantilevers and tips

While many types of structures have been used over the years as AFM cantilevers, for most modern AFM instruments, cantilevers are made using microfabrication techniques (Albrecht et al. 1990, Bhushan and Marti 2011); this enables them to be made very thin (0.5–1 μm) to provide for a low spring constant (0.1–50 N/m) and to be made fairly short (length <200 μm) to provide for a high resonance frequency (10 kHz to 1 MHz). Microfabricated cantilevers are either triangular in shape, as illustrated in Figure 8.5(b), or rectangular in shape, as illustrated in Figure 8.6.

Sharp tips are fabricated to protrude from the free end of the cantilever beam with ultra-small radii of curvature at the end (typically, $R < 10$ nm), as shown in the example in Figure 8.6. For AFM imaging, the tip sharpness ensures that, when the tip scans across the sample surface, the contact zone is only a few atoms across, resulting in a high—potentially sub-nanometer—lateral resolution. For force measurements, such a sharp tip means that only a small number of atoms and molecules at the end of the tip are involved in generating the forces at small separation distances. Indeed, the main advantage of using an AFM for studying contact forces is its ability to measure the forces operating when the first few atoms on the summit of a single asperity touch the atoms on the opposing surface, as illustrated in Figure 8.5(b). One disadvantage of these sharp tips is that force is much smaller than for blunter tips, making it difficult to measure. Another complication with working with such sharp tips is that they do not remain sharp after a few contacts with sample surfaces due to plastic deformation of the tip; it is not uncommon for a tip with an initial radius <10 nm to blunt to a few tens of nanometers.

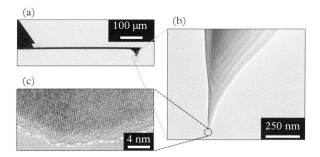

Figure 8.6 *Transmission electron microscopy (TEM) images of (a) the side-view of a rectangular shaped AFM cantilever and tip, (b) a magnified view of the tip, and (c) a high resolution image of the very end of the tip revealing the detailed shape of the tip apex, which is outlined by an added white dot-dashed line. In (c), the periodic pattern comes from the transmission of the electron beam through the crystalline atomic lattice of the silicon tip. Adapted with permission from Jacobs et al. (2016). Copyright 2016, AIP Publishing.*

One clever solution to this blunting of AFM tips is to attach a narrow carbon nanotube onto the end of the tip (Dai et al. 1996, Wong et al. 1998, Wilson and Macpherson 2009). Since these tubes are fairly flexible, they can bend when the repulsive force exceeds the small threshold needed to buckle the tube. This limits the stresses applied to the tip during contact, minimizing the damage to the end of the nanotube. Since the diameter of a carbon nanotube is typically <10 nm, high resolution imaging of surface forces can be achieved with little risk of damage; even if the end of the nanotube breaks off, the tube diameter, which determines the resolution, remains constant.

Commercially, AFM cantilevers have become available with a wide array of tip materials, including diamond and various metals, with particular utility for nanotribology studies.

8.2.2 Chemical force microscopy

One active area of research involves using tips coated with a well-defined molecular entity (Noy et al. 1997). Then, the interaction forces between this entity and that on the opposing sample surface can be measured and mapped over the sample surface. Force microscopy done with these functionalized tips is often referred to as *chemical force microscopy* (Noy et al. 1997, Noy 2006). Functionalized tips can be prepared in a research laboratory or purchased from commercial AFM cantilever vendors. The three most common ways of preparing tips with functionalized surfaces are:

1. silanization of a tip made of either silicon nitride or silicon oxide;
2. coating the tip first with gold then self-assembling thiols onto the gold film;
3. functionalizing the end of a carbon nanotube tip (Wong et al. 1998).

Another approach for obtaining a tip with a well-defined chemically active layer on its surface is to replace the standard sharp AFM tip with a smooth, spherical particle with a well-defined diameter and then to coat that particle with the molecular entity that one wants to study. Figure 8.5(c) shows a spherical particle at the end of an AFM cantilever coated with a functionalized monolayer. Using a spherical particle at the end of the cantilever has several advantages:

1. Since the sphere radius is much larger than a standard AFM tip (microns rather than nanometers), the force is much larger, and, hence, easier to measure. Despite the higher force, the larger contact area leads to lower stresses, allowing easier imaging of soft, easily-damaged materials.

2. Since the sphere's radius is known, the measured F vs. D curve can be converted into the energy per unit area versus separation distance curve by using the Derjaguin approximation (Section 6.1). For this method to work well at small separations, the spherical particles need to be as smooth as possible, as the Derjaguin approximation is only valid at separation distances significantly greater the surface roughness.

8.2.3 Dynamic atomic force microscopy

To achieve the highest resolution in an AFM force measurement, one has to measure the force acting between a single atom protruding from the end of the tip and a single atom sitting on a sample surface. Typically, this force is too small to be measured simply by measuring the small deflection of a cantilever. Instead, a more sensitive force detection method is used where the cantilever is oscillated at its resonance frequency and the shift of the resonance frequency is measured as a function of separation distance (Dürig 2000, Giessibl 2003). This method determines the derivative of the force acting on the tip, then integrating this derivative over separation distance yields the net force acting on the tip. Since a stiffer cantilever can be used, this oscillating method has the advantage that the attractive force can be measured for smaller separation distances without snap-in than with the static deflection method. In later Sections 8.3.1 and 8.3.2, we present an example where this oscillation method was used to measure the van der Waals force between a sharp silicon tip and a silicon surface.

8.2.4 Friction force microscopy

The lateral force detection capability for the AFM was first developed by Mate et al. (1987) and then adapted for the optical deflection method by Marti et al. (1990) where a quadrant photodiode is used to determine the lateral force, as shown in Figure 8.5(b); this has become the most popular method for measuring lateral forces and is used in nearly all commercial AFMs. In this design, a friction force exerted on the tip by the sample causes the cantilever to twist, which in turns deflects the reflected light beam either to the left or right along the photodiode by an amount proportional to the friction force. This deflection is independent of the deflection of the light beam upward or downward along the photodiode caused by the normal force. Thus, friction and normal forces can be simultaneously recorded. Since even angstrom deflections of the AFM tip result in measurable angular deflections, atomic-level resolution is possible of the lateral forces from static or kinetic friction.

8.3 Examples of forces acting on AFM tips

As discussed in Chapters 6 and 7, when two solids are brought in close proximity, numerous types of forces act between them. In the remaining sections of this chapter, we discuss a few examples of these forces as they act on AFM tips. The purpose here is

1. to illustrate the capabilities of AFM to measure and distinguish between the different surface forces;
2. to illustrate under what conditions these different surface forces become important;
3. to illustrate the magnitude of these forces.

The results discussed here represent only a miniscule fraction of the AFM results in the published literature. For a more extensive survey of what has been learned about surface forces using AFM, the reader is referred to several review papers (Carpick and Salmeron 1997, Cappella and Dietler 1999, Janshoff et al. 2000, Giessibl 2003, Butt et al. 2005, Cappella 2016). A similarly extensive literature also exists regarding surface force measurements by SFA. The literature for SFA up to 1991 was reviewed in the book by Israelachvili (1991), with more recent work reviewed by Ruths and Israelachvili (2011). The SFA literature on properties of lubricant films has been reviewed by Yamada (2010), and on biomolecular interactions by Ruths et al. (2014).

8.3.1 Van der Waals forces under vacuum conditions

As discussed in Chapter 7, van der Waals forces are always present between surfaces. Consequently, if an AFM tip is brought near a surface, it will always experience an attractive van der Waals force. For separation distances much smaller than the tip radius, we can approximate the tip–sample geometry as being equivalent to the sphere-above-flat geometry. Using the relationship for a sphere-above-flat geometry (Figure 7.6), the van der Waals force on an AFM tip can be estimated as

$$F_{\text{vdw sphere-flat}}(D) = -\frac{A_{\text{tip-v-sample}}R}{6D^2}. \tag{8.4}$$

For example, if an AFM tip contacts a flat surface where

- tip radius $R = 10$ nm,
- Hamaker constant $A_{\text{tip-v-sample}} = 10^{-19}$ J,
- separation distance in contact $D_0 = 0.165$ nm (Section 7.3.4),

then

$$F_{\text{vdw sphere-flat}}\left(D_{\text{contact}}\right) = -\frac{(10^{-19}\,\text{J})(10\,\text{nm})}{6(0.165\,\text{nm})^2}.$$
$$= 6.1\,\text{nN}$$
(8.5)

This magnitude of van der Waals force (6.1 nN) is readily measurable with a wide range of AFM cantilevers. It quickly drops off, however, as the tip is moved away from the sample: by $D = 1$ nm, the van der Waals force acting on this 10 nm tip is reduced to only 0.17 nN. Consequently, the van der Waals force at a few nanometers separation can be difficult to measure unless one uses a low spring constant cantilever, but this will make the AFM measurement susceptible to the snap-in and snap-out behavior discussed in Figure 8.2.

If a tip with a larger radius is used, the van der Waals force, which scales with radius, becomes easier to measure. Figure 8.7(a) shows the simulated F vs. D_{sample} curves for the van der Waals force acting on an AFM tip with radius $R = 30$ nm for two different cantilever spring constants. For the very weak cantilever spring constant of 0.01 N/m, the van der Waals force induces a substantial cantilever deflection (a few angstroms at a separation distance of 20 nm), but this weak cantilever also results on approach in a "snap-in" at a separation of 10 nm and on withdrawal a "snap-out" to ~1100 nm separation. Increasing the cantilever spring constant by a factor of ten reduces the snap-in/snap-out distances, but also decreases the force sensitivity by a factor of ten.

While eq. (8.4) provides a reasonable estimate of the van der Waals force at small separations, once contact occurs, the adhesive forces distort the surfaces somewhat away from the sphere-on-flat geometry. As discussed in Section 6.2.2, the coupling of adhesive forces with elasticity increases the contact zone area, changing the magnitude of the expected adhesive force. These effects can be analyzed using the JKR (Johnson et al. 1971), DMT (Derjaguin et al. 1975), Maugis (1992), or MYD (Muller et al. 1980) theories, with the choice of theory depending on the relative strengths of the adhesive force and elastic modulus as discussed in Section 6.2.2.2. Figure 8.7(b) shows the calculated force, as a function of separation distance, from the combination of the van der Waals interaction and the repulsive elastic contact, for silicon tips of different radii as they come into contact with a flat silicon sample. The effect of the elasticity is taken into account with an MYD analysis (Janshoff et al. 2000).

As an example of what has been achieved experimentally in measuring the weak van der Waals force on AFM tips, Figure 8.8 shows the F vs. D_{sample} curve for a sharp silicon AFM tip approaching a reconstructed silicon(111) surface under ultra-high vacuum conditions ($<10^{-9}$ Torr) (Lantz et al. 2001). Doing this experiment in ultra-high vacuum not only prevents the formation of a meniscus around the tip from capillary condensation, but also help ensures that the silicon surface remains free of contamination during the experiment.

Even though the force is very weak—only 40 pN at a distance of 20 Å and 3.8 nN at contact at $D \sim 2$ Å—the Dürig method (Dürig 2000) of oscillating the cantilever at its resonance frequency enables force to be measured with a good signal-to-noise ratio. Fitting eq. (8.4) to the data in Figure 8.8 enabled Lantz et al. to determine

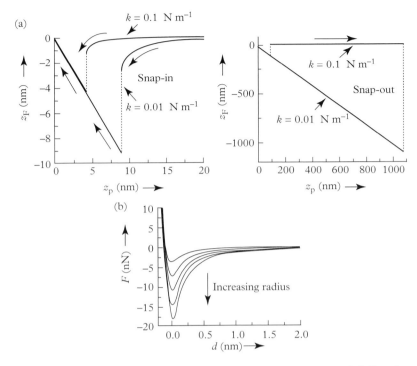

Figure 8.7 *(a) Simulated van der Waals F vs. D sample curves during approach (left) and separation (right) for a 30 nm radius AFM tip near a surface where the Hamaker constant $A_{tip-v-sample} = 10^{-19}$ J. Here z_F is the cantilever deflection ($D_{cantilever}$) and z_p is the distance that the sample moves (D_{sample}). (b) Simulated van der Waals force F vs. d (separation distance) curve for a silicon tip interacting with a silicon surface. From Janshoff et al. (2000), with permission from Wiley-VCH Verlag GmbH and Co, KG.*

that $A_{Si-v-Si}R = 9.1 \times 10^{-28}$ J m. Using the calculated Hamaker constant for silicon $A_{Si-v-Si} = 1.865 \times 10^{-19}$ J (Senden and Drummond 1995), the tip radius for this experiment is estimated to be 4.9 nm. Even though this is a very sharp radius, it only represents the average curvature at the end of the tip; for atomic lateral resolution in an AFM image, the atomic scale roughness at the end of the tip can often provide an individual atom protruding far enough to achieve atomic resolution.

8.3.2   Atomic resolution of short-range forces

Figures 8.9(d) and 8.9(e) show examples of the atomic resolution that can be achieved using dynamic atomic force microscopy with frequency modulation (FM-AFM) (Sugimoto et al. 2007). Figure 8.9 also illustrates that this type of AFM imaging method can distinguish between chemically different atoms on the surface. These chemical differences manifest themselves by changing the shape of the force-vs-distance curve,

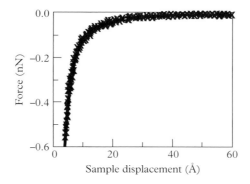

Figure 8.8 *F vs. D_{sample} curve measured under ultra-high vacuum conditions between a silicon AFM tip and a silicon (111) surface with a (7 × 7) reconstruction. The force is determined by measuring the shift in cantilever resonance frequency as the tip approaches the sample surface. For the F vs. D_{sample} curve shown, the tip was positioned over the corner atom of the unit cell of the (7 × 7) reconstructed surface, where the attractive force is expected to arise predominantly from the van der Waals force. A bias voltage of +1.16 V was applied to the sample to cancel the electrostatic attraction (discussed in Section 7.5.1). The solid line is the fit to eq. (8.4) to obtain $A_{Si-v-Si}R = 9.1 \times 10^{-28}$ J m. Reprinted with permission from Lantz et al. (2001). Copyright 2001 AAAS.*

which is dominated at small separations by the short-range forces between a single atom on the tip apex and the individual atoms on the sample surface. In order to accurately measure these short-range forces, these types of measurements have to done under ultra-high vacuum conditions, both to remove the influence of other molecules and to make contributions from long range forces (such as electrostatic) easier to separate out.

The short-range forces between two atoms come from the combination of van der Waals, chemical bonding, and Born repulsion. Figure 8.9(c) shows the calculated values for the van der Waals force, the short range chemical force, and the total force for the atomic geometry illustrated in Figure 8.9(b). As this plot shows, these short-range forces are only significant at separations less than 1 nm.

The topographic images in Figures 8.9(d) and 8.9(e) are of the samples consisting of a single atomic layer of either Sn or Pb grown on silicon substrates. Within this monolayer there are Si atomic defects where a Si atom has substituted for a Sn or Pb atom (characterized by a smaller topographical height in the FM-AFM image, that is, the darker atoms in the images). By measuring the force-vs-distance curves with the tip carefully positioned over individual Si, Sn, or Pb atoms, Sugimoto et al. were able to determine that the reason why the Si atoms appear to have smaller heights in the FM-AFM topography image is that the attractive component of the Sn's and Pb's short-range chemical force extends ~0.5 Å further away from the sample than that of the Si defect atoms.

8.3.3   Capillary condensation of contaminants and water vapor

Few tribological contacts occur between clean surfaces under vacuum conditions. More typically the surfaces are either immersed in a liquid (such as a lubricant or an aqueous

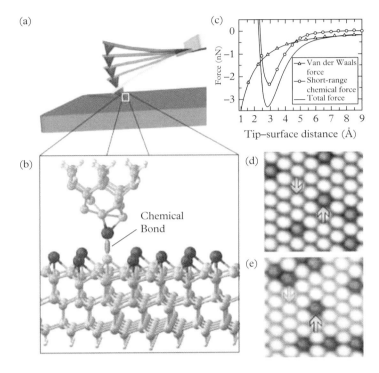

Figure 8.9 *Atomic resolution of short-range forces using dynamic atomic force microscopy. (a) Schematic showing the oscillation of the AFM cantilever (magnitude greatly exaggerated) at its mechanical resonance frequency. In dynamic atomic force microscopy, small changes in this resonance frequency are measured, which are proportional to the derivative of the force acting of the tip. (b) Enlarged area of tip apex interacting with the sample illustrating the short range interaction between the atom at the tip apex and an atomic defect on the sample. The sample is a Si(111) substrate covered with a monolayer of either Sn or Pb atoms (shown in black). (c) Calculated values for the van der Waals force, the short range chemical force, and the total force for the geometry in (b). (d, e) Topographic images of a single atomic layer of Sn (d) or Pb (e) on the silicon substrate obtained in ultra-high vacuum. The darker atoms correspond to silicon atoms defects within the Sn or Pb monolayer. Reprinted from Sugimoto et al. (2007) with permission from Springer Nature, copyright 2007.*

environment) or surrounded by a gaseous environment. One might think that surface forces in a gaseous environment should be fairly similar to those in a vacuum, as the gas itself does not have much influence on these forces. However, unless the gases are very pure, the small amounts of contaminant vapors present in most gases can have a significant impact on surface forces either by adsorbing onto the surfaces to form a contamination layer, or by condensing around the contact points.

For tribological contacts exposed to air, water vapor is the most common contaminant, either adsorbing directly onto surfaces or undergoing capillary condensation around the contacts. Consequently, the relative humidity tends to have a major impact on surface forces. As an example of the effect that humidity can have on forces acting on an AFM tip near a sample, Figure 8.10 shows the F vs. D curves as a tip first approaches and then withdraws from a silicon wafer in 0% and 92% relative humidity. At large tip–sample separations, the force is defined to be zero, and it stays near zero as the sample approaches the tip (the van der Waals force is too small to be measure). At point A, a sudden jump in the attractive force is observed due to the formation of a meniscus, as illustrated in Figure 8.11. This meniscus forms even at zero humidity, when no water is available to condense around the tip, indicating that instead a trace amount of some other mobile contaminant is still able to condense around an AFM tip at this small separation distance. As the humidity is increased from 0% to 92%, the separation distance A where the meniscus forms moves 30 Å further out as water vapor condenses more easily around the tip and swells the meniscus.

As the sample is withdrawn, the meniscus eventually breaks at point C in Figure 8.10, a process that is illustrated in Figure 8.11(c). Intuitively, the meniscus should break when the separation distance is somewhat less than twice the meniscus radius r_1 (Mate and Novotny 1991). A detailed analysis of the shape of the meniscus and the point of breakage can be found in the paper by Bowles et al. (2006).

Figure 8.11(a) illustrates the situation when the vapor that undergoes capillary condensation has a liquid surface tension γ_L that is higher the solid surface energies of the contacting surfaces, resulting in a finite contact angle θ. As discussed in Section 6.3.1, the meniscus force for this situation is given by eq. (6.52); when the contact angles equal

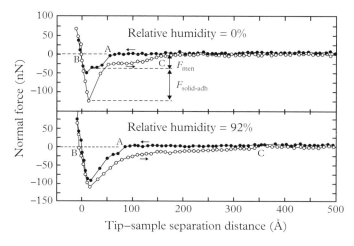

Figure 8.10 *Normal F vs. D curves for a AFM tip contacting a clean silicon wafer in a nitrogen atmosphere at low and high humidities, with a tip radius ~100 nm and a cantilever spring constant $K_{spring} = 50$ N/m. The zero force is defined to occur at large separations, and zero separation is defined to occur at point B when the repulsive force on the inward approach goes through zero. A indicates the separation where a meniscus forms around the tip during approach due to condensation of water vapor (high humidity case) or contaminants (low humidity case). C indicates the point where the meniscus breaks upon separation. Reprinted with permission from Binggeli and Mate (1995). Copyright 1995, American Institute of Physics.*

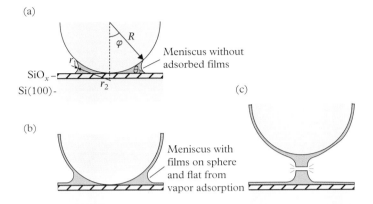

Figure 8.11 *(a) The meniscus that forms around an AFM tip–silicon wafer contact from capillary condensation, but with no adsorbed films of water or other contaminant on the tip and sample surfaces. (b) The meniscus that forms from capillary condensation when the vapor also condenses as a thin layer on both the tip and sample surfaces. (c) The breakage of the meniscus with the tip and sample are sufficiently separated.*

zero ($\theta = 0$), the meniscus force is $L_{men} = 4\pi R\gamma_L$. In contrast, Figure 8.11(b) illustrates the situation where the solid surface energies are higher than the liquid surface tension; this results in the vapor molecules adsorbing to form thin layers on the solid surfaces, with the thickness of the adsorbed layer determined by the adsorption isotherm for this vapor–sample combination. The presence of these films with finite thickness, however, results in larger values of the meniscus force than would be expected from $L_{men} = 4\pi R\gamma_L$. The reason for this is that, as the film thickness increases, the area subtended by the meniscus ($\pi R^2 \sin^2 \varphi$) over which the Laplace pressure acts becomes larger. Since the meniscus force = (meniscus area) × (Laplace pressure), the meniscus force is enhanced over the $4\pi R\gamma_L$ value that occurs in the absence of thick adsorbed layers.

Figure 8.12 shows an example of this effect. The solid lines in the plot show the values of the meniscus force calculated by Hsiao et al. (2010) for alcohol vapors condensing around an AFM tip on SiO_x surfaces, where the effect of the adsorbed layers' thickness is taken into account. The symbols in Figure 8.12 are the experimental values, which agree reasonably well with the calculated values. As shown in Figure 8.12, an adhesive force several times $4\pi R\gamma_L$ occurs at low relative vapor pressures, as the adsorbed layers initially cause the meniscus size to expand much more rapidly than would happen in their absence. At higher humidities, this effect diminishes as the meniscus becomes very large, and the adhesive force returns closer to the value of $4\pi R\gamma_L$ without the adsorbed layers.

8.3.4 Bonded and unbonded perfluoropolyether polymer films

Figure 8.13 shows examples of the normal force measured in air between a tungsten AFM tip and silicon wafers with either an unbonded, liquid polymer film or a bonded

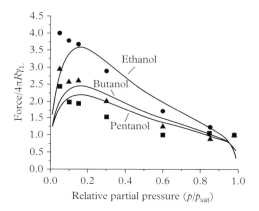

Figure 8.12 *The solid lines show the theoretical meniscus force for the situation illustrated in Figure 8.11(b) where the partial pressure p/p_{sat} results in capillary condensation of alcohol vapors around a sphere-on-flat geometry and in adsorbed films on the sphere and flat surfaces. The symbols show the experimental results for ethanol (circles), butanol (triangles), and pentanol (squares) condensing around a silicon AFM tip and a silicon oxide surface. Reprinted from Hsiao et al. (2010), copyright 2010, with permission from Elsevier.*

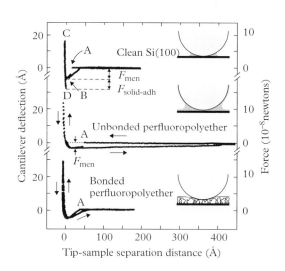

Figure 8.13 *Normal F vs. D curves comparing the behavior of a clean Si(100) wafer, a 40 Å thick unbonded liquid perfluoropolyether polymer film, and an 15 Å thick bonded perfluoropolyether polymer film deposited onto a silicon wafer. The insets show the type of meniscus that forms around the AFM tip and the sample, and, for the bonded lubricant, the penetration of the tip into the bonded polymer film. Adapted with permission from Blackman et al. (1990). Copyright 1990, American Physical Society.*

soft-solid polymer film. For comparison, the F vs. D curve is also shown for a nominally clean silicon wafer (upper plot). As before, the sudden increase in attractive force at point A during the approach to the clean silicon wafer is due to water and other vapors in the air condensing around the tip to form a meniscus.

When the sample is covered with a thin film of liquid polymer (the middle F vs. D curve), a sudden attractive force is observed at point A as the tip approaches the sample due to a liquid meniscus forming around the tip. In this case, the meniscus forms from liquid in the polymer film flowing across the sample surface to form the meniscus rather than from vapor condensation.

The lower F vs. D curve in Figure 8.13 is for a tip contacting a bonded polymer film on a surface. This sample was originally covered with the same type of perfluoropolyether polymer film as for the middle curve, but the polymer was then bonded to the silicon wafer by heating to chemically react the end groups of the polymer to the silicon oxide surface, followed by rinsing with a solvent to remove the unbonded portion. This process converts the liquid film into a soft, but robust solid film on the surface, the structure of which is sketched schematically in the inset in the lower right-hand side of Figure 8.13. This structure resembles spaghetti lying on a plate, with each end of the spaghetti strands attached to the plate surface. Since the bonded perfluoropolyether film has a low surface energy (17.9 mN/m), it also suppresses the condensation of contamination and water vapor. When the tip first contacts this bonded polymer film (point A), the attractive force does not increase sharply, indicating the absence of a liquid meniscus; instead, the attractive forces increases gradually as the tip penetrates into the film.

As the tip–sample separation distance nears zero, eventually the force turns repulsive when the molecules in the contact zone are squeezed out or compressed between the tip and sample surface. In this repulsive area, the slope of the force vs. separation curve gives the stiffness of the contact zone. For the clean silicon wafer, the stiffness is 100 N/m at point B, but rises quickly to 350 N/m when in hard-wall contact at point C. The stiffness S is related to the elastic properties and the size of the contact zone by

$$S = kaE_c, \tag{8.6}$$

where k is a geometric factor between 1.9 and 2.4, a is the radius of contact, and E_c is the composite elastic modulus for the tip and sample (Pethica and Oliver 1987). If we estimate the composite elastic modulus $E_c = 2 \times 10^{11}$ N/m^2 and if $S = 110$ N/m, then eq. (8.6) show that the contact radius is estimated at $a = 3$ Å, indicating that a contact zone a few atoms across forms when the tip first makes solid–solid contact.

To separate the tip and sample surfaces from contact, enough force needs to be applied to overcome the adhesive force. For the clean Si(100) surface (upper curve), the maximum solid–solid attractive force during retraction occurs at point D where tip and sample suddenly jump apart. For this tip–silicon wafer contact, the total adhesive force at pull-off is $F_{\text{tot-adh}} = -5 \times 10^{-8}$ N, of which -3×10^{-8} N is from the meniscus force and -2×10^{-8} N is from solid–solid adhesion:

$$F_{\text{adh-tot}} = F_{\text{men}} + F_{\text{solid-adh}}$$
$$-5 \times 10^{-8}\text{N} = \left(-3 \times 10^{-8}\right) + \left(-2 \times 10^{-8}\text{N}\right). \tag{8.7}$$

From the middle curve, we see that the addition of the perfluoropolyether polymer liquid film reduces the pull-off adhesive force to just the capillary force, indicating that, in the presence of this particular liquid, $W_{\text{tip-liquid-sample}} \sim 0$.

As the sample is further withdrawn in Figure 8.13, the attractive forces gradually decrease as the tip tries to break free from the molecules that have gathered in the contact zone. For the liquid perfluoropolyether, the tip has to be retracted 400 Å before it breaks free from the rather large liquid meniscus that forms around the contact zone, while for the other surfaces the break-free distance is 40–50 Å. Breaking free from menisci that form around contact points is a major contributor to the adhesion hysteresis (Section 5.6).

8.3.5   Electrostatic double-layer force

To eliminate the influence of capillary forces on AFM force measurements, the AFM cantilever can be completely immersed in a liquid. This immersion is often used to allow examination of the non-capillary forces that arise from a liquid sandwiched between solid surfaces (Section 7.4): solvation, electrostatic double-layer, hydration repulsion, hydrophobic attraction, etc. In this section, we provide an example of an AFM measurement of the electrostatic double-layer force, a force commonly encountered in aqueous solutions.

As discussed in Section 7.4.2.1, oxides covered with water develop a surface charge that depends on the water's pH. This surface charge is balanced by the counterions dissolved in the water, which segregate towards the surface. Bringing two of these surfaces into close proximity disrupts the counterion densities near the surfaces resulting in a repulsive force called the electrostatic-double layer force, which has a decay length given by the Debye length (eq. (7.45)).

Figure 8.14 plots the force divided by particle radius for a silica particle mounted on an AFM cantilever as it approaches a flat rutile titania crystal in different pH environments. According to the Derjaguin approximation (Section 6.1), normalizing the force on the particle by dividing it by the particle's radius makes the ordinate axis in Figure 8.14 equivalent to $2\pi W_{\text{silica-water-titania}}(D)$. Using a particle attached to the cantilever rather than a conventional sharp AFM tip makes this normalization easier as the particle's larger radius (2.5 µm) can be measured with a much smaller percent error than a sharp AFM tip (radius $\sim$ 10 nm).

The density of charge on both surfaces is mainly determined by pH. At high pH, both materials are negatively charged resulting in electrostatic repulsion. In Figure 8.14, this repulsion is readily observable for pH = 8.8 out to separations of 40 nm. As the pH is decreased below the isoelectric point of titania at pH 5.6, the titania surface becomes positively charged, and the electrostatic repulsion turns into an electrostatic attraction. The silica surface remains negatively charged until the pH is lowered below its isoelectric point at pH 3. The F vs. D curve in Figure 8.14 for pH 3 is clearly attractive for all separation distances prior to solid–solid contact near zero separation. These repulsive and attractive forces at finite separation can be adequately described with the DLVO

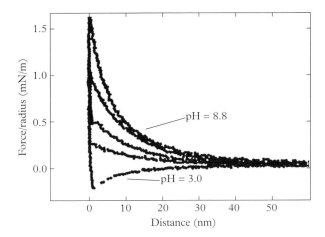

Figure 8.14 *F vs. D curve between a 5 μm diameter SiO₂ spherical particle and TiO₂ crystal surface. The different curves are, from top to bottom, at pH = 8.8, 7.2, 6.3, 5.3, and 3.0 with 1 mM KNO₃. Reprinted with permission from Larson et al. (1995). Copyright 1995, American Chemical Society.*

theory that accounts for both the electrostatic double-layer force and the van der Waals force (Larson et al. 1995), which was discussed in Chapter 7.

8.3.6   Friction force mapping

In Chapter 11, we will discuss several scientific research results concerning atomic-scale friction, many of which take advantage of the experimental capabilities of the AFM described in this chapter. To illustrate the instrumental capabilities of lateral force measurements with AFM, here we discuss an example of lateral force mapping that illustrates how variations in friction can be measured with nanometer-scale lateral resolution.

Figure 8.15 shows a friction force image of a polydiacetylene monolayer (Carpick et al. 1999). The monolayer is comprised of nearly vertical alkyl chains that are connected near their midway point by a diacetylene polymer backbone chain, as illustrated in Figure 8.15(a), which forces the molecules within a particular domain to all tilt a particular angle with respect to the surface normal. The image reveals multiple domains where friction is relatively uniform within each domain, but varies significantly from one domain to the next. The domains are all indistinguishable in height, but exhibit different friction forces against the tip, with the friction being lowest when sliding parallel to the backbones, and nearly three times larger when sliding perpendicular to the backbones. Carpick et al. (1999) hypothesize that the film is more compliant in shear if the shearing direction is perpendicular to the backbones compared to direction parallel to the backbones. In other words, the backbone-linked polymer chains are like a series of guitar strings: they can be deformed more readily by strumming your fingers across the strings, but little deformation occurs if you slide your fingers along the direction of the strings. This

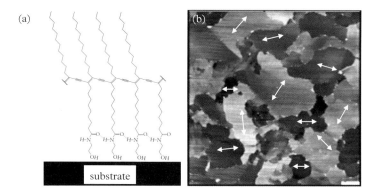

Figure 8.15 *(a) Side view schematic of polydiacetylene monolayer structure showing how the ethanol-amine head groups bond the molecules to the substrate and how the cross-linking between the polymer carbon backbones cause these molecules to all tilt in one direction within a structural domain. (b) Friction force microscopy image revealing the structural domains of a polymer monolayer revealed by friction force microscopy imaging. The $50 \times 50~\mu m^2$ friction image of a polydiacetylene monolayer film reveals different domains; the white arrows indicate the orientation of the tilt of the polymer backbones within each domain, as determined from higher resolution images (not shown). Adapted from Carpick et al. (1999) with permission of Springer Nature, copyright 1999.*

means that more elastic energy can be stored, and then released and dissipated, when sliding across the backbones, just as your fingers will encounter more resistance, and then produce more sound, when strumming across the guitar strings.

....

8.4 PROBLEMS

1. A system is considered to be in stable equilibrium if it is at a point of minimum potential energy. This can be used to discuss conditions for stability of an AFM probe experiencing an attractive force as it approaches a surface.

    Consider this quantitatively using the Lennard-Jones (LJ) force between materials (Figure 8.2). The potential energy in the LJ model is discussed briefly in Section 7.2; it is composed of an attractive van der Waals term and an empirically-based repulsive term. For a sphere above a plane, the LJ force $F_{LJ}(D)$ takes the form:

$$F_{LJ}(D) = \frac{2\pi WR}{3}\left[\left(\frac{\sigma}{D}\right)^8 - 4\left(\frac{\sigma}{D}\right)^2\right] \tag{8.8}$$

where D is the sphere-plane separation distance, R is the sphere radius, W is the work of adhesion, and σ represents a characteristic interaction distance. A negative force represents attraction. We shall assume this model is appropriate for a silicon AFM tip (the sphere) approaching a diamond surface (the plane), where the silicon tip has a radius of 10 nm. We shall use $\sigma = 0.30$ nm and $W = 0.060$ J/m^2.

(a) Find the value of the equilibrium separation distance D_0 where $F_{LJ}(D) = 0$.

Now assume that the AFM cantilever has a spring constant of 0.05 N/m and that the tip is 2.0 μm tall with a tapered conical shank that is 1.0 μm in radius at the "top" (i.e., where it is connected to the cantilever) and narrows to 10 nm radius at the bottom. Also, assume the tip and its shank are rigid.

(b) Find the tip–sample separation D and the cantilever–sample separation at which the tip will snap into contact with the surface.
(c) Find how stiff the cantilever must be to avoid snap-in altogether.

2. Consider the AFM measurements shown in Figure 8.10. In a separate measurement, the contact angle between water and the clean silicon wafer was measured to be 60°. By estimating appropriate values from the plot and using others given in the figure caption, estimate the solid–vapor surface energy of the clean silicon wafer. For simplicity, assume that the tip has the same surface energy as the sample.

. . . .

8.5 REFERENCES

Albrecht, T. R., S. Akamine, T. E. Carver and C. F. Quate (1990). "Microfabrication of cantilever styli for the atomic force microscope." *Journal of Vacuum Science & Technology A: Vacuum, Surfaces, and Films* **8**: 3386–96.

Bhushan, B. and O. Marti (2011). "Scanning probe microscopy – principle of operation, instrumentation, and probes." In: *Nanotribology and nanomechanics I: measurement techniques and nanomechanics*. B. Bhushan, Ed. Berlin: Springer, pp.37–110.

Binggeli, M. and C. M. Mate (1995). "Influence of water-vapor on nanotribology studied by friction force microscopy." *Journal of Vacuum Science & Technology B* **13**(3): 1312–5.

Binnig, G., C. F. Quate and C. Gerber (1986). "Atomic force microscope." *Physical Review Letters* **56**(9): 930–3.

Blackman, G. S., C. M. Mate and M. R. Philpott (1990). "Interaction forces of a sharp tungsten tip with molecular films on silicon surfaces." *Physical Review Letters* **65**(18): 2270–3.

Bowles, A. P., Y. T. Hsia, P. M. Jones, J. W. Schneider and L. R. White (2006). "Quasi-equilibrium AFM measurement of disjoining pressure in lubricant nano-films. I. Fomblin Z03 on silica." *Langmuir* **22**(26): 11436–46.

Butt, H. J., B. Cappella and M. Kappl (2005). "Force measurements with the atomic force microscope: technique, interpretation and applications." *Surface Science Reports* **59**(1–6): 1–152.

Cappella, B. (2016). "Physical principles of force–distance curves by atomic force microscopy." In: *Mechanical properties of polymers measured through AFM force-distance curves*. Cham, Switzerland: Springer International Publishing, pp.3–66.

Cappella, B. and G. Dietler (1999). "Force–distance curves by atomic force microscopy." *Surface Science Reports* **34**(1–3): 1–104.

Carpick, R. W. and M. Salmeron (1997). "Scratching the surface: fundamental investigations of tribology with atomic force microscopy." *Chemical Reviews* **97**(4): 1163–94.

Carpick, R. W., D. Y. Sasaki and A. R. Burns (1999). "Large friction anisotropy of a polydiacetylene monolayer." *Tribology Letters* **7**(2–3): 79–85.

Dai, H. J., J. H. Hafner, A. G. Rinzler, D. T. Colbert and R. E. Smalley (1996). "Nanotubes as nanoprobes in scanning probe microscopy." *Nature* **384**(6605): 147–50.

Derjaguin, B. V., V. M. Muller and Y. P. Toporov (1975). "Effect of contact deformations on adhesion of particles." *Journal of Colloid and Interface Science* **53**(2): 314–26.

Ducker, W. A., Z. Xu, D. R. Clarke and J. N. Israelachvili (1994). "Forces between alumina surfaces in salt-solutions: non-DLVO forces and the implications for colloidal processing." *Journal of the American Ceramic Society* **77**(2): 437–43.

Dürig, U. (2000). "Extracting interaction forces and complementary observables in dynamic probe microscopy." *Applied Physics Letters* **76**(9): 1203–5.

Frantz, P., N. Agrait and M. Salmeron (1996). "Use of capacitance to measure surface forces. 1. Measuring distance of separation with enhanced spatial and time resolution." *Langmuir* **12**(13): 3289–94.

Frantz, P., F. Wolf, X. D. Xiao, Y. Chen, S. Bosch and M. Salmeron (1997). "Design of surface forces apparatus for tribology studies combined with nonlinear optical spectroscopy." *Review of Scientific Instruments* **68**(6): 2499–504.

Gao, J., W. Luedtke, D. Gourdon, M. Ruths, J. Israelachvili and U. Landman (2004). *Frictional forces and Amontons' law: from the molecular to the macroscopic scale*. ACS Publications.

Giessibl, F. J. (2003). "Advances in atomic force microscopy." *Reviews of Modern Physics* **75**(3): 949–83.

Hirz, S. J., A. M. Homola, G. Hadziioannou and C. W. Frank (1992). "Effect of substrate on shearing properties of ultrathin polymer-films." *Langmuir* **8**(1): 328–33.

Homola, A. M., J. N. Israelachvili, M. L. Gee and P. M. McGuiggan (1989). "Measurement of and relation between the adhesion and friction of two surfaces separated by molecularly thin liquid films." *Journal of Tribology* **111**: 675.

Horn, R. G. and D. T. Smith (1991). "Analytic solution for the 3-layer multiple beam interferometer." *Applied Optics* **30**(1): 59–65.

Horn, R. G., D. T. Smith and W. Haller (1989). "Surface forces and viscosity of water measured between silica sheets." *Chemical Physics Letters* **162**(4–5): 404–8.

Houston, J. E. and T. A. Michalske (1992). "The interfacial-force microscope." *Nature* **356**(6366): 266–7.

Hsiao, E., M. J. Marino and S. H. Kim (2010). "Effects of gas adsorption isotherm and liquid contact angle on capillary force for sphere-on-flat and cone-on-flat geometries." *Journal of Colloid and Interface Science* **352**(2): 549–57.

Israelachvili, J., Y. Min, M. Akbulut, A. Alig, G. Carver, W. Greene, K. Kristiansen, E. Meyer, N. Pesika, K. Rosenberg and H. Zeng (2010). "Recent advances in the surface forces apparatus (SFA) technique." *Reports on Progress in Physics* **73**(3): 036601.

Israelachvili, J. N. (1973). "Thin film studies using multiple-beam interferometry." *Journal of Colloid and Interface Science* **44**: 259–72.

Israelachvili, J. N. (1991). *Intermolecular and surface forces* (2nd ed.). London: Academic Press.

Israelachvili, J. N. and G. E. Adams (1978). "Measurement of forces between two mica surfaces in aqueous electrolyte solutions in the range 0–100 nm." *Journal of the Chemical Society, Faraday Transactions* **174**: 975–1001.

Israelachvili, J. N. and P. M. McGuiggan (1990). "Adhesion and short-range forces between surfaces. Part I. New apparatus for surface force measurements." *Journal of Materials Research* **5**(10): 2223–31.

Jacobs, T. D. B., G. E. Wabiszewski, A. J. Goodman and R. W. Carpick (2016). "Characterizing nanoscale scanning probes using electron microscopy: a novel fixture and a practical guide." *Review of Scientific Instruments* **87**(1): 013703.

Janshoff, A., M. Neitzert, Y. Oberdorfer and H. Fuchs (2000). "Force spectroscopy of molecular systems – single molecule spectroscopy of polymers and biomolecules." *Angewandte Chemie, International Edition* **39**(18): 3212–37.

Johnson, K. L., K. Kendall and A. D. Roberts (1971). "Surface energy and contact of elastic solids." *Proceedings of the Royal Society of London, Series A: Mathematical and Physical Sciences* **324**(1558): 301–13.

Klein, J. and E. Kumacheva (1995). "Confinement-induced phase transitions in simple liquids." *Science* **269**(5225): 816–9.

Lantz, M. A., H. J. Hug, R. Hoffmann, P. J. A. van Schendel, P. Kappenberger, S. Martin, A. Baratoff and H. J. Guntherodt (2001). "Quantitative measurement of short-range chemical bonding forces." *Science* **291**(5513): 2580–3.

Larson, I., C. J. Drummond, D. Y. C. Chan and F. Grieser (1995). "Direct force measurements between dissimilar metal-oxides." *Journal of Physical Chemistry* **99**(7): 2114–18.

Levins, J. M. and T. K. Vanderlick (1995). "Impact of roughness on the deformation and adhesion of a rough metal and smooth mica in contact." *Journal of Physical Chemistry* **99**(14): 5067–76.

Marti, O., J. Colchero and J. Mlynek (1990). "Combined scanning force and friction microscopy of mica." *Nanotechnology* **1**(2): 141–4.

Mate, C. M. and V. J. Novotny (1991). "Molecular-conformation and disjoining pressure of polymeric liquid-films." *Journal of Chemical Physics* **94**(12): 8420–7.

Mate, C. M., G. M. McClelland, R. Erlandsson and S. Chiang (1987). "Atomic-scale friction of a tungsten tip on a graphite surface." *Physical Review Letters* **59**: 1942.

Maugis, D. (1992). "Adhesion of spheres: the JKR–DMT transition using a Dugdale model." *Journal of Colloid and Interface Science* **150**(1): 243–69.

Mukhopadhyay, A., J. Zhao, S. C. Bae and S. Granick (2003). "An integrated platform for surface forces measurements and fluorescence correlation spectroscopy." *Review of Scientific Instruments* **74**(6): 3067–72.

Muller, V. M., V. S. Yushchenko and B. V. Derjaguin (1980). "On the influence of molecular forces on the deformation of an elastic sphere and its sticking to a rigid plane." *Journal of Colloid and Interface Science* **77**(1): 91–101.

Noy, A. (2006). "Chemical force microscopy of chemical and biological interactions." *Surface and Interface Analysis* **38**(11): 1429–41.

Noy, A., D. V. Vezenov and C. M. Lieber (1997). "Chemical force microscopy." *Annual Review of Materials Science* **27**: 381–421.

Parker, J. L. and H. K. Christenson (1988). "Measurements of the forces between a metal-surface and mica across liquids." *Journal of Chemical Physics* **88**(12): 8013–4.

Parker, J. L., H. K. Christenson and B. W. Ninham (1989). "Device for measuring the force and separation between two surfaces down to molecular separations." *Review of Scientific Instruments* **60**(10): 3135–8.

Pethica, J. B. and W. C. Oliver (1987). "Tip surface interactions in STM and AFM." *Physica Scripta* **T19A**: 61–6.

Ruths, M. and J. N. Israelachvili (2011). "Surface forces and nanorheology of molecularly thin films." In: *Nanotribology and nanomechanics II*, B. Bhushan, Ed. Berlin: Springer, pp.107–202.

Ruths, M., N. Alcantar and J. Israelachvili (2003). "Boundary friction of aromatic silane self-assembled monolayers measured with the surface forces apparatus and friction force microscopy." *The Journal of Physical Chemistry B* **107**(40): 11149–57.

Ruths, M., C. Drummond and J. N. Israelachvili (2014). "Characterization of biomolecular interactions with the surface forces apparatus." In: *Handbook of imaging in biological mechanics*, C. P. Neu and G. M. Genin, Eds. Boca Raton, FL: CRC Press, pp.457–80.

Senden, T. J. and C. J. Drummond (1995). "Surface-chemistry and tip sample interactions in atomic-force microscopy." *Colloids and Surfaces A: Physicochemical and Engineering Aspects* **94**(1): 29–51.

Seror, J., L. Zhu, R. Goldberg, A. J. Day and J. Klein (2015). "Supramolecular synergy in the boundary lubrication of synovial joints." *Nature Communications* **6**: 6497.

Sugimoto, Y., P. Pou, M. Abe, P. Jelinek, R. Pérez, S. Morita and O. Custance (2007). "Chemical identification of individual surface atoms by atomic force microscopy." *Nature* **446**(7131): 64–7.

Tonck, A., J. M. Georges and J. L. Loubet (1988). "Measurements of intermolecular forces and the rheology of dodecane between alumina surfaces." *Journal of Colloid and Interface Science* **126**(1): 150–63.

Vigil, G., Z. Xu, S. Steinberg and J. Israelachvili (1994). "Interactions of silica surfaces." *Journal of Colloid and Interface Science* **165**(2): 367–85.

Wilson, N. R. and J. V. Macpherson (2009). "Carbon nanotube tips for atomic force microscopy." *Nature Nanotechnology* **4**(8): 483–91.

Wong, S. S., A. T. Woolley, E. Joselevich, C. L. Cheung and C. M. Lieber (1998). "Covalently-functionalized single-walled carbon nanotube probe tips for chemical force microscopy." *Journal of the American Chemical Society* **120**(33): 8557–8.

Yamada, S. (2010). "Nanotribology of symmetric and asymmetric liquid lubricants." *Symmetry* **2**(1): 320–45.

9

Lubrication

When dry surfaces of metals, ceramics, and most polymers slide against each other, friction coefficients μ are typically 0.5 or higher. In many practical situations, however, such high values of μ can imply intolerably high frictional forces, energy losses, and wear. For those situations where high friction is not explicitly needed (like the brakes on your car), *lubricants* are used to reduce friction and wear to acceptable levels in sliding or rolling element bearings.

The benefits of applying liquids and greases to surfaces to reduce friction and wear have been known since antiquity (Dowson 1978). Traces of animal fats have been found on ancient wheel axles, indicating that a practical knowledge of lubrication goes back as far as the invention of the wheel. Mural paintings in ancient Egyptian tombs illustrate the method of applying oils, greases, and water to wooden boards over which large stones, statues, and obelisks could then be slid with relative ease (pulled by only a few hundred slaves) during the construction of these magnificent monuments.

Only with the work of Tower and Reynolds in the late 1880s did scientists and engineers begin to sort out how lubricants alleviate friction and wear, and only in the past few decades has progress been made in understanding lubrication mechanisms at the molecular level.

9.1 Lubrication regimes

The main way lubricants function is by introducing a layer between sliding surfaces that lowers the shear strength. The different ways that these lubricating layers work can be categorized as follows:

- *Hydrostatic lubrication*—The sliding surfaces are separated by a fluid film pressurized by an external supply to provide a lifting force that separates the surfaces. The thickness of the fluid film is great enough for the asperities on the opposing surfaces not to make contact.

- *Hydrodynamic lubrication*—In this case, the fluid film separating the sliding surfaces is pressurized by viscous dragging of the lubricant into a narrowing gap, which

Tribology on the Small Scale: A Modern Textbook on Friction, Lubrication and Wear. Second edition. C. Mathew Mate and Robert W. Carpick. © Oxford University Press 2019. Published in 2019 by Oxford University Press.
DOI: 10.1093/oso/ 9780199609802.001.0001

provides a lifting force great enough so that the opposing surfaces do not make contact, particularly at the trailing edge of the wedge.

- *Elastohydrodynamic lubrication (EHL)*—Highly localized pressures within the squeezed lubricant film result in elastic distortion of the surfaces. If the materials are sufficiently stiff, the pressures can be high enough to increase the lubricant's viscosity. Both effects help prevent asperity–asperity contact from occurring between the sliding surfaces by interposing a thin lubricant film.

- *Mixed lubrication*—Within a hydrodynamic or elastohydrodynamic (EHD) lubricated contact, if the film thickness becomes comparable to the surface roughness, some amount of solid–solid contact occurs leading to a mixture of hydrodynamic lubrication and solid–solid friction.

- *Boundary lubrication*—Asperity–asperity contact occurs between the sliding surfaces, but the asperity surfaces are covered with a strongly adhered, lubricant film that lowers the shear strength.

- *Solid lubrication*—A solid film with low shear strength separates the sliding surfaces.

9.1.1   Stribeck curve

In the most common lubrication system, a liquid lubricant is used to provide hydrodynamic or EHD lubrication without externally supplied pressure. This type of lubrication is found in most bearings and gears throughout all types of machinery.

As illustrated in Figure 9.1, the mechanism of lubrication within such a system can change as sliding or rolling conditions change. The solid line is known as the *Stribeck curve* (Stribeck 1902), and illustrates how the friction coefficient of a lubricated bearing changes with the sliding parameters. The Stribeck curve is generally representative of the

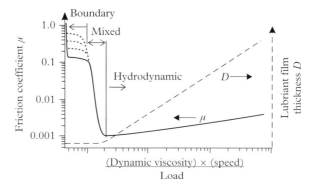

Figure 9.1 *The solid line shows the Stribeck curve, which represents the typical behavior of the friction coefficient μ for an oil lubricated journal bearing. The friction coefficient at the far left of the curve (low speeds and/or high loads) depends on the condition of the surface and the type of boundary lubricant present (dotted lines). The dashed line shows how the film thickness D changes for different parts of the Stribeck curve.*

(a)　　　　　　　　　　　(b)

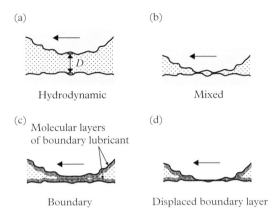

Hydrodynamic　　　　　　Mixed

(c) Molecular layers　　　　(d)
of boundary lubricant

Boundary　　　　　　Displaced boundary layer

Figure 9.2 *Different lubrication regimes between two sliding or rolling surfaces separated by a liquid lubricant.*

behavior when the contact is *conformal*, such as a journal inside a sleeve, or other convex part nested inside a concave part.

Figure 9.2 illustrates the section of the bearing closest to contact for the different lubrication regimes of the Stribeck curve. Moving from right to left along the Stribeck curve:

- *Hydrodynamic lubrication* (Figure 9.2(a))—Under conditions of light loads and/or high sliding speeds, shearing of the liquid lubricant film provides enough hydro-dynamic lift for the surfaces to sufficiently separate so that the asperities on the opposing surfaces do not make contact. This is the *hydrodynamic* lubrication regime. As one moves right to left along the Stribeck curve with increasing load or decreasing sliding speed, the separation distance D decreases, until a point is reached where the friction coefficient is a minimum. Moving further left along the curve transitions the bearing into the *mixed lubrication* regime.
- *Mixed lubrication* (Figure 9.2(b))—To the left of the point of minimum friction, the separation distance is small enough for solid–solid contact to start to occur over a fraction of the lubricated zone. These solid-solid contact regions have higher friction than the regions still separated by the hydrodynamic film. As the separation decreases, the fraction of the interface involving solid-solid contact increases, and so the friction coefficient increases as well. This is termed the *mixed* lubrication regime since a mixture exists of both liquid-lubricated and solid-solid contacts.
- *Boundary lubrication* (Figure 9.2(c))—For even higher loads or lower speeds, high pressures squeeze out the remaining liquid lubricant from the contacting asperities, resulting in a plateauing of the friction coefficient at a somewhat high value. Within a hydrodynamic bearing, this is sometimes referred to as *inadequate lubrication*. If a layer of molecules from the lubricant bonds to the solid surfaces, the friction remains lower than for sliding between bare surfaces. This *boundary* lubrication

layer reduces the shear strength for these solid–solid contacts. Typically, commercial lubricants are blended with a small percentage of additives generally known as *friction modifiers* that serve as good boundary lubricants for the rubbing surfaces.

- *Displaced boundary layer* (Figure 9.2(d))—At extremely high loads and slow siding speeds, the resulting high contact pressures displace the boundary lubricant layer, and friction becomes comparable to the unlubricated situation.

As mentioned above, the Stribeck curve shown is typical for conformal contacts. Many contacts are *nonconformal* (or sometimes called *counterformal*). Examples include gear teeth and the cam-follower contact in an engine. In these cases, the contact pressures are high enough that they will deform the materials in contact, even if made of high elastic modulus materials such as steel, putting the contact in the EHD regime.

In this chapter, we introduce the physics for hydrodynamic and EHD lubrication, while the next chapter covers boundary lubrication and other topics associated with lubrication in confined spaces.

9.2 Viscosity

For those lubrication regimes—hydrostatic, hydrodynamic, and EHD lubrication—where no solid–solid contact occurs and only a fluid film is present, *viscosity* is the principal fluid characteristic that determines how well it functions as a lubricant.

9.2.1 Definition and units of viscosity

As discussed in Chapter 3, when a shearing force is applied to a linear elastic solid, as illustrated in Figure 9.3(a), it deforms elastically so that the top surface moves sideways a distance x, which is proportional to the shear stress τ_s, then stops:

$$\gamma = \frac{x}{d} = \frac{\tau_s}{G}$$
$$x = \frac{d}{G}\tau_s \tag{9.1}$$

where γ is the shear strain and G is the shear modulus. In other words, the solid materials display the property of *rigidity*: They can withstand a moderate shear stress for indefinite period of time. If the shear stress exceeds the threshold for *plasticity* or *fracture*, the solid flows or breaks.

A fluid behaves completely differently in that it has no rigidity at all: if you apply a shearing force, a fluid gives way because it flows, and it continues to flow as long you continue to push on it. Viscosity describes the shear forces that exist in the fluid and, consequently, quantifies the fluid's resistance to shear.

Suppose we have the situation illustrated in Figure 9.3(b) where a fluid is being sheared between two parallel plates. The bottom plate is held stationary, and a force F_f is

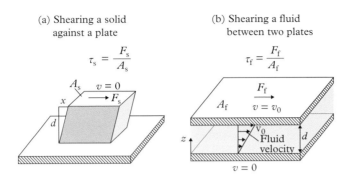

Figure 9.3 *(a) Shearing a solid against a plate. (b) Shearing a fluid between two plates.*

applied to the top plate so that it moves at a velocity v_0. For most experimental situations where v_0 is not too great, F_f is found to be proportional to the area of the plate A_f and to the shear rate or strain rate $d\gamma/dt = v_0/d$, where d is the distance between the plates:

$$\frac{F_f}{A_f} = \tau_f = \eta \frac{v_0}{d}. \tag{9.2}$$

The *dynamic* (or *absolute*) *viscosity* η is the proportionality constant between the shear stress τ_f and the shear rate $d\gamma/dt$. We can rewrite eq. (9.2) to obtain a definition of the dynamic viscosity:

$$\eta = \frac{\tau_f}{v_0/d}$$

$$\text{dynamic viscosity} = \frac{\text{shear stress}}{\text{shear rate}} \tag{9.3}$$

Within Systeme Internationale (SI) units, the shear stress has units N/m^2 (or *pascal*, Pa) and the shear rate has units $(m/s)/m$ or s^{-1}, so the dynamic viscosity η has units of $N\,s/m^2$ or Pa s. In the c.g.s. unit system, the units of dynamic viscosity are dyne s/cm^2; this c.g.s. unit of viscosity has been given the name *poise* (P) after the French doctor Jean Poiseuille (1797–1869) who studied the movement of water through glass tubing of very-small-diameter in order to understand how blood flows through capillaries in the human body. Since water has a viscosity close to 0.01 P, the centipoise (cP) is the unit most commonly used for dynamic viscosity. These units for dynamic viscosity are related to each other by

$$1\,\text{cP} = 10^{-2}\,\text{P} \equiv 10^{-3}\,\text{Pa s} = 1\,\text{mPa s}. \tag{9.4}$$

Typical viscosities of lubricating oils at room temperature range from 2 to 1000 cP, while air (the most common gaseous lubricant) has a viscosity of 0.019 cP or 53 times less than water. Table 9.1 lists the dynamic viscosities for some common fluids at

Table 9.1 *Typical viscosities of some common fluids at room temperature.*

Fluid	Viscosity (mPa s)
Air	0.019
Ether	0.24
Benzene	0.64
Water	1.0
Mercury	1.5
Mineral oils	2 to >1500
Castor oil	650
Glycerin	1410

room temperature; note that these viscosities typically rapidly decrease with increasing temperature as discussed in Section 9.2.4.

For eq. (9.2) to be valid, the "no-slip" boundary condition needs to hold: When fluid molecules come into contact with one of the plate surfaces in Figure 9.3(b), they stick long enough for the average velocity of the molecules next to the plate to be treated as being equal to the plate velocity. To some extent, molecules always slip a small amount while in contact, so the "no-slip" condition is not strictly true, but for most situations the slip is negligible. (A later section covers slip in more detail.) With the no-slip boundary condition, the fluid velocity $v(y)$ between the plates in Figure 9.3(b) increases linearly from the bottom plate ($v(0) = 0$) to the top plate, ($v(d) = v_0$) with the illustrated velocity profile.

While the dynamic viscosity is the most relevant parameter for lubrication problems, *kinematic viscosity* is often the one tabulated, and is defined by the relation

$$\text{kinematic viscosity} = \frac{\text{dynamic viscosity}}{\text{density}}$$

$$v = \frac{\eta}{\rho} \tag{9.5}$$

In situations where flow is driven by gravity, the kinematic viscosity turns out to be the more relevant expression for the viscous behavior of the fluid. Since viscosity is usually measured using gravity-driven flow, the kinematic rather than the dynamic viscosity is measured more directly and ends up being the viscosity property most often reported.

In SI units, kinematic viscosity has units of m^2/s. The c.g.s. unit is cm^2/s, which is called a *stoke* (St), after George Gabriel Stokes (1819–1903), a professor at Cambridge University, who developed much of the mathematical analysis for fluid flow. The most common unit for kinematic viscosity is centistokes (cSt); for example, water has a

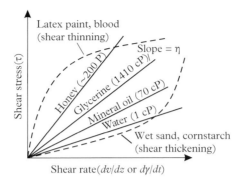

Figure 9.4 *Shear stress is linear with shear rate for Newtonian fluids (solid lines) and nonlinear for non-Newtonian fluids (dashed lines).*

kinematic viscosity of 1.00 cSt at room temperature. The units for kinematic viscosity are related to each other by

$$1 \text{ cSt} = 10^{-2} \text{ St} = 1 \text{ mm}^2/\text{s}. \tag{9.6}$$

9.2.2 Non-Newtonian behavior and shear degradation

In general, the relationship in eq. (9.3) between the shear stress τ and shear rate $d\gamma/dt$ can be written as

$$\tau = \frac{d\gamma}{dt} \eta_{\text{eff}} \left(\frac{d\gamma}{dt} \right), \tag{9.7}$$

where the effective viscosity η_{eff} is a function of shear rate $d\gamma/dt$.

For low and moderate shear rates ($<10 \text{ s}^{-1}$), the shear stress is a linear function of shear rate for most fluids; these fluids are classified as *Newtonian fluids* and are characterized by a constant viscosity. The solid lines in Figure 9.4 illustrate this Newtonian behavior for several liquids. When the shear stress is not linear with shear rate, the fluid is called *non-Newtonian*. The dashed lines in Figure 9.4 illustrate two types of non-Newtonian behavior: shear thickening and shear thinning.

For *shear thickening* fluids, it is fairly easily to move them slowly, but it becomes harder at higher shear rates. An everyday example of this is a mixture of cornstarch with a modest amount of water. If you tilt the mixture, it flows with the consistency of whipping cream, but if you stir it with a spoon, the mixture becomes remarkably stiff in the region next to the stirring spoon. What is happening? A cornstarch mixture is a concentrated suspension of spherical particles that repel one another. When left unperturbed, the particles arrange in an ordered structure that keeps them reasonably well separated from each other when sheared at a low shear rate. At higher shear rates, the order becomes disrupted, bringing some of the particles into contact resulting in higher friction forces. The repulsive interaction pushes them back so, on average, the spacing between the particles becomes larger. This thickening of the material is called *dilatancy*. Due to this thickening, the material becomes noticeably drier (as the fraction of water decreases) and stiffer (due to the higher meniscus forces between the particles as well as the contact friction between the particles).

Another example of dilatancy is when you stand on a damp beach, the sand around your feet becomes noticeably drier due to shear thickening.

For *shear thinning* fluids, it becomes easier to shear the fluid at higher shear rates. Shear thinning is common in suspensions, such as blood and latex paint, and in liquids composed of long, randomly oriented polymers. For paint, shear thinning is a desirable feature, as the paint spreads easily when sheared by a brush but resists flowing down the vertical wall being painted when the brush is removed. For blood, the application of shear breaks up the aggregates of blood cells, platelets, etc. suspended in the blood plasma while aligning these constituents so that they flow over one another more easily. With the shear thinning of polymers, the application of shear helps to align the polymer chains, making it easier for them to slide past each other, reducing their viscosity.

Shear thinning is sometimes referred to as *pseudoplasticity*, particularly if it takes a small amount of shear stress to initiate flow. Lubricating greases are an example of this type of pseudoplastic behavior. These greases are used to lubricate mechanisms where a liquid lubricant would flow out, whereas the semisolid character at low shear stresses of the grease enables it to stay in place. Greases typically consist of a mineral or synthetic base lubricant and a soap that acts as a matrix, which provides some rigidity at low shear stresses. When a slight shear stress is applied to the grease, the viscosity drops to that of the base lubricant.

A number of models have been developed for characterizing the shear thinning behavior of lubricants. Figure 9.5 illustrates the relationship between the shear stress and the shear rate for the following three commonly used models:

- *Eyring model*—This model is derived in the next section. The relationship between shear stress and shear rate ($\dot{\gamma} = d\gamma/dt$) is

$$\tau = \tau_{\text{Eyring}} \sinh^{-1}\left(\frac{\dot{\gamma}\,\eta_{\text{Newtonian}}}{\tau_{\text{Eyring}}}\right) \qquad (9.8)$$

  where $\eta_{\text{Newtonian}}$ is the viscosity at low shear rates where the fluid undergoes Newtonian flow, that is, $\tau = \eta_{\text{Newtonian}}\dot{\gamma}$. This model works well for describing lubricant shear thinning behavior in many EHD contacts.

- *Carreau model*—The effective viscosity η_{eff} follows a power law relationship

$$\eta_{\text{eff}} = \frac{\eta_{\text{Newtonian}}}{\left[1 + (\dot{\gamma}/\dot{\gamma}_c)^{n/2}\right]} \qquad (9.9)$$

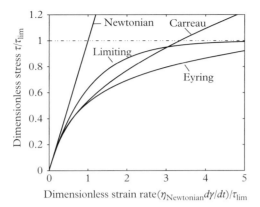

Figure 9.5 *The stress versus shear rate relationships for three different models for shear thinning of lubricants and for Newtonian flow ($\tau = \eta_{Newtonian} d\gamma/dt$). In the Eyring model, the Eyring stress is set to $\tau_{Eyring} = \tau_{lim}/4$; the Eyring stress represents the stress above which the shear behavior deviates from Newtonian flow.*

and the relationship between shear stress and shear rate is

$$\tau = \dot{\gamma}\eta_{\text{eff}}$$
$$= \frac{\dot{\gamma}\eta_{\text{Newtonian}}}{\left[1 + (\dot{\gamma}/\dot{\gamma}_c)^2\right]^{n/2}} \tag{9.10}$$

where $\dot{\gamma}_c$ is the critical shear rate above which shear thinning becomes significant. The exponent n is typically between ½ and 1. This model has been found to work well for fitting shear thinning data from high shear rate viscometry measurements and molecular dynamic simulations (Kioupis and Maginn 2000, Bair 2001, Martini et al. 2006). As can be seen in Figure 9.5, the Carreau and Eyring models predict similar shear stress/shear rate relationships, and the merits of applying either model to lubricated EHD contacts are discussed by Spikes and Jie (2014).

- *Limiting stress model*—At very high shear rates, or at moderate shear rates and high pressures, liquid often transition from liquid-like behavior to solid-like behavior characterized by a limiting shear stress τ_{lim} (Bair and Winer 1979, 1992, Luengo et al. 1996). While various models have been proposed for how materials approach this transition, the simplest rheological model leads to the shear stress approaching its limiting value in an exponential fashion:

$$\frac{\tau}{\tau_{\text{lim}}} = 1 - \exp\left(\frac{\eta_{\text{Newtonian}}\dot{\gamma}}{\tau_{\text{lim}}}\right). \tag{9.11}$$

Shear thinning due to the slippage of lubricant molecules against each other is reversible in that the viscosity returns to its original value when the shear rate is reduced back to the linear region. However, some lubricants undergo irreversible degradation at high shear rates. For these lubricants, the molecules are so entangled with one another that high shear rates cannot disentangle them to the extent needed to relieve the stress through slippage alone. Instead, the stress breaks the bonds within the lubricant molecule; the lubricant becomes permanently degraded, and the viscosity no longer returns to its original value at low shear rates. This shear degradation can be thermally promoted by frictional heating and affected by impurities in the lubricant or on the sliding surfaces.

9.2.3   Eyring model for viscosity and shear thinning

In 1936, Henry Eyring (1901–81) introduced a simple model for viscous flow of liquids based on the transition state theory for chemical reaction rates that he had previously developed (Eyring 1936, Kauzmann and Eyring 1940). While more complex and more accurate theories for viscosity have been developed since then, it is useful to go through Eyring's theory in order to introduce how applying a stress to a material changes the energy activation barriers for the motion and reaction of atoms and molecules in that material. This is an important concept for understanding the molecular processes responsible for the friction and wear (Spikes and Tysoe 2015), as will be discussed in later chapters.

Figure 9.6 illustrates Eyring's model for flow, which considers the probability that an individual liquid molecule moves to a neighboring void or cavity a distance Δx away large enough to accommodate the molecule. This molecule can be thought of as undergoing a transition from one equilibrium state to another over an energy barrier E_a, as illustrated in Figure 9.6(b). When no external stress is applied, the rate k at which the molecule goes over the barrier (moves to the neighboring cavity) is given by

$$k = r_0 e^{-E_a/k_B T} \tag{9.12}$$

where E_a is the activation energy barrier for flow, r_0 the pre-exponential factor, k_B the Boltzmann constant, and T the temperature. The effect of applying a shear stress τ is to lower the activation barrier for flow in the direction of the applied stress and to increase the barrier for flow against the shear stress, as illustrated by the dashed line in Figure 9.6(b). In his model, Eyring assumed that energy barrier E_a was raised or lower by an amount equal to the work done by moving the molecule half way to the neighboring cavity, so as to reach the top of the energy barrier:

$$\Delta E_a = \frac{f \Delta x}{2} = \frac{\tau a_2 a_3 \Delta x}{2}, \tag{9.13}$$

where a_2 and a_3 are the dimensions of the molecule, respectively, in the direction of the applied stress and the transverse direction; $f = \tau a_2 a_3$ then corresponds to the net force acting on an individual molecule in the direction the applied stress.

(a)

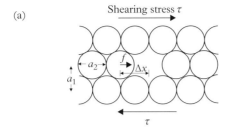

(b)

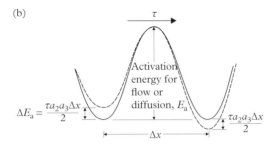

Figure 9.6 *(a) In Eyring's model for viscosity, a shear stress τ is applied to a liquid, exerting a force f on an individual molecule causing it to move a distance Δx to a neighboring hole in the liquid. (b) The solid line shows the energy potential in the direction of motion without the applied shear stress and the dashed line with the shear stress τ. Adapted from Kauzmann and Eyring (1940) and Spikes and Tysoe (2015) with permission of the American Chemical Society. Copyright 1940, American Chemical Society.*

Applying a shear stress results in a flow rate k_f in the forward direction that is higher than the rate k_b in the backward direction. These rates are given by

$$k_f = Be^{-(E_a - \tau a_2 a_3 \Delta x/2)/k_B T} = ke^{\tau a_2 a_3 \Delta x/2k_B T} \tag{9.14}$$

and

$$k_b = Be^{-(E_a + \tau a_2 a_3 \Delta x/2)/k_B T} = ke^{-\tau a_2 a_3 \Delta x/2k_B T}. \tag{9.15}$$

The net speed v at which this molecule moves relative to the layer below it is then given by the net rate constant $(k_f - k_b)$ times the distance moved Δx:

$$\begin{aligned}
v &= (k_f - k_b)\,\Delta x \\
&= k\Delta x \left(e^{\tau a_2 a_3 \Delta x/2k_B T} - e^{-\tau a_2 a_3 \Delta x/2k_B T} \right) \\
&= 2k\Delta x \sinh\left(\frac{\tau a_2 a_3 \Delta x}{2k_B T} \right).
\end{aligned} \tag{9.16}$$

From the definition of dynamic viscosity in eq. (9.3) we obtain

$$\eta\left(\tau, T\right) = \frac{\tau}{v/a_1} = \frac{\tau a_1}{2k \sinh\left(\frac{\tau a_2 a_3 \Delta x}{2k_B T}\right)}. \tag{9.17}$$

At low enough shear stresses, we would expect Newtonian flow with a constant value for the viscosity η; this occurs when $\tau << k_B T/a_2 a_3 \Delta x$ and results in

$$\eta_{\text{Newtonian}} = \frac{k_B T a_1}{k a_2 a_3 \Delta x^2}. \tag{9.18}$$

Eyring's model works reasonably well in predicting the viscosities of molecules that are approximately spherical in shape. However, for polymeric molecules it predicts viscosities that are too low; this led Eyring to speculate that the flow process for these molecules is dominated by individual polymer segments.

Eyring's model also provides a simple molecular model for shear thinning. If we set $\tau_{\text{Eyring}} = 2k_B T/a_2 a_3 \Delta x$ where is τ_{Eyring} is called the "Eyring stress," then the shear rate $d\gamma/dt$ is related to the shear stress τ by

$$d\gamma/dt = \frac{2k\Delta x}{a_1} \sinh\left(\frac{\tau a_2 a_3 \Delta x}{2k_B T}\right)$$
$$= \frac{\tau_{\text{Eyring}}}{\eta_{\text{Newtonian}}} \sinh\left(\frac{\tau}{\tau_{\text{Eyring}}}\right), \tag{9.19}$$

and eq. (9.8) results when inverted to solve for how the shear stress depends on shear rate. At high shear rates ($\dot{\gamma}\eta_N/\tau_e > 2$), eq. (9.8) reduces to

$$\tau \simeq \tau_{\text{Eyring}} \log\left(\frac{2\dot{\gamma}\eta_{\text{Newtonian}}}{\tau_{\text{Eyring}}}\right). \tag{9.20}$$

Eyring applied his theory with some success to explain the shear thinning of polymer melts, solutions, and colloids (Ree et al. 1958). More recent studies have shown that many lubricants in EHD contacts display a shear thinning behavior well described by the sinh relationships in eqs. (9.8) and (9.19) (Tevaarwerk and Johnson 1975, Johnson and Tevaarwerk 1977, Jadhao and Robbins 2017).

9.2.4   Temperature dependence

Liquid viscosities depend strongly on temperature. For example, the viscosity of water near freezing is 1.8 times higher than at 20°C. So, it is not just because it so cold that it is more difficult to swim in freezing water, it indeed takes more physical effort. For mineral oils—a common type of lubricant—the dependence of viscosity on temperature has empirically been found to follow the relationship:

$$\log \log (\eta/\rho + A) = b - c \log T \tag{9.21}$$

where b and c are constants. Equation (9.21) indicates that a lubricant's viscosity drops precipitously with increasing temperature. As the lift generated by a hydrodynamic bearing is proportional to viscosity, it is greatly reduced as the temperature increases. Also, frictional heating within the bearing can increase the bearing temperature and thereby lower the lubricant's viscosity, leading to a bearing that was initially operating in the hydrodynamic regime to move to the left along the Stribeck curve in Figure 9.1, perhaps into the boundary lubrication regime, resulting in higher friction and wear and early failure of the bearing.

9.3   Fluid film flow between parallel surfaces

Let us now look more closely at the fluid confined between two parallel plates, but this time considering the more general case where, in addition to the shear stress applied to the fluid by the plate surface, a pressure gradient exists in the x direction, as illustrated in Figure 9.7. Sketched within the fluid in Figure 9.7(a) is an infinitesimal-size fluid element, and the pressures and stresses acting on this element causing it to flow. For steady-state flow these forces sum up to zero, leading to the relationship

$$\frac{\partial p}{\partial x} = \frac{\partial \tau}{\partial z}. \tag{9.22}$$

Extending eq. (9.2) to this element:

$$\tau = \eta \frac{\partial v_{\mathrm{x}}}{\partial z}. \tag{9.23}$$

This relationship defines the dynamic viscosity at the microscopic level as the ratio of the local shear stress τ to the local shear strain rate $\partial v_{\mathrm{x}}/\partial z$.
   Combining eqs. (9.22) and (9.23), we obtain

$$\frac{\partial p}{\partial x} = \eta \frac{\partial^2 v_{\mathrm{x}}}{\partial z^2}, \tag{9.24}$$

which is the simplified Navier–Stokes equation describing the motion of the fluid element. For the geometries in Figure 9.7, $\partial p/\partial x$ does not depend on z, so integrating eq. (9.24) twice over the z variable obtains the expression for the fluid element's velocity in the x direction as a function of z:

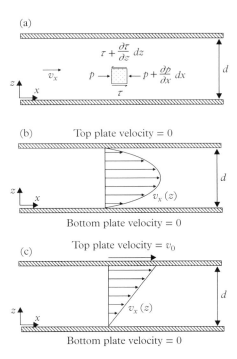

Figure 9.7 *A fluid between two parallel plates that extend much greater distances in the lateral x and y directions than their separation distance d. (a) The pressures and shear forces acting on a fluid element between the plates. (b) Poiseuille flow driven by a pressure gradient dp/dx. (c) Couette flow driven by the shear forces generated by the relative motion of the top and bottom plates.*

$$v_x(z) = \frac{1}{2\eta}\left(-\frac{dp}{dx}\right)z\,(d-z) + v_0\frac{z}{d} \tag{9.25}$$

where the no-slip boundary conditions are used: $v_x = 0$ at $z = 0$, and $v_x = v_0$ at $z = d$.

The first term on the right-hand side of eq. (9.25) represents the flow velocity driven by the pressure gradient dp/dx, with a negative pressure gradient driving the flow to the right. This pressure-induced flow is called *Poiseuille flow*, as this was the type of flow found in Jean Léonard Marie Poiseuille's original studies for water through the glass capillaries. The Poiseuille flow velocity has a parabolic dependence on z, as illustrated in Figure 9.7(b).

The second term in eq. (9.25) represents the flow velocity driven by the shearing motion of the two plates. This shear-induced flow is called *Couette flow* after Maurice Couette who described this type of flow in his 1890 PhD thesis for the first successful viscometer based on shearing a liquid between coaxial cylinders (Couette 1890). The Couette flow depends linearly on z, as illustrated in Figure 9.7(c).

Equation (9.25) can be integrated again over z to obtain the volumetric flow rate per unit width of the plate:

$$q = \int_0^d v_x(z)dz = \frac{d^3}{12\eta}\left(-\frac{dp}{dx}\right) + \frac{v_0 d}{2}. \tag{9.26}$$

Again, the first term on the right is the pressure-driven (Poiseuille) volumetric flow rate, and the second term the shear-driven (Couette) volumetric flow rate.

9.4 Slippage at liquid–solid interfaces

If one keeps increasing the shear rate, say for example by factors of ten, it is unreasonable to expect that the resistance to flow, as expressed by eqs. (9.2) and (9.26), keeps rising forever by factors of ten: eventually something gives, either at the walls or within the fluid. Shear thinning at high shear rates is an example of a breakdown occurring within the liquid; in this case, slippage occurs between the molecules and acts to lower the liquid's viscosity. Another example is slippage at the liquid–solid interface.

The equations in the previous section assumed the *no-slip boundary condition* where the fluid adjacent to a solid surface is assumed to move at the same velocity as the solid. For most ordinary fluids in most ordinary situations, this assumption works quite well. However, if we look closely at how the fluid molecules interact with solid surfaces, we realize that these molecules are going to slip to some degree across the surface. For ordinary liquids flowing in macro-sized channels and gaps, slippage has an imperceptible influence on flow. But, as dimensions become smaller, slip effects become more important.

9.4.1 Definition of slip length

While the no slip boundary condition has been successfully used since scientists began analyzing fluid flows at the beginning of the nineteenth century, doubt has always existed about it and the possibility of slip continuously debated. The standard way of characterizing slip was introduced early on by Navier (1823), who posited that the slip velocity v_s of the fluid contacting the surface is proportional to the shear rate at the surface:

$$v_s = b\frac{dv_x(z)}{dz}\bigg|_{z=0}. \tag{9.27}$$

Since the shear rate dv_x/dz is also proportional to the shear stress (eq. (9.23)), this approach is equivalent to the slip velocity v_s being proportional to the shear stress τ_s at the surface:

$$v_s = \frac{b}{\eta}\tau_s. \tag{9.28}$$

From eqs. (9.27) and (9.28), we see that the parameter b characterizes the slippage of fluid at a surface. b is called the *slip length* as it has the unit of length. Generally, b is

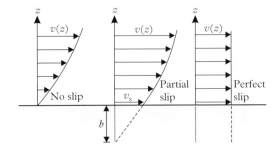

Figure 9.8 *Profiles of the fluid flow velocity next to a surface as a function of the distance z away from the surface, where the extrapolation length b characterizes the amount of slippage. Three types of slip are illustrated: no slip (b = 0), partial slip (0 < b < ∞), and perfect slip (b = ∞).*

assumed to be constant, making this a linear boundary condition. If b is a function of v_s or τ_s, the boundary condition is nonlinear.

Figure 9.8 illustrates the velocity profiles of a fluid flowing next to a surface with various degrees of slip: no slip, partial slip, and perfect slip. From Figure 9.8 we see that the slip length b corresponds to the distance below the surface where flow velocity extrapolates to zero.

9.4.2   Example: shear stress in the presence of slip

For a liquid film between two parallel surfaces with no slippage at the surfaces, eq. (9.2) indicates how much shear stress τ_f needs to be applied to move one surface at a velocity v_0 relative to the other surface. If instead slippage occurs with a slip length b at both plate surfaces, eq. (9.2) becomes

$$\tau_f = \frac{F_f}{A_f} = \eta \frac{v_0}{d + 2b}$$

$$= \eta \frac{v_0}{d} \left(1 + \frac{2b}{d}\right)^{-1} \tag{9.29}$$

So, slippage reduces the shear stress needed to generate a shear rate v_0/d by a factor $(1+2b/d)^{-1}$. Figure 9.9 shows how the shear stress is reduced by various amount of slip for a 100 cP liquid being sheared between two parallel surfaces separated by $d = 1\ \mu$m. Only when b becomes comparable to the separation distance d does the shear stress become significantly less than that for the no-slip case.

9.4.3   Example: measuring slip lengths using viscous force during drainage

When a curved surface is pushed toward a flat or another curved surface while immersed in a liquid, the liquid drains via Poiseuille flow, generating a viscous force that opposes

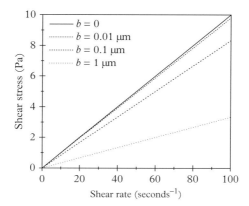

Figure 9.9 *Shear stress needed to shear a 1 μm thick liquid film between two parallel surfaces with a slip length b at both the top and bottom surfaces.*

the change in separation distance D. For a sphere of radius R near a flat surface or for two crossed cylinders with mean radius R, this viscous force F can be expressed as

$$F = -f^* \frac{6\pi \eta R^2 V}{D} \tag{9.30}$$

where $V = dD/dt$ is the velocity of the change in separation and f^* is the slip factor that quantifies how the slippage affects the viscous force: $f^* = 1$ if the no-slip condition occurs, and $f^* < 1$ if slip occurs. For the case where the slip length b is the same on the two opposing surfaces, the slip factor f^* can be expressed (Hocking 1973, Vinogradova 1996) as

$$f^* = \frac{D}{3b} \left[\left(1 + \frac{D}{6b} \right) \ln \left(1 + \frac{6b}{D} \right) - 1 \right]. \tag{9.31}$$

In the limit of small separations where $D \ll b$, then f^* goes to zero as $f^* \simeq D \ln (6b/D) / 3b$. Therefore, as the separation distance goes to zero, the viscous force in eq. (9.30) varies logarithmically with D due to slippage.

Numerous studies have exploited this drainage technique for measuring slip lengths, using either the surface force apparatus (SFA) or atomic force microscopy (AFM); for a review see Neto et al. (2005). The SFA is particularly effective for these measurements, as not only can D be made as small as a single layer of liquid molecules, but also the absolute value of D can be measured with the precision of a few angstroms. With R in an SFA typically being a few centimeters, the viscous force is easily measurable, either as a function of separation velocity V or separation distance D.

Zhu and Granick have conducted the most extensive series of slip length measurements with the SFA technique, and have found that the no-slip condition generally holds for hydrophilic surfaces and for surfaces with any significant amount of roughness, while slip lengths a few tens of nanometers to a few microns occur for molecularly smooth, hydrophobic surfaces (Zhu and Granick 2001, 2002a, 2002b, 2002c, Granick et al. 2003). Zhu and Granick have also observed that a critical shear stress needs to be exceeded for a significant slip length to be observed and the slip length increases with shear rate, indicating that b is not a constant as typically assumed for a linear boundary condition. Channel experiments have also observed slip lengths increasing with shear stress (Choi et al. 2003).

9.4.4 Mechanisms for slip at liquid–solid interfaces

Equations (9.27) and (9.28) characterize the slippage within the continuum description for fluid flow, but do not tell us how the slip occurs or what determines the slip length b. In our discussion, we will divide the possible mechanisms for slippage into two categories:

1. *Molecular slip* where the liquid molecules remain in intimate contact with the solid surface. For molecular slip, values for slip length typically range from 0 to 100 nm (Neto et al. 2005, Lauga et al. 2007). Following Martini et al. (2008), molecular slip can be further subdivided into two subcategories:

   - *Defect slip*—Most liquid molecules obey the zero slip condition but a few individual molecules hop to nearby vacancies preferentially in the direction of the shear stress (and the vacancies or defects migrate in the opposite direction).
   - *Global slip*—While defect slip dominates at lower forcing shear rates, above a critical shear rate (or critical shear stress) the forcing becomes sufficient for all the liquid molecules adjacent to the solid surface to slide together in the direction of applied shear stress.

2. *Apparent slip* where the liquid is not in intimate contact with the solid, but instead a thin medium with much lower viscosity separates the liquid from the solid.

9.4.4.1 Theories for defect slip

Most models for defect slip are conceptually similar to Eyring's model for liquid viscosity. In these models, most molecules remain stuck in their equilibrium sites at the solid-liquid interface, while isolated molecules hop from one equilibrium site to another at a rate described by Arrhenius kinetics.

Tolstoi was one of the first to extend Eyring's theory to explain the mobility of molecules at the liquid–solid interface (Tolstoi 1952, Blake 1990). With his theory, Tolstoi was trying to demonstrate how slip lengths should depend on surface energies. In Tolstoi's analysis, an individual molecule's mobility is determined by the probability of a void being created via thermal excitation large enough for the molecule to diffuse into this void. This probability is related to the energy required to form the void: If a is the characteristic molecular dimension and γ_L the liquid surface tension, then the energy to create a molecular size hole is roughly $\gamma_L a^2$. When a liquid wets the solid, the work of adhesion of the liquid with the solid is greater than the liquid's internal work of cohesion (Section 5.4.1), and it takes more energy to create a hole within the liquid monolayer adjacent to the solid; this suppresses the mobility within the adjacent layer and results in a slip length less than the size of the molecule. When the liquid has a finite contact angle with the solid (partial wetting and non-wetting), the liquid has a lower affinity for the solid than for itself. In this case, more holes are thermally excited in the molecular layer adjacent to a non-wetting surface, resulting in a finite slip length. One shortcoming of the Tolstoi analysis is that it predicts that the slip length increases exponentially with increasing contact angles, with the slip length eventually approaching many millimeters, many orders of magnitude higher than observed in experiments.

A number of research groups have measured slip lengths experimentally on both wetting and non-wetting surfaces, and have systematically found larger slip lengths for the non-wetting surfaces. As pointed out by Lauga et al. (2007), however, when results from different groups are compared with each other, slip lengths correlate poorly with contact angles. The poor correlation between slip and contact angles among research groups (as well as the occasional observation of no-slip on non-wetting surfaces) may result from other parameters that influence slip, such as roughness, that are varying from one experimental setup to the next and masking the underlying relationship between surface energy and slip.

More recently, Lichter et al. (2007) have developed a transition rate theory for defect slip where the activation barrier is assumed to be proportional to solvation pressure rather than surface energy. This theory has been found to be in good agreement with molecular dynamic simulations (Martini et al. 2008).

9.4.4.2 *Global slip of polymer melts*

At sufficiently high shear rates, the shear stress at the solid wall becomes high enough to force all the liquid molecules adjacent to the solid out of their equilibrium sites and to move downstream at the same speed. At this critical shear stress the slippage transitions from defect slip to global slip (Lichter et al. 2004, Martini et al. 2008).

An example of where global slip can occur at moderate to low shear rates is polymer melts where high viscous forces lead to slippage. Since these polymers are entangled, they strongly oppose being sheared as this tends to disentangle them; the entanglement results in viscosities that are enormously enhanced compared with normal liquids and non-entangled polymers. With such high viscosities, it becomes easier to concentrate the shear at the interface with the solid rather than spread the shear rate out over the polymer melt. The basic model for slippage at a polymer–solid interface was proposed by de Gennes (1979) and predicts that the slip length b is related to the bulk viscosity η of the polymer and to the molecular size a by

$$b = a\frac{\eta}{\eta_1}, \tag{9.32}$$

where η_1 is the viscosity of a liquid consisting of monomer units of the polymer (i.e., a liquid with the same interactions, but no entanglements). Since entanglement greatly

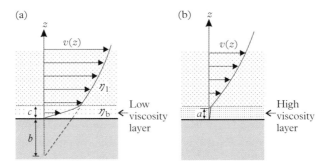

Figure 9.10 *Profiles of the liquid flow velocity next to a solid surface. (a) Apparent slip occurs due to a lower viscosity layer forming next to the surface. (b) A higher viscosity boundary layer forms next to the surface shifting the no-slip boundary condition a distance a away from the surface.*

enhances the viscosity of the polymer melt, η can be many orders of magnitude greater than η_1, resulting in slip lengths of many microns to millimeters.

The slippage of polymer melts is a well-studied field in polymer processing, as slippage often dominates the flow properties of polymers through the pipes used in their processing, and processing additives that promote slip are often added to plastics to facilitate their flow through molds (Achilleos et al. 2002).

9.4.4.3 Apparent slip

When a lower viscosity component segregates out between a liquid and a solid surface, it shears more easily leading to *apparent slip* at the liquid–solid interface. Figure 9.10(a) illustrates the velocity profile for flow in this situation. Notice that the velocity profile extrapolates through the low viscosity boundary film and below the solid surface to give an apparent slip length b, even though a no-slip boundary actually exists where the low viscosity fluid meets the solid surface. Since the shear stress is continuous across the boundary between the high and low viscosity fluids, the velocity gradient in the boundary layer must be higher than the rest of the liquid by the ratio of the two viscosities [eq. (9.23)]. This leads to an apparent slip length

$$b = c\,(\eta_1/\eta_b - 1), \tag{9.33}$$

where c is the boundary layer thickness, η_l is the liquid viscosity, and η_b is the boundary layer viscosity.

Some of the ways apparent slip is generated are as follows:

- *Lotus effect of trapped air around surface roughness*—This mechanism occurs for water flowing over hydrophobic surfaces where the slopes of the surface roughness are very high and the roughness features have very high aspect ratios, as illustrated in Figure 9.11(a). These types of surfaces are referred to as *superhydrophobic*. The high water contact angle on the sides of these surface features prevents the water

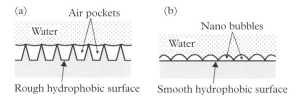

Figure 9.11 *For water-hydrophobic interfaces, a low viscosity gaseous layer causes apparent slip. (a) Pockets of air trapped between the peaks of the surface roughness. (b) Nanobubbles formed by gas dissolved in the water segregating to the surface.*

from penetrating below the roughness summits, so only air or vapor exists around the roughness features. Since the aspect ratio is high, only a small fraction of the wall-water interface consists of a water-solid interface with a zero slip condition, and slippage over the trapped air pockets dominates. Since the ratio of water viscosity to air viscosity equals 53, then eq. (9.33) indicates that the slip length b should be 52 times the thickness of the trapped air layer. This mechanism is often referred to as the *lotus effect*, since it is responsible for water droplets quickly running off lotus leaves, which are covered with tall, thin hairs of a hydrophobic, wax-like material and provide the leaves with the right roughness for this effect (Barthlott and Neinhuis 1997, Samaha and Gad-el-Hak 2014). Several research groups have demonstrated that large slip lengths—tens of microns—can be engineered onto surfaces by nano-fabricating them to have micron high surface features made or covered with low surface energy films (Martines et al. 2005, Choi and Kim 2006).

- *Nanobubbles*—For smoother hydrophobic surfaces, the air pockets prevalent on the superhydrophobic surfaces become thermodynamically unstable and collapse. Even though air bubbles are not supposed to be stable on these surfaces, numerous recent studies have found evidence that bubbles, only a few nanometers in height (*nanobubbles*) form at these water–hydrophobic surface interfaces. While the mechanism for nanobubble formation is not well understood, they are suspected to form from gases previously dissolved in water segregating to the hydrophobic surface (Figure 9.11(b)). (The evidence for nanobubbles and possible explanations have been reviewed by Neto et al. (2005) and Lauga et al. (2007).)

- *Polymer solution and particle suspensions*—If a polymer is dissolved in a solvent with lower viscosity and the solvent molecules preferentially adsorb on the solid surface, a low viscosity layer forms to provide for apparent slip. Similarly, if the particles in a suspension are repelled by a nearby surface, apparent slippage occurs due to the reduced density of particles at the suspension–solid interface (Barnes 1995).

Figure 9.10(b) illustrates the opposite effect to apparent slip, where a layer of higher viscosity material segregates to the liquid–solid interface, moving the plane for the no-slip boundary condition into the liquid flow a distance a, where a is often on the order of the size of a molecule. For lubricated bearings, this high viscosity layer is often formed by the additives blended into the base lubricant, which segregates out to serve as a boundary lubricant layer. Often these are polymers with end-groups with an affinity for the solid surface, anchoring them to the surface with a high effective viscosity.

9.4.5 Why does the no-slip boundary condition work so well?

Before leaving this discussion of slip, it is useful to return to the question: Why does the no-slip boundary condition work so well for most ordinary liquid–solid interfaces?

One reason why has already been touched on. When the liquid molecules make contact with the walls, the intermolecular forces cause them to stick and resist their movement to the adjacent surface sites. Only when these interactions are reduced, for example by lowering the interface energy, are slip lengths increased above a few molecular diameters.

Another reason is that the roughness prevalent on most surfaces suppresses the potential for slip. In essence, the viscous dissipation of flowing over surface roughness brings the liquid to rest, regardless of the strength of the molecular interactions with the surface (Granick et al. 2003). For example, Zhu and Granick (2002b) have shown an rms roughness of 6 nm is sufficient to suppress slippage of water at a hydrophobic surface. In general, any surface heterogeneity, of which roughness is just one type, can pin the flowing liquid, suppressing slip. The situation is physically similar to the pinning of liquid droplets on surfaces, due to chemical heterogeneity and roughness of the surface that leads to contact angle hysteresis (Section 5.4.2.1). Increasing the shear rate of the liquid flowing over these surface irregularities eventually exerts enough shear stress to overcome this pinning and induces slip. For most practical surfaces, where the roughness is typically more than the few nanometers needed to suppress slip, the critical shear stress is never reached for ordinary flow conditions, and the no-slip boundary condition holds. Superhydrophobic surfaces represent a special case where a certain kind of surface roughness is combined with a hydrophobic surface to induce slip.

9.5 Fluid film lubrication

Over the past century, perhaps the biggest achievement of tribological engineering has been the ability to design bearings with lubricating fluid films. For a well-designed fluid film bearing, the sliding surfaces almost never come into solid–solid contact, as shearing takes place entirely across the fluid film. Such a bearing can operate within its designed operating range practically forever with low friction and minimal wear, so long as the fluid is not lost or degraded. Virtually every machine, where components move relative to each other for long periods of time, now relies on fluid bearings; some examples are automobile engines, electric motors, turbines, pumps, gearboxes, and computer hard drives.

While lubricated bearings have been around for centuries, scientific understanding of what it takes to achieve effective fluid film lubrication only started in the late 1880s, with the work of Beauchamp Tower and Osborne Reynolds.

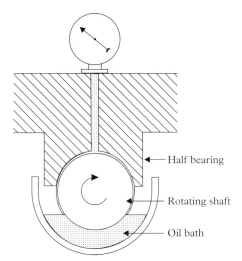

Figure 9.12 *Schematic of the partial journal bearing used by Beauchamp Tower for simulating the bearing tribology of railway axles. In the later stages of his experiments, he added the pressure gauge shown at the top of the bearing for determining the oil pressure.*

First came Tower's experimental finding of hydrodynamic lubrication. Tower was employed by a British railway company to investigate friction in railroad journal bearings. During the course of his work, he had drilled a hole in the top of a bearing, which was lubricated from the bottom by an oil bath, with the intention of adding more lubricant from the top (Figure 9.12). When the shaft rotated against the bearing, he found that, not only was the oil pulled into the bearing, but, to his amazement, it also rose up the hole with enough pressure to pop out the wooden plug he had used to stop up the hole. His subsequent investigations showed that this high oil pressure was correlated with very low friction coefficients for the bearing, from $\mu = 0.001$ to 0.01.

After hearing of Tower's intriguing result, Osborne Reynolds analyzed how a journal could drag oil into a bearing and how a sufficient pressure builds up to support the loading force. In 1886, he published a detailed analysis of fluid film bearings— including a differential equation now called the *Reynolds equation*—that still forms the theoretical basis for fluid film lubrication. This equation is used to determine the pressure distribution within a fluid film bearing from the geometry of the surfaces, the relative sliding velocities, and the fluid's viscosity and density. The Reynolds equation can be derived from the general equations of motion for viscous fluids, the *Navier–Stokes equations*, by assuming that the fluid flows in a laminar manner (not turbulent), that the fluid film thickness is small compared to the lateral dimensions, and that mainly viscous forces act within the fluid. The reader is referred to specialized texts on fluid flow and lubrication theory for the derivation of Reynolds equation and more detailed discussion of fluid film lubrication (Williams 1994, Szeri 1998, Hamrock et al. 2004).

Only in a few simple cases can the Reynolds equation be solved analytically (some of which are discussed below). These days, most analyses of fluid film lubrication problems are done numerically using computer software specifically written for this task. The coupling over the past few decades of computer aided design (CAD) of bearing geometries with new precision engineering methods for fabricating surface topographies and new specialty lubricants has led to a renaissance of high performance bearing engineering. The greatly improved lubrication performance of modern bearing designs has led to key advances for numerous technologies such as disk drives, gyroscopes, ultracentrifuges, MRI machines, and dental drills.

9.5.1 Hydrodynamic lubrication

In hydrodynamic lubrication, fluid is dragged into a bearing by the relative motion of the two sliding surfaces. For hydrodynamic lubrication to work, a portion of the bearing needs to converge so that the outlet for the entrained fluid is smaller than the inlet. This convergence causes the fluid to be squeezed as the relative motion of the two surfaces pulls lubricant into the gap; this pressurizes the fluid film, generating a lifting force that pushes the surfaces apart. Figure 9.13 shows three such convergent geometries often encountered in hydrodynamic bearings: (a) an inclined plane or tilted pad bearing, (b) a Rayleigh step bearing, and (c) a non-concentric journal bearing.

It should be kept in mind that the wedge angles in Figure 9.13 are greatly exaggerated; the actual convergence angles in hydrodynamic bearings are quite small, typically only about a quarter of a degree. Also, the lift force generated does not depend so much on the precise shape of the converging geometry as on the ratio of inlet to outlet height h_i/h_o.

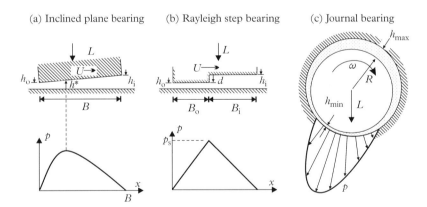

Figure 9.13 *Convergent bearing geometries (top) frequently encountered in hydrodynamic lubrication. The bearings are sufficiently wide in the direction out of the page so that the fluid flow can be treated as being entirely parallel to the plane of the page. The distributions of pressure p above ambient along the bearings (bottom) that support the load L per unit width, where the width is the length of the bearing perpendicular to the page. Note that the pressure distributions are not symmetrical about the bearing midpoint.*

In order for hydrodynamic lubrication to occur without the sliding surfaces making contact, the film thickness needs to be thick enough to prevent machining irregularities or load-induced surface deformations from contacting, particularly at the trailing edge of the contact. The film thickness in typical oil bearings ranges from 1 μm to 1 mm.

9.5.1.1   Inclined plane bearing

First let's discuss the pressure distribution that develops for one of the simplest and most common types of fluid bearing geometries: the inclined plane geometry shown in Figure 9.13(a). To simplify the problem, we consider an incompressible fluid (most liquids can be treated as incompressible) and assume that viscosity is constant throughout the bearing. For this situation, the Reynolds equation for determining the variation in pressure along the bearing can be expressed as

$$dp/dx = -6\eta U \left(h - h^*\right)/h^3, \tag{9.34}$$

where h^* is the separation distance at the x location where the pressure is maximum ($dp/dx = 0$). If we integrate eq. (9.34) with the boundary conditions $p = 0$ at $x = 0$ and $x = B$, we obtain the pressure distribution shown in the lower part of Figure 9.13(a). Integrating again, we obtain an expression for how much load L the bearing supports per unit width (i.e., the lift generated by the bearing):

$$L = 6\eta K U \left(\frac{B}{h_{\mathrm{o}}}\right)^2, \tag{9.35}$$

where

$$K = \frac{\ln\left(1 + n\right)}{n^2} - \frac{2}{n\left(2 + n\right)} \tag{9.36}$$

and $n = h_{\mathrm{i}}/h_{\mathrm{o}} - 1$. An inclined plane bearing generates maximum lift at $h_{\mathrm{i}}/h_{\mathrm{o}} = 2.2$ where $K = 0.027$.

9.5.1.2   Rayleigh step bearing

In 1918 Lord Rayleigh was able to show, using variational calculus, that the bearing shape that provides maximum lift consists of two parallel sections, such as that shown in Figure 9.13(b). (An optimal Rayleigh step bearing generates a factor of 1.29 more lift than an optimal inclined plane bearing.)

The parallel sections of a Rayleigh step bearing make the analysis more straightforward than for other bearing types. First, the Reynolds equation predicts a constant pressure gradient in the parallel sections of the step bearing:

$$\left\{\frac{dp}{dx}\right\}_{\mathrm{i}} = \frac{p_{\mathrm{s}}}{B_{\mathrm{i}}} \quad \text{and} \quad \left\{\frac{dp}{dx}\right\}_{\mathrm{o}} = -\frac{p_{\mathrm{s}}}{B_{\mathrm{o}}}, \tag{9.37}$$

where p_s is the peak pressure, which occurs at the step, and B_1 and B_2 are the lengths of the two parallel sections. This leads to the pressure profile shown in the bottom of Figure 9.13(b).

As before, we assume that the fluid is incompressible (i.e., a liquid) and that the viscosity is constant throughout the bearing. Since the volume flow rate is the same in both sections, from eq. (9.26) we can write

$$q = -\frac{h_i^3}{12\eta}\left\{\frac{dp}{dx}\right\}_i + \frac{Uh_i}{2} = -\frac{h_o^3}{12\eta}\left\{\frac{dp}{dx}\right\}_o + \frac{Uh_o}{2}. \tag{9.38}$$

Combining this equation with eq. (9.37) and rearranging to solve for p_s gives

$$p_s = \frac{P(H-1)}{H^3+P}\frac{6U\eta B_o}{h_o^2}, \tag{9.39}$$

where $H = h_i/h_o$ and $P = B_i/B_o$.

The load L per unit width that can be carried by a Rayleigh step bearing is given by

$$L = \int_0^B pdx$$
$$= \frac{1}{2}p_s(B_i+B_o)$$
$$= \frac{U\eta B^2}{h_o^2}\frac{3P(H-1)}{(1+P)(H^3+P)} \tag{9.40}$$

For a given viscosity η and sliding speed U, maximum lift occurs when $H = 1.87$ and $P = 2.59$ (Williams 1994: p.248).

9.5.1.3 *Journal bearing*

Figure 9.13(c) illustrates a journal bearing where a shaft (journal) rotates within a sleeve (bearing or bushing), dragging lubricant into the clearance gap. Since the loading force shifts the shaft axis off center from the bearing sleeve, a convergent wedge of lubricant forms, which generates the hydrodynamic pressure that counterbalances the load during steady state hydrodynamic lubrication.

At first glance, a journal bearing looks like an inclined plane bearing that has been wrapped round on itself. The solution for the Reynolds equation, however, is complicated by the presence of the divergent portion of the lubricant film in the journal bearing. If the Reynolds equation is simply integrated around the bearing, one obtains the plot labeled *full Sommerfeld solution* in Figure 9.14 for the pressure as function of angle. The problem with this solution is that the areas underneath the positive and negative portions of the pressure curve cancel, so this solution provides no net load support. That journal bearings do support loads suggests that liquid lubricants can only sustain

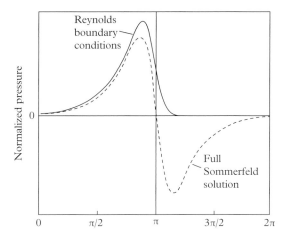

Figure 9.14 *Pressure distribution as function of angle in a journal bearing; 0 is at the location of h_{max} and π at the location of h_{min}.*

a limited amount of negative pressure (i.e., below the ambient pressure). When a liquid is subjected to a negative pressure, it is under a tensile stress which causes the liquid to rupture when a threshold is exceeded. This rupture process is known as *cavitation* as it creates voids in the liquid.

In the Reynolds equation analysis, the phenomenon of cavitation is typically handled by requiring that the pressure in the liquid be greater than or equal to the ambient pressure, which provides solutions close to what is measured in experiments. In Reynolds' original 1886 paper, he used the boundary condition that the pressure and its first derivative with respect to position around the bearing go to zero simultaneously as the liquid exits the convergent portion of the journal bearing; this Reynolds solution is plotted in Figure 9.14.

The expression for the load bearing capacity of a journal bearing, even though the analysis is more complicated, ends up being similar in form to that for an inclined plane bearing (eq. (9.35)); the load L per unit width that a journal bearing supports is given by:

$$L = \eta SU\left(\frac{R}{h}\right)^2,\qquad(9.41)$$

where h is the mean thickness of the lubricant film, R the shaft radius, and $U = \omega R$ the peripheral speed of the shaft. Here h is also the maximum distance that the shaft axis can be displaced from the bearing center before contacting the bearing surface. For typical journal bearings, R/h is usually in the range 1000–10,000.

In eq. (9.41), S is a dimensionless number, called the *Sommerfeld number*, determined by the eccentricity of the shaft in its journal housing and by the ratio of the bearing width to its diameter. So, it expresses how this ratio and the ratio of the maximum to the minimum lubricant film thickness (h_{max}/h_{min}) impacts the load bearing capacity. S can be obtained by numerically solving the Reynolds equation for any eccentricity and width to diameter ratio, and is tabulated in many standard texts on bearing lubrication. An obvious case is when the shaft is concentric with journal housing: $S = 0$, as no convergent wedge forms to generate a load bearing capacity. For a more typical case when $h_{max}/h_{min} = 4$, S is about 2 when bearing width is the same as the diameter, and $S \approx 7$ for an infinitely long bearing. As the load force L increases, the bearing becomes more eccentric, increasing h_{max}/h_{min} and the pressures in the convergent film. The maximum load bearing capacity is determined by the smallest h_{min} that can be achieved before the shaft roughness starts to contact the roughness of the journal housing.

The viscous friction or tangential force F per unit width on an infinitely wide journal bearing is

$$F = 2\pi \eta U R / h. \tag{9.42}$$

This also provides a good estimate for the friction experienced by a journal bearing of finite width. Therefore, the friction coefficient μ is estimated using

$$\mu = F/L = \left(\frac{2\pi}{S}\right)\left(\frac{h}{R}\right). \tag{9.43}$$

From eq. (9.43), we can see that very low values of μ can be obtained with hydrodynamic lubrication; for example, if $S = 7$ and $h/R = 1/1000$, then $\mu = 0.0009$.

9.5.1.4   *Cavitation*

As mentioned above, lowering the pressure of a liquid below atmospheric pressure induces tensile stresses within it that will cause rupture at fairly small pressures below atmosphere, creating voids or cavities. Cavitation typically nucleates at particles in the liquid or at pits, cracks, and other surface defects on walls of the bearing, as these lower the activation energy for the nucleation of voids thereby promoting cavitation at low negative pressures.

Three different mechanisms for cavitation in lubricated bearings have been identified (Braun and Hannon 2010):

1. *Gaseous cavitation*—Dissolved gases in the lubricant segregate out to form the voids. Churning of a lubricant within a bearing typically aerates it, creating a plentiful source of dissolved gases and small bubbles for cavitation.

2. *Pseudo-cavitation*—A form of gaseous cavitation where existing gas bubbles expand without any further diffusion or migration of gas molecules through the liquid to the voids.

3. *Vaporous cavitation*—Voids that form when the absolute pressure in the liquid falls below the vapor pressure of the liquid, causing the liquid to evaporate and creating voids. Unlike gaseous cavitation, bubbles formed by vaporous cavitation are

susceptible to sudden collapse, a process that can generate very high compressive stresses (as high as 0.5 GPa.) For bubbles collapsing next to the bearing surfaces, these high stresses can result in significant damage called *cavitation wear*.

9.5.2   Gas bearings

While liquid lubricants are used within most fluid film bearings, for some bearings gas is a more desirable lubrication fluid. Air is the most common gas used due to its inherent advantages of ease of supply, non-degradability, and little environmental risk when exhausted. This is particularly true for most hydrostatic gas bearings (or more properly *aerostatic* bearings) where they are continually exhausted into the environment. Aerostatic bearings are frequently used on assembly lines to guide parts through with low friction without contaminating the surfaces of the moving parts. An air hockey game is another example where an aerostatic bearing provides the puck with ultra-low friction motion.

Since gas viscosities are typically 0.1% to 0.01% that of oil lubricants, sliding speeds for an *aerodynamic* bearing need to be many times that of a comparable liquid hydrodynamic bearing to generate a similar load bearing capacity. For high speed bearings, however, gas lubrication provides some distinct advantages:

- Gases are not susceptible to the degradation that plagues oil lubricants at high shear rates.
- The low gas viscosity provides for low friction.
- Gases can lubricate over a wide temperature range without the risk of solidifying at low temperatures or boiling at high temperatures.

Therefore, gas journal bearings are commonly used in high rotation speed devices, such as turbomachinery, machine tool spindles, gyroscopes, and dental drills.

Since the lower viscosity provides less lift, to compensate gas bearings typically have a smaller separation distances than liquid bearings. The smaller gap sizes mean that cleanliness is a more critical issue for gas bearings, since hard contaminant particles with dimensions smaller than the clearance can cause serious damage if ingested into the bearing. The smaller gap size also means that the manufacturing tolerances and surface roughness of gas bearings have to be much more tightly controlled than for oil bearings. An extreme example of tight tolerances for an aerodynamic bearing occurs in a disk drive, where the recording head slider is designed to fly over a rotating disk with a gap less than ten nanometers at its trailing edge. To ensure that a few nanometers of clearance is achieved between the slider and disk, these surfaces are manufactured with an rms roughness less than a nanometer.

For gas bearings, the compressibility of the gas needs to be taken into account. One way of doing this is to define a dimensionless *compressibility* or *bearing number* when analyzing the bearing. For the inclined plane geometry in Figure 9.15, the bearing number is

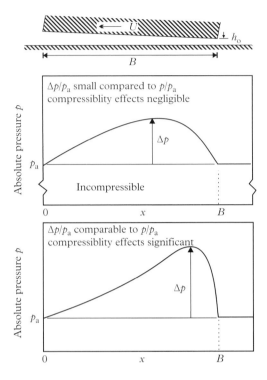

Figure 9.15 *Pressure distributions for a gas inclined plane bearing for the incompressible and compressible regimes of the gas. When the changes in pressure are small relative to the ambient pressure, the gas can be treated as incompressible. When the changes in pressure are comparable to the ambient pressure, the compressibility of the gas needs to be considered.*

$$\Lambda = \frac{6U\eta_a B}{p_a h_o^2},$$

(9.44)

where p_a is the ambient pressure and η_a is the gas viscosity at ambient pressure. For low values of the ratio U/h_o, the bearing number is small, the variation in pressure along the bearing is small relative to the ambient pressure, and the gas can be treated as incompressible. At high values of U/h_o corresponding to large bearing numbers, changes in pressure along the bearing become comparable to the ambient pressure, and compressibility effects need to be taken into account when analyzing the Reynolds equation for the bearing. Figure 9.15 illustrates how compressibility affects the resulting pressure distribution for an inclined plane bearing, with the center of pressure shifting toward the bearing's trailing edge for a compressible fluid compared to an incompressible fluid.

9.5.2.1 Slip flow in gas bearings

Just as for liquids, molecular slippage also occurs for gases flowing over a solid surface. When the thickness h of the gas film is much greater than the mean free path λ of the gas molecules, this slip has a negligible impact on the gas flow, and the no-slip boundary condition applies. When h is comparable to λ, gas flow through the bearing increases significantly above what would be expected with the no-slip boundary condition. This is quantified through the Knudsen number:

$$Kn = \frac{\lambda}{h}. \tag{9.45}$$

For $Kn < 0.01$, the gas can be treated as undergoing continuum or laminar flow with a no-slip boundary condition; for $0.01 < Kn < 15$, slip effects become important; and, for $Kn > 15$, slip flow is fully developed.

Slippage tends to have a much more pronounced effect on the hydrodynamic performance of gas bearings than on liquid lubricated bearings. The reason for this is that, since the viscosity of gas is much less than for a liquid, the fluid film thickness h needs to be much smaller for gas bearing compared to a liquid bearing to generate the same lift. This means that gas bearings are typically designed to have very narrow gaps, with a height often comparable to the mean free path of the gas (e.g., $\lambda = 65$ nm of air at 1 atm).

When flows are analyzed with the Reynolds equation in continuum theory, the effect of slip is often treated as an effective rarefaction of the gas rather than as a slip length. Burgdorfer (1959) introduced an approach for estimating how rarefaction reduces the effective viscosity of a gas in a narrow gap by applying the kinetic theory of gases to gas film lubrication:

$$\eta_{\text{eff}} = \frac{\eta_0}{1 + (6a\lambda/h)}, \tag{9.46}$$

where η_0 is the unrarefied gas viscosity and a is the surface correction coefficient for λ. This approach is often referred to as the first-order slip model and has been extended over the years to include higher order correction terms in a second-order slip model (Hsia and Domoto 1983, Colin 2005) and a 1.5-order slip model (Mitsuya 1993). Another approach, introduced by Fukui and Kaneko (1988), is to use a generalized lubrication equation based on the Boltzmann equation, which describes the behavior of the gas molecules statistically. In this method, the corrections to the Poiseuille flow rates are calculated in advance and a lookup table generated for future numerical simulations (Fukui and Kaneko 1990).

9.5.3 Elastohydrodynamic lubrication (EHL)

As previously noted, to achieve the most effective hydrodynamic lubrication, a fluid film bearing should generate enough lift to ensure that the asperities on the opposing sliding surfaces do not contact. For many years, one of the puzzles of hydrodynamic lubrication was that many commonly used lubricants provide effective lubrication even in situations where the standard hydrodynamic lubrication theory predicted insufficient lift to prevent asperity–asperity contact. This is particularly true in applications where one part has to roll against another, which necessitates that surfaces be curved and non-conforming and results in very high pressures in the contact zones. Standard hydrodynamic theory predicts that such high local pressures should squeeze out the lubricant film and result in solid–solid contact. Some common examples include the following:

- gear teeth rolling and sliding against each other;
- cam sliding against a follower;
- balls in a ball bearing rolling against the inner and outer races.

Beginning in the 1940s, the theory of *elastohydrodynamic lubrication (EHL)* was developed to explain why effective fluid film lubrication still occurs in these situations. This theory incorporates two additional effects into the standard hydrodynamic theory which manifest themselves at higher contact pressures and help maintain a continuous lubricant film between the sliding surfaces:

1. the exponential increase in a lubricant's viscosity with increasing pressure;
2. the elastic deformation of surfaces due to the high local pressures within a lubricant film.

Contacts where this occurs are referred to as *elastohydrodynamic (EHD) contacts*.

9.5.3.1 *Pressure dependence of viscosity*

In addition to their viscosity decreasing with temperature (as described by eq. (9.21)), lubricants also have the characteristic that their viscosity often increases steeply with pressure. This is particularly true for oil-based lubricants, and helps account for their effectiveness in many situations. The viscosity of an oil can usually be modeled as having exponential dependence on pressure, described by the *Barus equation*:

$$\eta = \eta' e^{\alpha p}, \tag{9.47}$$

where η' is the viscosity at one atmosphere, p is the hydrostatic pressure relative to one atmosphere, and α is a constant for the particular lubricant. For example, for a typical mineral oil where $\alpha \sim 2 \times 10^{-8}$ Pa^{-1}, a peak pressure of 230 MPa results in the peak local viscosity η being 100 times higher than η'.

So, while the high pressure is trying to push the lubricant out from between the approaching sliding surfaces, the exponential rise in viscosity works to resist the flow of lubricant out of the bearing, countering the effect of the high pressure and thus maintaining effective hydrodynamic lubrication. Since the lift generated by a bearing scales with the lubricant viscosity (eqs. (9.35), (9.40), and (9.41)), the higher viscosity due to the increasing pressure works to increase the separation distance between the surfaces.

It is not unusual for the peak pressure in EHD contacts to exceed 1 GPa, which for mineral oil, the Barus equation would predict a viscosity enhancement in excess of 10^8.

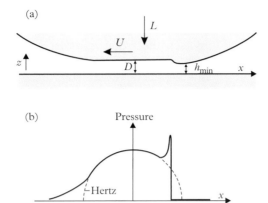

Figure 9.16 *(a) How elastohydrodynamic lubrication causes a cylindrical roller to deform as it slides or rolls over a flat surface (z scale exaggerated for clarity). The lubricant film over most of the Hertzian contact zone has nearly constant thickness D, except near the outlet where a constriction reduces the separation to h_{min}. This constriction leads to a spike in the pressure distribution (b). The dashed semi-ellipse indicates what the pressure would be for a Hertzian contact.*

In practice, such extreme pressure enhancements typically do not occur as shear thinning tends to suppress the viscosity enhancement, or the lubricant might solidify above a particular pressure and display a limiting shear stress.

9.5.3.2 *Pressure induced elastic deformation*

Another essential factor in keeping sliding surfaces out of solid–solid contact is the elastic deformation away from contact due to the high pressures within lubricant films. Even for light loading conditions, small elastic deformations can have a major impact in maintaining a continuous film of lubricant within a contact zone.

Analyzing EHD contacts is necessarily complex as the Reynolds equation must be solved while taking into account the elastic distortion of surfaces due to the hydrodynamically generated pressure along with how the viscosity varies with pressure.

Figure 9.16 illustrates the essential features of a line contact distorted by EHL. For the most part, the elastic deformation and the pressure distribution follow closely to that of an unlubricated Hertzian contact. At the lubricant inlet, the surfaces become flatter so that they are nearly parallel, with a lubricant film thickness D. Near the outlet, the elastic deformation deviates substantially from a Hertzian contact as a constriction or "nip" develops, reducing the film thickness by typically about 25%. This constriction causes a sharp spike in pressure before it returns back to ambient pressure as the lubricant exits.

It may initially seem odd that the pressure can vary as much as shown in Figure 9.16, where the surfaces are nearly parallel. According to the Barus equation, when $e^{\alpha p} \gg 1$ the local viscosity becomes very pressure sensitive. Thus, minute changes in film thickness produce large variations in pressure. Consequently, the mean separation between the surfaces has a fairly weak dependence on load, since the pressure of the liquid in the gap rises rapidly to counter an increasing load as the separation is slightly reduced.

When analyzing an EHD contact, the typical approach is to solve the Reynolds equation numerically and then fit power laws to the numerical solutions to determine the dependence on various parameters. One such example for a line contact (Figure 9.16) is the power law expression for the minimum film thickness $h_{\min}$ (Dowson and Higginson 1959, 1966):

$$\frac{h_{\min}}{R} = 2.65\{2\alpha E_c\}^{0.54}\left\{\frac{\overline{U}\eta'}{2E_cR}\right\}^{0.7}\left\{\frac{L}{2E_cR}\right\}^{-0.13}, \qquad (9.48)$$

where L is the load per unit length, R is the equivalent radius (eq. (3.13), η' and α are defined by eq. (9.47), E_c is the composite elastic modulus, and $\overline{U}$ is the entraining velocity, which is the mean velocity of the two surfaces. The final term $\{L/2E_cR\}^{-0.13}$ is a non-dimensional parameter that describes how the minimum film thickness depends on the loading force. From this term, we can see that the minimum film thickness has a very weak dependence on the load $(L^{-0.13})$; this is a critical factor in successful EHL, as the film thickness can only be reduced to the point of solid–solid contact by applying an exceptionally high load.

Another consideration in EHL is that the fluid film thickness throughout the bearing needs to be much larger than the surface roughness, so that the asperities on the opposing surfaces do not rub against each other. The ratio of film thickness to composite roughness is often called the "lambda ratio" or denoted Λ and as a rule of thumb, should be greater than 1 for an EHL contact.

9.5.3.3   *Experimental measurements of EHL*

Developing experimental techniques for characterizing EHD lubricant has long been challenging since these EHD films are, by nature, localized to small volumes that are difficult to access. As a consequence, the initial development of EHL theory was done without the aid of experimental input. Fortunately, advances in scientific techniques for characterizing nanoscale phenomena over the past few decades have led to a number of valuable approaches being developed specifically for characterizing thin EHD films. With these new techniques, it is now possible to determine, almost down to molecular dimensions, the physical and chemical properties of EHD films, as well as the complex rheology of films at high shear rates and contact pressures (Spikes 1999, Spikes and Jie 2014).

The thickness of the lubricant film under EHD conditions has long been considered to be the most basic parameter to be characterized by experiment. This thickness can now be measured by a variety of techniques, such as optical interferometry, capacitance, and laser fluorescence. One of the principal uses of these thickness measurement techniques has been to confirm how accurately EHL theory predicts the lubricant film thickness profile for a wide variety of EHD conditions.

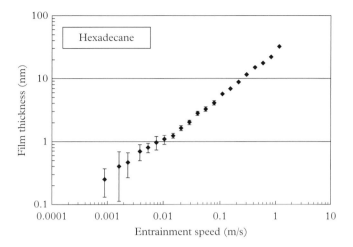

Figure 9.17 *Average lubricant film thickness for a steel ball rolling on a glass disk lubricated with hexadecane with an applied load of 20 N. Film thickness is measured at the center of the contact zone using optical interferometry. The bars correspond to the standard deviation. For this experiment, the ball had an rms roughness of 9.5 nm and the disk 8.2 nm. Reproduced from Glovnea et al. (2003) with permission of Springer Nature, copyright 2003.*

Figure 9.17 shows an example from Glovnea et al. (2003), where optical interferometry is used to measure the thickness of a hexadecane lubricant film as a function of entraining speed for a steel ball rolling on a glass disk. The remarkable aspect of the data in Figure 9.17 is that the film thickness continues to follow the power law dependence expected from EHL theory for a spherical rolling contact down to a thickness of less than 1 nm, which is comparable to the size of the hexadecane molecule. For this sphere-on-flat geometry, the theory of Hamrock and Dowson (1981) predicts that the EHD film thickness D follows

$$D = k\left(\overline{U}\eta\right)^x, \tag{9.49}$$

where x is a constant with a value between 0.6 and 0.75; the slope 0.72 of the log–log plot in Figure 9.17 falls in this expected range. This is even more remarkable when you realize that the composite surface roughness is 12.5 nm. So, the contact pressures are high enough to flatten any asperity roughness (in this situation, the glass surface elastically conforms to the steel surface), and a molecularly thin film of lubricant is maintained between the rolling surfaces with the predicted thickness, instead of being squeezed out. For thicknesses less than 1 nm, the film thickness in Figure 9.17 deviates slightly from the expected power law, which is attributed by Glovnea et al. (2003) to a thin layer—1–2 Å thick—of hexadecane becoming attached to one of the surfaces.

One of the principal reasons why the data in Figure 9.17 agrees well with theory is that the lubricant is entrained into the EHD interface by the rolling action of the ball on

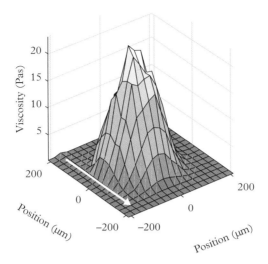

Figure 9.18 *Map of local viscosity measured for a rolling point contact lubricated with poly (oxyethylene)nonylphenyl ether (IGEPAL). The local viscosity is determined by measuring the fluorescent lifetimes of Thioflavin-T molecules dissolved in the lubricant. The arrow shows the direction of lubricant flow. Reproduced from Dench et al. (2016) with permission of Springer Nature, copyright 2016.*

the disk. Since the surfaces roll past each other without sliding, there is no shearing of the lubricant film in the EHD contact (very high shear rates, however, exist in the inlet to the bearing, generating the high pressures that lead to EHL).

Reducing the degree of rolling and increasing the relative sliding speed increases the shear rate applied to the lubricant film, with shear rates of 10^5–10^7 s^{-1} not being uncommon in the inlet to EHD contacts. But the shear stress cannot keep increasing forever with shear rate: eventually something has to give, either through slippage at the lubricant–solid interfaces, shear thinning, or from the lubricant's limiting shear strength. In these situations, EHD theory overestimates the shear stress expected for purely Newtonian behavior with no-slip boundary conditions; indeed, EHD experiments with sliding interfaces commonly observe a weaker than expected shear stress. Consequently, a major area of EHD research is to determine the appropriate rheological models for incorporating non-Newtonian and slippage effects into EHL theory (Bair and Winer 1992, Ehret et al. 1998, Spikes and Jie 2014).

If one of the EHD surfaces is transparent, optical spectroscopies such as infrared and Raman can be used to study the molecular character of the lubricant film under EHD conditions. These spectroscopies can be useful in establishing the film's molecular composition, the degree of molecular alignment, and whether molecular degradation is occurring. Optical techniques have also been developed for measuring the local viscosities and velocity profiles of lubricant films within EHD contacts (Ponjavic and Wong 2014, Ponjavic et al. 2015, Dench et al. 2016). Figure 9.18 shows an example where such a technique was used to measure and map the local viscosity over an EHD contact zone. This map confirms that the viscosity of this lubricant rises dramatically within a rolling contact zone in a manner consistent with the exponential dependence of viscosity on pressure expected from the Barus equation (eq. (9.47)): at the peak pressure of 380 MPa, the viscosity is 175 higher than the viscosity at ambient pressure.

9.6   Important physical and chemical properties of lubricants

As discussed at the beginning of this chapter, viscosity is the most important physical property for determining the lubricating properties of hydrodynamic and EHD bearings, as borne out by the numerous expressions in the chapter containing viscosity η as a parameter. Here, we briefly discuss a few other physical and chemical properties of lubricants that are often important for determining their success or failure.

9.6.1   Surface tension

Generally, it is desirable for lubricants to have low surface tensions for two reasons:

- As discussed in Chapter 5, low surface tension corresponds to low cohesion energy. Generally, the weaker the forces between the molecules, the easier they slide over each other, and the lower the lubricant shear strength. So, low surface tension tends to correlate with low shear strengths and low friction in the boundary lubrication regime.

- In order for a liquid to provide lubrication, it must first cover the solid sliding surfaces. This is not a problem if the sliding surfaces are immersed in the liquid, such as for a sealed bearing filled with lubricant. In many situations however, either the bearing is only partially immersed or a thin film of lubricant is applied topically to the surfaces to provide the necessary lubrication. In both cases, the liquid lubricant needs to wet the surfaces to ensure that sliding does not occur on an uncovered portion of the surfaces. As most polymer lubricants have surface tensions in the range 20–35 mN/m, wetting is typically not a problem with metal or oxide surfaces as these have surface energies above this range. Lubrication for plastics is more problematic, as their surface energies overlap the high end of the range for lubricant surface tensions. So, either lubricant with very low surface tension should be used or the plastic parts should be immersed in the lubricant.

Also, as mentioned in Section 9.4.4.1, if walls that confined the lubricant film have surface energies much lower than the liquid lubricant, then the potential for significant slippage exists at these walls. It has been suggested this phenomenon could be used to design liquid-lubricated bearings with significantly reduced hydrodynamic friction (Spikes 2003, Choo et al. 2007). Indeed, experimental studies by Kalin and Polajnar (2013) have shown that properly tailoring the surface energies of diamond-like carbon coatings on steel surfaces to promote slippage can reduce friction by more than 30% in the EHL regime.

9.6.2   Thermal properties

One of the principal drawbacks of polymer lubricants is the limited range of temperatures over which they can provide hydrodynamic lubrication. The exponential dependence of viscosity on temperature indicated by eq. (9.21) means that the viscosity declines rapidly with increasing temperature. If temperatures are too low, the viscosity becomes so high that the lubricant effectively solidifies. At too high a temperature, the viscosity becomes too low to provide sufficient lift.

Further, at a few hundred degrees Celsius many lubricant molecules oxidize or break down chemically (for most polymer lubricants, this typically occurs well below their boiling temperature). For example, mineral oils are among the least stable with oxidation occurring above 135°C, while perfluoropolyethers (thermal stability $\sim$ 370°C) and polyphenyl ethers (thermal stability $\sim$ 430°C) are among the most thermally stable. Consequently, care must be taken when choosing a lubricant to ensure that it has suitable viscosity and is thermally and oxidatively stable over the expected range of temperatures.

For hydrocarbon-based lubricants, the main way degradation takes place is through oxidation in the presence of air or oxygen. This oxidation is detrimental as the degradation products form smears or coalesce into solid-like deposits on the sliding surfaces hindering lubrication, or they can form organic acids that corrode the surfaces and promote further lubricant decomposition. Consequently, most hydrocarbon based lubricants are blended with anti-oxidation agents that react preferentially with any oxygen dissolved in the lubricant.

Even if the operating temperatures are within acceptable limits, thermal degradation of the lubricant can still be a problem if high local temperatures occur from frictional heating at contacting asperities. This can be alleviated somewhat if lubricants have high thermal conductivity to help dissipate heat away from these flash points, but this is often difficult to achieve.

9.7   PROBLEMS

1. In the Eyring model, the energy activation barrier E_a in eq. (9.12) corresponds to the energy needed to create a void in the liquid. At one atmosphere pressure this can be estimated as $E_a = \gamma a^2$ where γ is the surface tension of the liquid and a is an appropriate molecular dimension. If the pressure is increased above one atmosphere by an amount p, then an additional work $p\delta V$ needs to be provided to create the void, where δV is the volume of the void large enough the accommodate motion of the molecule. Re-derive Eyring's equation (9.17) for how viscosity depends on the pressure p by including a $p\delta V$ term in the activation energy and show that this equation reduces to the Barus equation (9.47) at low shear stresses.

2. Consider a line contact as illustrated in Figure 9.16. Assume this describes a steel cylinder of radius $R = 5$ cm sliding with an applied load per unit length of 1000 N/mm at a speed of 10 m/s against a steel flat that is stationary. The sliding takes place in a mineral oil with a viscosity of 0.04 Pa s at ambient pressure and $\alpha = 3 \times 10^{-8}$ Pa^{-1}. The composite roughness of the steel surfaces is measured to be 80 nm. Does the system fall into the EHL regime? What is the minimum thickness of the lubricating film?

9.8 REFERENCES

Achilleos, E. C., G. Georgiou and S. G. Hatzikiriakos (2002). "Role of processing aids in the extrusion of molten polymers." *Journal of Vinyl and Additive Technology* **8**(1): 7–24.

Bair, S. (2001). "Measurements of real non-Newtonian response for liquid lubricants under moderate pressures." *Proceedings of the Institution of Mechanical Engineers, Part J: Journal of Engineering Tribology* **215**(3): 223–33.

Bair, S. and W. O. Winer (1979). "Shear strength measurements of lubricants at high pressure." *Journal of Lubrication Technology* **101**(3): 251–7.

Bair, S. and W. O. Winer (1992). "The high-pressure high shear-stress rheology of liquid lubricants." *Journal of Tribology, Transactions of the ASME* **114**(1): 1–13.

Barnes, H. A. (1995). "A review of the slip (wall depletion) of polymer solutions, emulsions and particle suspensions in viscometers: its cause, character, and cure." *Journal of Non-Newtonian Fluid Mechanics* **56**(3): 221–51.

Barthlott, W. and C. Neinhuis (1997). "Purity of the sacred lotus, or escape from contamination in biological surfaces." *Planta* **202**(1): 1–8.

Blake, T. D. (1990). "Slip between a liquid and a solid: D.M. Tolstoi's (1952) theory reconsidered." *Colloids and Surfaces* **47**(Supplement C): 135–45.

Braun, M. and W. Hannon (2010). "Cavitation formation and modelling for fluid film bearings: a review." *Proceedings of the Institution of Mechanical Engineers, Part J: Journal of Engineering Tribology* **224**(9): 839–63.

Burgdorfer, A. (1959). "The influence of molecular mean free path on the performance of hydrodynamic gas lubricated bearings." *ASME Journal of Basic Engineering* **81**: 94–100.

Choi, C. H. and C. J. Kim (2006). "Large slip of aqueous liquid flow over a nanoengineered superhydrophobic surface." *Physical Review Letters* **96**(6): 066001.

Choi, C. H., K. J. A. Westin and K. S. Breuer (2003). "Apparent slip flows in hydrophilic and hydrophobic microchannels." *Physics of Fluids* **15**(10): 2897–902.

Choo, J. H., H. A. Spikes, M. Ratoi, R. Glovnea and A. Forrest (2007). "Friction reduction in low-load hydrodynamic lubrication with a hydrophobic surface." *Tribology International* **40**(2): 154–9.

Colin, S. (2005). "Rarefaction and compressibility effects on steady and transient gas flows in microchannels." *Microfluidics and Nanofluidics* **1**(3): 268–79.

Couette, M. F. A. (1890). "Études sur le frottement des liquides." *Annales de Chimie et de Physique* **21**: 433–510.

de Gennes, P. G. (1979). "Viscometric flows of tangled polymers." *Comptes Rendus Hebdomadaires Des Seances De L Academie Des Sciences Serie B* **288**(14): 219–20.

Dench, J., N. Morgan and J. S. S. Wong (2016). "Quantitative viscosity mapping using fluorescence lifetime measurements." *Tribology Letters* **65**(1): 25.

Dowson, D. (1978). *History of tribology*. London: Longman.

Dowson, D. and G. R. Higginson (1959). "A numerical solution to the elastohydrodynamic problem." *Journal of Mechanical Engineering Science* **1**: 6–20.

Dowson, D. and G. R. Higginson (1966). *Elasto-hydrodynamic lubrication: the fundamentals of roller and gear lubrication* (1st ed.). Oxford: Pergamon Press.

Ehret, P., D. Dowson and C. M. Taylor (1998). "On lubricant transport conditions in elastohydrodynamic conjunctions." *Proceedings of the Royal Society of London, Series A: Mathematical, Physical and Engineering Sciences* **454**(1971): 763–87.

Eyring, H. (1936). "Viscosity, plasticity, and diffusion as examples of absolute reaction rates." *The Journal of Chemical Physics* **4**(4): 283–91.

Fukui, S. and R. Kaneko (1988). "Analysis of ultra-thin gas film lubrication based on linearized Boltzmann equation. 1. Derivation of a generalized lubrication equation including thermal creep flow." *Journal of Tribology, Transactions of the ASME* **110**(2): 253–62.

Fukui, S. and R. Kaneko (1990). "A database for interpolation of Poiseuille flow-rates for high Knudsen number lubrication problems." *Journal of Tribology, Transactions of the ASME* **112**(1): 78–83.

Glovnea, R. P., A. K. Forrest, A. V. Olver and H. A. Spikes (2003). "Measurement of sub-nanometer lubricant films using ultra-thin film interferometry." *Tribology Letters* **15**(3): 217–30.

Granick, S., Y. X. Zhu and H. Lee (2003). "Slippery questions about complex fluids flowing past solids." *Nature Materials* **2**(4): 221–7.

Hamrock, B. J. and D. Dowson (1981). *Ball bearing lubrication: the elastohydrodynamics of elliptical contacts*. New York: John Wiley & Sons.

Hamrock, B. J., S. R. Schmid and B. O. Jacobson (2004). *Fundamentals of fluid film lubrication*. Boca Raton, FL: CRC Press.

Hocking, L. M. (1973). "The effect of slip on the motion of a sphere close to a wall and of two adjacent spheres." *Journal of Engineering Mathematics* **7**(3): 207–21.

Hsia, Y. T. and G. A. Domoto (1983). "An experimental investigation of molecular rarefaction effects in gas lubricated bearings at ultralow clearances." *Journal of Lubrication Technology, Transactions of the ASME* **105**(1): 120–30.

Jadhao, V. and M. O. Robbins (2017). "Probing large viscosities in glass-formers with nonequilibrium simulations." *Proceedings of the National Academy of Sciences* **114**(30): 7952–7.

Johnson, K. L. and J. L. Tevaarwerk (1977). "Shear behavior of elastohydrodynamic oil films." *Proceedings of the Royal Society of London, Series A: Mathematical, Physical and Engineering Sciences* **356**(1685): 215–36.

Kalin, M. and M. Polajnar (2013). "The effect of wetting and surface energy on the friction and slip in oil-lubricated contacts." *Tribology Letters* **52**: 185–94.

Kauzmann, W. and H. Eyring (1940). "The viscous flow of large molecules." *Journal of the American Chemical Society* **62**(11): 3113–25.

Kioupis, L. I. and E. J. Maginn (2000). "Impact of molecular architecture on the high-pressure rheology of hydrocarbon fluids." *Journal of Physical Chemistry B* **104**(32): 7774–83.

Lauga, E., M. Brenner and H. Stone (2007). "Microfluidics: the no-slip boundary condition." In: *Springer handbook of experimental fluid mechanics*, C. Tropea, A. L. Yarin and J. F. Foss, Eds. Berlin: Springer, pp.1219–40.

Lichter, S., A. Martini, R. Q. Snurr and Q. Wang (2007). "Liquid slip in nanoscale channels as a rate process." *Physical Review Letters* **98**(22): 226001.

Lichter, S., A. Roxin and S. Mandre (2004). "Mechanisms for liquid slip at solid surfaces." *Physical Review Letters* **93**(8): 086001.

Luengo, G., J. Israelachvili and S. Granick (1996). "Generalized effects in confined fluids: new friction map for boundary lubrication." *Wear* **200**(1–2): 328–35.

Martines, E., K. Seunarine, H. Morgan, N. Gadegaard, C. D. W. Wilkinson and M. O. Riehle (2005). "Superhydrophobicity and superhydrophilicity of regular nanopatterns." *Nano Letters* **5**(10): 2097–103.

Martini, A., Y. Liu, R. Q. Snurr and Q. J. Wang (2006). "Molecular dynamics characterization of thin film viscosity for EHL simulation." *Tribology Letters* **21**(3): 217–25.

Martini, A., A. Roxin, R. Q. Snurr, Q. Wang and S. Lichter (2008). "Molecular mechanisms of liquid slip." *Journal of Fluid Mechanics* **600**: 257–69.

Mitsuya, Y. (1993). "Modified Reynolds equation for ultra-thin film gas lubrication using 1.5-order slip-flow model and considering surface accommodation coefficient." *Journal of Tribology, Transactions of the ASME* **115**(2): 289–94.

Navier, C. L. M. H. (1823). "Memoire sur les lois du mouvement des fluides." *Memoires de l'Academie Royale des Sciences de l'Institut de France* **VI**: 389–440.

Neto, C., D. R. Evans, E. Bonaccurso, H. J. Butt and V. S. J. Craig (2005). "Boundary slip in Newtonian liquids: a review of experimental studies." *Reports on Progress in Physics* **68**(12): 2859–97.

Ponjavic, A. and J. S. S. Wong (2014). "The effect of boundary slip on elastohydrodynamic lubrication." *RSC Advances* **4**(40): 20821–9.

Ponjavic, A., J. Dench, N. Morgan and J. S. S. Wong (2015). "In situ viscosity measurement of confined liquids." *RSC Advances* **5**(121): 99585–93.

Ree, F., T. Ree and H. Eyring (1958). "Relaxation theory of transport problems in condensed systems." *Industrial & Engineering Chemistry* **50**(7): 1036–40.

Samaha, M. and M. Gad-el-Hak (2014). "Polymeric slippery coatings: nature and applications." *Polymers* **6**(5): 1266–311.

Spikes, H. and Z. Jie (2014). "History, origins and prediction of elastohydrodynamic friction." *Tribology Letters* **56**(1): 1–25.

Spikes, H. and W. Tysoe (2015). "On the commonality between theoretical models for fluid and solid friction, wear and tribochemistry." *Tribology Letters* **59**(1): 21.

Spikes, H. A. (1999). "Thin films in elastohydrodynamic lubrication: the contribution of experiment." *Proceedings of the Institution of Mechanical Engineers Part J: Journal of Engineering Tribology* **213**(J5): 335–52.

Spikes, H. A. (2003). "The half-wetted bearing. Part 2. Potential application in low load contacts." *Proceedings of the Institution of Mechanical Engineers Part J: Journal of Engineering Tribology* **217**(J1): 15–26.

Stribeck, R. (1902). "Characteristics of plain and roller bearings." *Zeitschrift Verein Deutscher Ingenieure* **46**: 1341–8.

Szeri, A. Z. (1998). *Fluid film lubrication: theory and design.* Cambridge: Cambridge University Press.

Tevaarwerk, J. and K. L. Johnson (1975). "A simple non-linear constitutive equation for elastohydrodynamic oil films." *Wear* **35**(2): 345–56.

Tolstoi, D. M. (1952). "Molecular theory for slippage of liquids over solid surfaces." *Doklady Akademiya Nauk SSSR* **85**: 1089.

Vinogradova, O. I. (1996). "Hydrodynamic interaction of curved bodies allowing slip on their surfaces." *Langmuir* **12**(24): 5963–8.

Williams, J. A. (1994). *Engineering tribology*. Oxford: Oxford University Press.

Zhu, Y. X. and S. Granick (2001). "Rate-dependent slip of Newtonian liquid at smooth surfaces." *Physical Review Letters* 87(9): 096105.

Zhu, Y. X. and S. Granick (2002a). "Apparent slip of Newtonian fluids past adsorbed polymer layers." *Macromolecules* 35(12): 4658–63.

Zhu, Y. X. and S. Granick (2002b). "Limits of the hydrodynamic no-slip boundary condition." *Physical Review Letters* 88(10): 106102.

Zhu, Y. X. and S. Granick (2002c). "No-slip boundary condition switches to partial slip when fluid contains surfactant." *Langmuir* 18(26): 10058–63.

10

Lubrication in Tight Spots

At the beginning of the last chapter, we discussed how the different lubrication regimes for a lubricated bearing correspond to different portions of the Stribeck curve (Figure 9.1). In the hydrodynamic lubrication regime, which occurs at high sliding speeds and light loads, the sliding surfaces are separated by a thick oil film, and friction depends on the oil's hydrodynamic properties, particularly viscosity, and not on the nature of the solid surfaces. Increasing the load or decreasing the relative speed thins the lubricating film within the bearing and eventually contact occurs between the sliding surfaces. With this contact, friction and wear not only increase dramatically, but also start to depend strongly on the specific nature of the solid surfaces and the molecular films adsorbed on them.

In this chapter, we look more closely at how the properties of the lubricating films change as the space between the sliding surfaces becomes very small—down to molecular dimensions. As we shall see, several interesting phenomena dominate the lubricating properties of this nanoscale regime:

- *Confined liquids*—enhanced viscosity and solidification of nanoscale liquid films;
- *Boundary lubrication*—thin films, chemically distinct from the starting materials, formed by adsorption and/or reactive processes to provide protection that reduces friction and wear;
- *Capillary and disjoining pressures*—mechanisms responsible for meniscus formation and meniscus forces from lubricant films.

10.1   Confined liquids

If we consider the example of a smooth ball of radius R being pushed with a loading force L against a liquid droplet on a flat surface, the liquid between the two surfaces undergoes viscous flow as it is squeezed out (Granick 1991, 1999). When the thickness of this liquid film is macroscopic, the viscous flow is well described by Reynolds theory of hydrodynamic lubrication. However, when the ball–flat separation distance D becomes

Tribology on the Small Scale: A Modern Textbook on Friction, Lubrication and Wear. Second edition. C. Mathew Mate and Robert W. Carpick. © Oxford University Press 2019. Published in 2019 by Oxford University Press.
DOI: 10.1093/oso/ 9780199609802.001.0001

less than about 10 molecular diameters, interesting phenomena begin to happen due to the confinement of the liquid molecules in this narrow gap.

One of these phenomena, which was discussed in Section 7.4.1, is that between two solid surfaces liquid molecules tend to order into a layered structure. This tendency to form layers manifests itself as oscillations in the local density of the liquid, as illustrated in Figure 7.14; the period of this oscillation corresponds to a molecular dimension, and the oscillation amplitude decays away from the solid surfaces. This ordered structure gives rise to solvation or structural forces that oppose the squeezing out of individual layers and remarkably enable the liquid film to support the loading force acting on the ball (Figure 7.15.) As you might expect, the larger the loading force is, the smaller the final number of molecular layers at equilibrium. Quite a substantial pressure, however, is needed to squeeze out the last molecular layer; for example, estimates range from 300 MPa for short alkanes to 800 MPa for long alkanes (Sivebaek et al. 2004).

Another interesting effect for liquids confined in small gaps is that they exhibit different viscoelastic properties than the bulk liquid. A common way of determining the confined liquid's viscoelastic properties is to oscillate one surface parallel to the other so as to shear the liquid film. A convenient way to measure this is to use the surface force apparatus (SFA) discussed in Section 8.1, which has the advantage that the solvation forces elastically deform the mica surfaces to form two parallel surfaces, which sandwich between them a thin film of liquid of uniform thickness D, as illustrated in Figure 10.1(c).

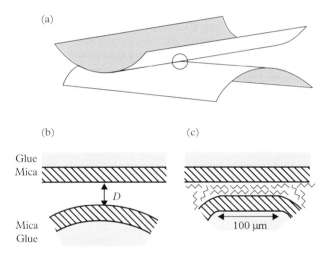

Figure 10.1 *(a) Cross cylinder geometry of an SFA experiment. (b, c) Enlarged cross-sectional views of the encircled contact zone for (b) undistorted mica surfaces when D is large or (c) elastically distorted mica surfaces when D is small with a few layers of molecules. The repulsive solvation forces from the molecular layers sandwiched between the mica surfaces elastically deform the mica sheets and the glue attaching them to the cylindrical lenses to form two parallel surfaces. In the figure, the scale normal to the surface is greatly exaggerated: the parallel region typically extends up to 100 μm across, while the gap between the surfaces is only a few nanometers.*

This parallel geometry simplifies the expression of the force F needed to shear the liquid film to (eq. (9.2))

$$F = \frac{\eta A v}{D},$$

(10.1)

where η is the liquid's viscosity, A the area of the parallel surfaces being sheared, and v the velocity that the surfaces move relative to each other. Another advantage of using an SFA is that the mica surfaces are atomically smooth, allowing the film thickness D to be made arbitrarily small if enough pressure can be applied to squeeze out a sufficient number of molecular layers. With the SFA technique, the area A over which the surfaces are parallel (typically tens of microns in diameter) is measured by optical microscopy; the separation D is measured by optical interference; and the lateral sliding velocity v is measured using a displacement sensor for lateral deflections. Then, by applying a known lateral force F to one of the mica surfaces, the effective viscosity of the liquid in the gap is determined from

$$\eta = \frac{DF}{Av}$$
$$= \frac{F}{A\,(d\gamma/dt)}$$

(10.2)

where $d\gamma/dt$ is the shear rate.

In a technique pioneered by Prof. Granick's group (Peachey et al. 1991), a known sinusoidal force is applied in the lateral direction to one mica surface while measuring the resulting amplitude and phase of the induced velocity of that surface. With this sinusoidal excitation, eq. (10.2) determines the real and imaginary parts of η, which correspond to the viscous loss and elastic modulus of the liquid film at that frequency and corresponding shear rate $d\gamma/dt$.

The general result of these viscoelastic experiments is that the more a liquid is confined, the more sluggish it becomes. This slowing of the molecular motion in the liquid with decreasing film thickness can be divided into three regimes:

1. *Bulk liquid*—For thick liquid films, the viscosity and other dynamic properties remain close to that of the bulk liquid.
2. *Enhanced viscosity*—When the liquid film thickness is reduced below four to ten molecular diameters (depending on the liquid), the liquid begins to transition to a more solid-like phase as manifested by a rapidly increasing viscosity and elastic modulus. The effective viscosity for a liquid confined between two mica surfaces in an SFA can be up to 10^6 times greater than the bulk viscosity.
3. *Solid-like*—When the film thickness becomes less than two to four molecular diameters (again depending on the liquid), a solid-like phase forms, which deforms elastically when sheared until its yield stress is exceeded. The solid-like behavior originates either from crystalline structure where the liquid molecules form

ordered layers between the wall surfaces (Section 7.4.1), or from the liquid transitioning to an amorphous glass when confined.

This transition with decreasing film thickness, that is, bulk liquid → enhanced viscosity → solid-like, has been observed for many kinds of liquids, such as linear and branched alkanes, polymers, glass formers, liquid crystals, ionic liquids, and aqueous solutions, and so it seems to be a generic feature of confinement rather than related to the molecular structure or chemical composition.

10.1.1   Example: enhanced viscosity of dodecane

n-Dodecane is a linear alkane liquid whose shear properties have been extensively studied when confined in a thin film with the experiments using the SFA (Hu et al. 1991) and by molecular dynamics simulations (Cui et al. 2003, Jabbarzadeh et al. 2006a, 2006b, 2007). Since dodecane manifests many of the general shear characteristics of liquid films confined in molecularly thin films, we will use it illustrate the behavior and origins of the enhanced viscosity when liquids are confined in thin films.

Dodecane has the chemical structure $CH_3(C_2H_2)_{10}CH_3$, so can be thought as a flexible linear chain of twelve carbon atoms, with a length of 18 Å and a diameter of 4 Å. When confined between mica surfaces in the SFA, the normal force exhibits oscillatory behavior when the film thickness is less than 45 Å, with the oscillation period equal to the dodecane chain diameter. Molecular dynamics simulations indicate that the dodecane molecules form layers in these thin films with their chains oriented predominately parallel to the mica surfaces. Figure 10.2 shows the viscosity of dodecane measured by SFA, and indicates that the effective viscosity is already greatly enhanced over the bulk viscosity when the film thickness is reduced to 40 Å film, and that it increases rapidly with decreasing film thickness.

In general, when a liquid is confined, not only does the liquid viscosity become strongly enhanced, but it also becomes highly non-Newtonian. Figure 10.3 shows how the non-Newtonian behavior is manifested in a six-layer film of dodecane (thickness = 27 Å). At very low shear rates, this confined film has an effective viscosity six orders greater than bulk. But, above a critical shear rate of $\dot{\gamma}_c = 20$ s^{-1}, the effective viscosity exhibits shear thinning: with increasing shear rates, the viscosity decreases back to the bulk viscosity behavior.

The non-Newtonian behavior exhibited in Figure 10.3 is fairly typical for liquids confined to a few molecular layers. Generally, it has been found that, when the shear rate exceeds a critical shear rate $\dot{\gamma}_c$ in SFA experiments and molecular dynamics simulations, the shear thinning behavior of these liquids can be described by the power law relationship:

$$\eta_{\text{eff}} \propto \left(\frac{d\gamma}{dt}\right)^n,$$ (10.3)

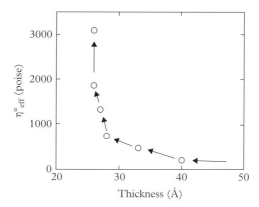

Figure 10.2 *How the effective viscosity increases with decreasing film thickness for dodecane being sheared between mica surfaces in an SFA at 28° C. At 28° C, dodecane's bulk viscosity is 0.01 poise and is independent of shear rate. Reprinted with permission from Hu et al. (1991). Copyright 1991, American Physical Society.*

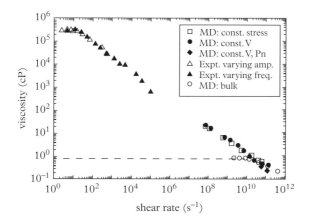

Figure 10.3 *Comparison of the effective viscosity versus shear rate for a six layer film of n-dodecane as determined in SFA experiments (triangles) and in molecular dynamics (MD) simulations. Above a critical shear rate of 20 s^{-1}, these films undergo shear thinning described by the power law relationship in eq. (10.3) with n ~ −2/3. These results are compared with viscosity for bulk n-dodecane (open circles) determined by MD simulations, with the dashed line showing the extrapolation of the constant bulk viscosity to low shear rates. The SFA experimental results are from Hu et al. (1991). Reprinted from Cui et al. (2003) with permission from AIP Publishing.*

where the exponent n ranges from −0.5 to −1 (Jabbarzadeh and Tanner 2006). With this range of exponents, this power law relationship is equivalent to the Carreau shear thinning model described by eq. (9.9).

10.1.2   Molecular origins of enhanced viscosity

So how does confinement lead to enhanced viscosity? The enhanced viscosity indicates that the molecular relaxation times in confined films are much longer than in the bulk liquid. For dodecane confined in a 27 Å gap, a relaxation time of 0.05 seconds can be estimated from the onset of shear thinning at a shear rate of 20 s^{-1} in Figure 10.3; this relaxation time is about 10^8 times slower than in the bulk (Hu et al. 1991). The likely explanation for the slower relaxation times during confinement is that, when one molecule diffuses through the liquid, its neighbors have to be displaced sufficiently to make room for this motion (Section 9.2.3), and these molecules in turn have to displace their neighbors. When the liquid is confined between two stiff surfaces, however, the liquid molecules quickly bump up against these hard walls, severely limiting their displacements in the direction normal to the film. So, the thinner the film, the greater these displacements have to extend out laterally before a molecule can diffuse in the liquid film. This leads to higher activation volumes and activation energies for molecular diffusion through the liquid film as it becomes thinner (Hu et al. 1991). Attractive interactions between the walls and the liquid molecules further slow molecular motion by pinning molecules to the walls (Alba-Simionesco et al. 2006). Another way of looking at this is that confinement lowers the entropy of the liquid film by reducing the number configuration states available. This lowering of entropy shifts the film's freezing point to higher temperatures increasing the liquid's viscosity and eventually inducing a liquid-to-solid transition (Thompson et al. 1992, Weinstein and Safran 1998, Braun and Peyrard 2003, Gao et al. 2004).

In the case of dodecane, molecular dynamics simulations by Jabbarzadeh et al. (2006a, 2007) indicate that the enhancement of viscosity with decreasing shear rate also comes from the formation of crystalline structures that bridge between the shearing surfaces, as shown in Figure 10.4. At shear rates below the critical shear rate $\dot{\gamma}_c$ for shear thinning, the confined dodecane films are purely crystalline, and above $\dot{\gamma}_c$ they consist of a mixture of disordered and crystalline regions, with the crystalline regions consisting of stacks of dodecane molecules layered with their backbone chains oriented parallel to the surface and with these stacks bridging between the two surfaces, as shown in Figure 10.4. For these crystalline structures, displacements within the crystalline structure are severely restricted and the viscous frictional force comes from the shear stress needed to slide one layer of molecules over another. At sufficiently high shear rates, the dodecane molecule structure transitions to a completely disordered state with bulk-like liquid viscosity behavior. Interestingly, these molecular dynamics simulations also indicate that with prolonged shearing, the dodecane molecules within the crystalline stacks become aligned with the flow direction, greatly reducing the shear stress needed to slide one plane over another (Jabbarzadeh et al. 2007).

Consistent with a confined liquid transitioning to a crystalline structure, SFA experiments with thin liquid films also find a measurable yield stress when sheared, an indicator of solid-like behavior. For example, this liquid-to-solid transition occurs with the silicone oil octamethylcyclotetrasiloxane (OMCTS) when the film thickness is reduced from 62 to 54 Å (seven to six molecular layers) (Gee et al. 1990, Klein and Kumacheva 1998,

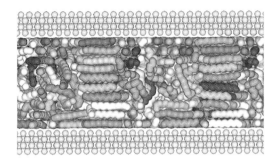

Figure 10.4 *Cross-section view obtained from a molecular dynamics simulation of a 3.95 nm thick film of dodecane undergoing shear above the critical shear rate for shear thinning ($\dot{\gamma}_c > \dot{\gamma}_c$). Isolated stacks of layered dodecane molecules bridge the top and bottom surfaces and are surrounded by disordered regions of dodecane. In the crystalline stacks, the dodecane molecules form a mosaic pattern with their backbone chains oriented parallel to the wall surfaces. Reprinted from Jabbarzadeh et al. (2006a). Copyright 2006 by the American Physical Society.*

Kumacheva and Klein 1998). This yield stress ~0.1 MPa is several orders of magnitude smaller than that needed to shear a bulk OMCTS crystal, indicating that this "solid-like" film is a long way from matching the crystalline solid. Molecular dynamics simulations also indicate that, while diffusion rates are much lower in these crystalline liquid films, they are still much greater than in a solid (Jabbarzadeh and Tanner 2006). So, while these thin liquid films exhibit some solid-like features when confined—that is, static friction, tendency to order, and high viscosity—they should not be thought of as a crystalline solid film: the ordering is more diffuse than in a solid, and the molecules move about within the film much faster than in a solid or glass.

Figure 10.5 shows how this yield stress or static friction force increases as the individual liquid OMCTS layers are squeezed out between mica surfaces in an SFA. The friction versus time trace in Figure 10.5 also exhibits a sawtooth fine structure due to stick-slip motion of the sliding surfaces (Section 4.3.1), indicating that static friction is larger than kinetic friction. For these thin liquid films, the transition from stick to slip is often interpreted as being due to the shear stress inducing a melting transition within the solid film to initiate sliding (Thompson and Robbins 1990). Figure 10.6 illustrates how this might occur for spherically shaped molecules such as OMCTS. In Figure 10.6, several possible disordered or slip processes are depicted as it is generally unclear how the molecules slide past each other in these narrow gaps, unless molecular dynamics simulations are used to elucidate the mechanism.

Often, when shearing polymers and other long chain molecules, the frictional force is observed to slowly decrease over time before plateauing at a stable value. For these type of molecules, this drop in friction after the initiation of sliding is attributed to the shearing process inducing alignment of the molecules, thereby enabling them to slide more easily past each other (Israelachvili 2005, Jabbarzadeh et al. 2007). Figure 10.7 shows this process schematically: when the shearing starts, the film first dilates to accommodate the

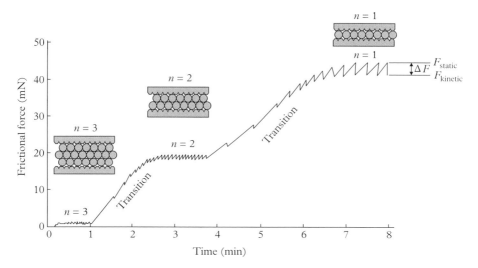

Figure 10.5 *How the friction force increases as the individual layers of octamethylcyclotetrasiloxane (OMCTS) are squeezed out from the sliding mica surfaces. Reprinted with permission from Israelachvili (2005). Copyright 2005 Materials Research Society.*

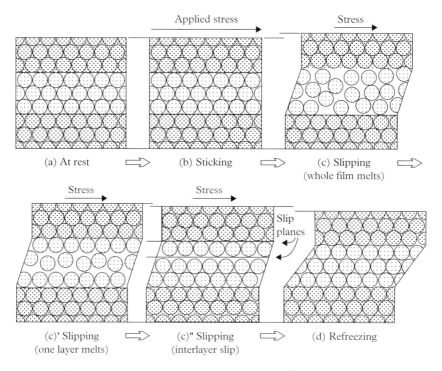

Figure 10.6 *Idealization of the number of ways that an ordered film of spherical molecules can rearrange in response to a shear that induced slippage: (c) total disorder of liquid molecules, (c)′ disorder of the internal layers, and (c)″ slippage between ordered layers. Reprinted with permission from Israelachvili (2005). Copyright 2005 Materials Research Society.*

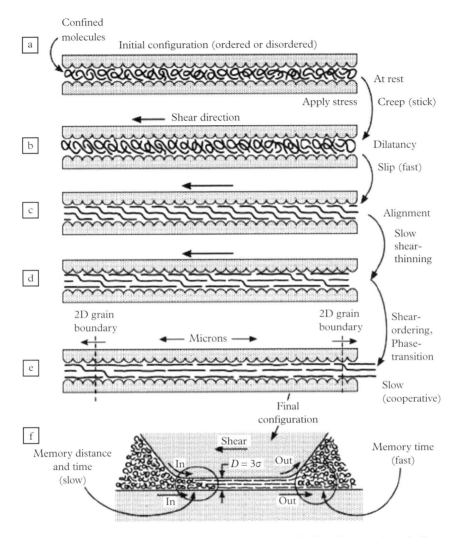

Figure 10.7 *(a–e) Schematic of how shearing causes a polymer liquid to disentangle and align to the shear direction; (f) how the polymer changes from disordered at the film inlet to aligned at the film exit during steady state sliding. Reproduced with permission from Drummond and Israelachvili (2000). Copyright 2000, American Chemical Society.*

molecular movement (Figure 10.7(b)). With continued shear, the molecules disentangle and start to orient with the shear direction enabling them to slide more easily past each other (Figure 10.7(c)–(e)]. One's first expectation might be that this alignment process takes place on the timescale of an individual molecular relaxation time (pico- to nanoseconds); however, in SFA experiments the timescale is found to be minutes or longer, during which time the sliding surfaces move many microns. The reason for this is that, when confined to these small gaps, the molecular movement becomes a highly cooperative affair, with many molecules needing to move to accomplish the molecular realignment. As a consequence, the frictional behavior of these molecules often exhibits a "memory" of how it has been previously sheared (Gao et al. 2004).

It should be noted that the SFA geometry provides a highly idealized contact environment for studying the enhanced viscosity and solidification during confinement. This prompts the question: to what extent do these phenomena persist in less idealized contact geometries? For example, molecular dynamics simulations have found that, if the surfaces have sufficient roughness to prevent ordering of the liquid molecules into layers between them, this also inhibits the solidification process (Gao et al. 2000, Jabbarzadeh et al. 2006b, 2007, Sivebaek and Persson 2016).

Also, fluorescence correlation spectroscopy experiments of Mukhopadhyay et al. (2002) found that the molecules become less sluggish toward the edges of a SFA contact zone, which indicates that the enhanced viscosity diminishes as the lateral extent of contact diminishes. This is further supported by atomic force microscope (AFM) experiments—where the contact zone is orders of magnitude smaller in area than an SFA—which have also shown much less enhancement of viscosity (Mate 1992b, Friedenberg and Mate 1996, O'Shea and Welland 1998, O'Shea et al. 2010). Molecular dynamics simulations do indicate, however, that liquid films between a rough asperity and a flat surface do have enhanced viscosity at small separations, though not as great as observed for parallel flat surfaces, and that very high pressures are still needed to squeeze out liquids from pockets between the contacting roughness summits (Sivebaek and Persson 2016). This has been backed up by AFM experiments where evidence has been observed for the formation of molecular layers (e.g., see Figure 7.17) and enhanced viscosity (Krass et al. 2016).

While enhanced viscosity and solidification occur in confined contact zones, this sensitivity to roughness and to the lateral extent of the contact zone makes it difficult to predict how these phenomena will influence the tribological performance of a particular contacting interface.

10.2   Boundary lubrication

A boundary lubricant is a thin film of material that coats a surface and provides for a low coefficient of friction (typically $\mu < 0.2$) and often greater resistance to wear than the original uncoated surface. These films are typically 1–10 nm in thickness, and can be composed of layers of molecules or a solid film. Since boundary lubricant films are so thin, their bulk properties (like the viscosity) are relatively unimportant; instead the lubrication characteristics are determined more by the film's chemical composition and structure and by the underlying substrate.

As we saw with the discussion of the Stribeck curve in Chapter 9, a bearing designed to run with hydrodynamic lubrication will operate in the boundary lubrication regime when the fluid film thickness between the sliding surfaces becomes small enough for solid–solid contact to occur (Figures 9.1 and 9.2). The fluid film thickness is reduced either by increasing the loading force on the bearing, decreasing the sliding speed, or using a fluid with a low viscosity. Commercial lubricants typically include molecular additives—including friction modifiers, anti-wear additives, and extreme pressure additives—that help form boundary lubricant films under sliding action. If a boundary lubricant layer is not present, often very high friction, as well as severe wear or scuffing, occurs when operating in the boundary lubrication regime.

For surfaces sliding against each other in dry conditions—that is, without a liquid lubricant film—a good boundary lubricant film is often essential for achieving low friction and wear.

10.2.1 Boundary lubricant materials

Since the film thickness of a boundary lubricant is typically much less than the height of the surface roughness, the true area of contact A_r is little modified when a boundary lubricant film exists on one of the contacting surfaces. As discussed in Section 4.2.1, when two solid surfaces slide over each other in the absence of hydrodynamic lubrication, the tangential resistance in the absence of plowing comes from *adhesive friction* F_{adh}, which equals the real area of contact times the shear strength s of the contact junction (eq. (4.2)):

$$F_{adh} = A_r s. \tag{10.4}$$

It then follows that the main attribute that provides for low friction values of a boundary lubricant is a low shear strength s. This mechanism is illustrated in Figure 10.8. Furthermore, these films reduce the wear of the materials by preventing direct metal-metal contact (which are susceptible to adhesive wear), and by reducing the contact stresses.

Some categories of commonly used low shear strength materials for providing good boundary lubrication:

- *Organic films*—While many types of molecules can provide good boundary lubrication, some of the best belong to the category of long chain hydrocarbons that have an active end-group such as a fatty acid, alcohol, or amine end-group. These types of end-groups readily adhere to metals and oxides, causing the molecules to form close packed structures with their long chains oriented away from the surface, as illustrated in Figure 10.9. The other end of the hydrocarbon chain is typically terminated by a neutral, non-polar end-group, such as the $-CH_3$ methyl group, that provides a low shear strength surface for sliding.

- *Soaps and greases*—When an organic lubricant with a fatty acid end-group reacts with a metal, it forms a *metallic soap*, with a new polar end-group consisting of an acidic anion and metallic cation. Metallic soaps tend to provide better boundary lubrication than their fatty acid precursors. Metallic soaps can either form naturally when a fatty acid molecule is adsorbed onto a bare metal surface (though an elevated temperature may be required for the reaction to occur) or be premanufactured and deliberately deposited onto surfaces to serve as a boundary lubricant. Metallic soaps

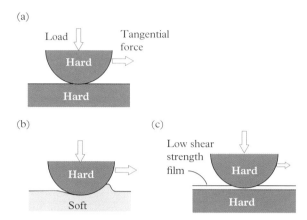

Figure 10.8 *(a) When two hard materials are slid across each other with a loading force, the contact area A_r will be relatively small as the hard materials have high elastic moduli. For ideal metals, the shear strength is 1/6 the hardness; the coefficient of friction from adhesive friction is then $\mu_{adh} = 1/6$, as explained in Section 4.2.1. (b) When a hard metal is slid against a soft metal, even though the contact area goes up, shear strength goes down proportionally and friction coefficient remains $\mu_{adh} = 1/6$. (c) Placing a low shear strength film between two hard materials lowers the tangential force for inducing sliding, as the contact area A_r remains similar to the hard-on-hard case shown in (a), but the lower value of the shear strength s means a much lower value of adhesive friction $F_{adh} = A_r s$. Moreover, the film can also protect the materials from wear by preventing direct metal–metal adhesive contact, and reducing normal and shear stresses at the contact points.*

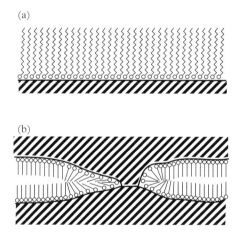

Figure 10.9 *(a) Self-assembled or Langmuir–Blodgett monolayer with the polar or reactive end-group adhering to a solid surface, and the closely packed alkane chains standing upright. (b) Bowden and Tabor's concept of how a boundary layer lubricant is displaced from an asperity contact (Bowden and Tabor 1950).*

(a) Structure of graphite

(b) Structure of molybdenum disulfide (MoS$_2$)

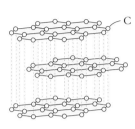
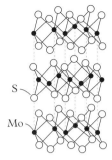

Figure 10.10 *Atomic crystal structure of (a) graphite and (b) molybdenum disulfide (MoS2). These two lamellar solids are effective solid lubricants since the basal crystal planes consist of strong covalent bonding within the planes (represented by solid lines) and weak van der Waals bonding between the planes (represented by the dashed gray lines), providing a low shear strength when these crystal planes slide over each other.*

are also commonly used as thickening agents for lubricant greases, where the soap's semi-solid character provides the grease with a slight rigidity to hold it in place; the metallic soap within the grease also serves as a boundary lubricant additive for the grease's base oil.

- *PTFE (polytetrafluoroethylene)*—PTFE or Teflon is a polymeric material that is widely used as a dry boundary lubricant. Due to its low surface energy, contacting surfaces coated with PTFE exhibit a very low interfacial adhesion, which makes it easier to slide them against each other. When an uncoated surface contacts with a PFTE coated surface, PTFE molecules are immediately transferred to the coated surface, and consequently the slip plane always occurs with PTFE molecules sliding over each other. During the sliding, the axes of the PTFE molecules become aligned in the direction of sliding resulting in very low coefficients of friction ($\mu < 0.05$).

- *Lamellar solids*—Graphite and molybdenum disulfide (MoS$_2$) are lamellar solids that are widely used as solid lubricants. The layered structure of these compounds is illustrated in Figure 10.10. The reason why these two particular lamellar solids are able to provide low friction is that the interatomic bonding between the layers is very weak (mainly van der Waals bonding), while the bonding within individual layers is much stronger, as it comes principally from covalent bonding. Consequently, these compounds have a low shear strength when the layers are slid over each other.

- *Thin metallic films*—While metal bearings are typically designed to operate with hydrodynamic lubrication, there can be times when operating conditions dictate that the bearing be run in the boundary lubrication regime of the Stribeck curve for short periods of time. For example, this occurs when relative motion with the bearing is first initiated or terminated during the start-up or shut-down of the machine incorporating the bearing. Plating or coating one of the bearing surfaces with a thin film of a low shear strength metal or alloy can help in these situations to reduce friction and wear. Since a shear strength of a metal is about 1/6 its hardness, soft metals—like lead, tin, indium, silver, and gold—are typically used. Not only should the coating metal be chosen to have low hardness, but it should also adhere well to the surface it is being coated onto and have a weak interaction with the opposing surface.

- *Anti-wear and extreme pressure additives and friction modifiers*—Certain additives to engine and gear oils are designed to react with metal surfaces to form thin, protective, solid-like films that improve the load-carrying capacity of the contacts by reducing friction and wear. These kinds of additives include:

  o *anti-wear additives*—if they reduce wear at moderate loads and temperatures,

  o *extreme pressure additives*—if they reduce wear at high loads and temperatures.

  o *friction modifiers*—if they reduce friction under at least some conditions or reduce the tendency for stick-slip behavior.

Some of these additives—such as the fatty acid ester glycerol monooleate (GMO), which is an effective friction modifier—also belong to the organic film-forming category described above. More typically though, rather than forming a film of intact molecules, they form a *tribofilm*, where tribochemical reactions dramatically modify the chemical structure of the additive molecules in the process of forming these films. In other words, when molecules of a tribofilm additive are confined between sliding asperities and subjected to compressive and shear stresses, they undergo chemical reactions that lead to the formation of films bound to the surfaces. Two common examples of such additives that form tribofilms are molybdenum dithiocarbamate (MoDTC), a friction modifier which forms a MoS_2-like film, and zinc dialkyldithiophosphate (ZDDP), an anti-wear additive which forms a Zn- and S-containing phosphate glass that protects steel surfaces against wear. These films are often soft enough to allow for low shear-strength sliding; the worn areas of the film are quickly replenished by new additive molecules arriving at the surface and undergoing tribochemical reactions.

10.2.2 Molecular mechanisms of boundary lubrication

For hydrodynamically-lubricated bearings, a small concentration (0.1–1 wt. %) of boundary lubricant molecules is typically added to the base lubricant oil as a friction modifier to provide a source for a molecular boundary layer. For situations where the contacting surfaces cannot be immersed in a liquid lubricant as in a hydrodynamic bearing, a thin film of the lubricant molecules can be applied to the sliding surfaces in advance, in order to minimize friction, adhesion, and wear during use.

When organic molecules are used as a boundary lubricant, they are often a long chain hydrocarbon with a reactive or polar end-group that typically adsorbs on a surface with a structure illustrated in Figure 10.9(a). As mentioned previously: these long chain molecules adsorb with their reactive or polar end-group adhering strongly to the solid surface; close packing of the chains causes them to line up perpendicularly to the surface; and neutral, non-polar terminal end-groups present a surface with low surface energy and low shear strength. When opposing surfaces of such layers are brought into contact (Figure 10.9(b)), both the molecular adhesion between the chains from van der Waals forces and the adhesion from the bonding of the end-group to the surface provide substantial resistance to penetrating asperities, forcing most of the shearing to take place at the interface between the boundary lubricant layers, except for those asperity summits where the lubricant layer has been displaced by the high contact pressures.

This concept of boundary lubrication originates from William Hardy (1864–1933) who in the 1920s studied lubrication by long chain hydrocarbons with various reactive end-groups (Hardy and Bircumshaw 1925); some of his results are shown in Figure 10.11. He found that, on metal surfaces, the hydrocarbons with reactive end-groups had lower friction coefficients than those without them, and that the friction coefficients decreased with increasing molecular weight. Friction coefficients as low as 0.05 could be obtained for the longest chain molecules, much lower than for unlubricated surfaces.

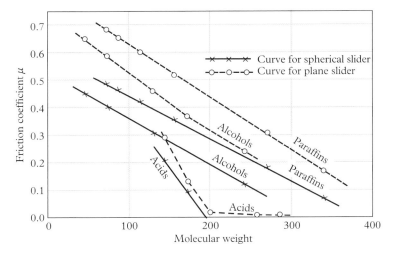

Figure 10.11 *Dependence of the friction coefficient on the molecular weight of a boundary lubricant. Reprinted from Hardy and Bircumshaw (1925).*

In more recent years, the molecular details of the type of boundary lubrication illustrated in Figure 10.9 (self-assembled or Langmuir–Blodgett monolayers with upright alkane chains) have become much better established (Carpick and Salmeron 1997, Barrena et al. 1999, Tutein et al. 2000, Perry et al. 2001, Salmeron 2001, Chandross et al. 2004, Mikulski et al. 2005). These studies confirm that low friction and strong resistance to penetration are maximized when the alkane chains are densely packed, as this packing maximizes the van der Waals cohesion between the chains. This effect is seen for chain lengths up to approximately 12–18 carbon atoms (Lio et al. 1997, Major et al. 2003), at which point the inter-chain van der Waals interactions that drive the packing saturate; thus, beyond this length, the chains do not exhibit increased ordering. Higher friction occurs for situations with lower packing densities, such as for short chain lengths or mixtures of different chain lengths; in this case, the higher friction originates from the chain segments having more room to adopt configurations away from a straight linear chain, providing more channels for energy dissipation.

It was originally suggested by Bowden and Tabor that molecules with long alkyl chains remain as rigid rods when squeezed between contacting surfaces (Figure 10.9); now the current evidence indicates that the chains first bend at their terminal $-CH_3$ methyl groups when compressive contact pressure is applied. This is seen in the image from the molecular dynamics simulation by Mikulski et al. (2005) shown in Figure 10.12; in this simulation the terminal methyl groups are found to bend away from the sliding interface at moderate contact pressures. Higher friction has been reported for methyl-terminated alkyl chains with odd numbers of carbon atoms compared to even numbers in these simulations (Mikulski et al. 2005) and in other molecular dynamic simulations (Ramin and Jabbarzadeh 2012b, 2012a). This difference is attributed to the terminal group for odd-numbered chains having a more upright configuration, which allows the terminal group to undergo more severe deformations when loaded, producing more gauche defects. AFM friction measurements for odd- and even-numbered chains, however, have produced contradictory results, with two papers (Kim and Houston 2000, Yang et al. 2010) reporting higher friction for odd-number chains and one paper (Wong et al. 1998) reporting lower friction.

When a sharp asperity contacts one of these monolayer films consisting of upright linear chains, additional types of structural changes occur within the molecular film around the edges of the contact zone. Molecular simulations contribute to an overall physical picture for this type of single asperity contact where the alkyl chains exhibit two types of responses to contact with an asperity:

- The alkyl chains directly underneath the asperity are compressed, which produces gauche defects and other structural changes to accommodate the deformation and which leads to increased friction.

- In contrast, chains near the perimeter of the contact area are splayed away from the excluded volume. At low loads, this produces an increased local packing density which can reduce friction locally, while at high loads, these chains can be displaced and removed from the surface, leading to high friction and to wear of the monolayer itself.

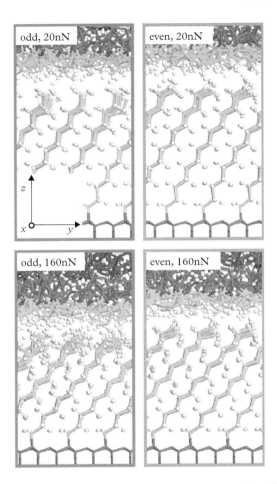

Figure 10.12 *Snapshots from molecular dynamics simulations of odd (left) and even (right) numbered alkyl chains bonded to a diamond surface and compressed from above by an amorphous carbon layer at two different loads. The wireframe format represents carbon atoms. Hydrogen atoms in the alkyl chains and the upper surface are small white spheres. Sliding is in the positive y direction. At the higher loads, a greater number of gauche defects occur for the terminal groups for the odd-numbered chains, which is attributed to the more upright configuration for odd-numbered chains which allows for severe deformations away from the surface normal. Reprinted from Mikulski et al. (2005).*

Figure 10.13 shows an example of the structural changes induced in an alkylsilane self-assembled monolayer due to a sliding sharp asperity as determined from a molecular dynamics simulation.

The molecular dynamics simulation results are consistent with the results of AFM and sum-frequency generation spectroscopy experiments of Carpick and Salmeron (1997), Barrena et al. (1999), Perry et al. (2001), and Salmeron (2001). Figure 10.14 shows schematically a few of the molecular structures adopted at different levels of compression

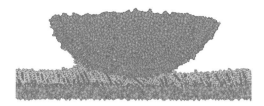

Figure 10.13 *Snapshot from a molecular dynamics simulation of a silicon oxide asperity, 10 nm in radius, in contact with an alkylsilane self-assembled monolayer (SAM) sliding to the left for 2 ns at a speed of 2 m/s under a normal load of 10 nN. The image shows a 4 nm slice from the center of the SAM under the asperity. Directly under the tip, alkyl chains are compressed and exhibit kinks and gauche defects. At the edges of the contact, particularly on the left side (which is the leading edge), the SAM molecules are splayed laterally but are still mostly linear and are compressed together to a higher packing density. Reprinted with permission from Chandross et al. (2008), copyright 2008, American Chemical Society.*

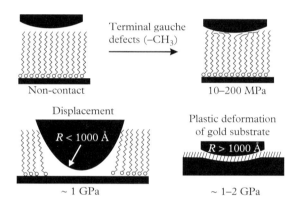

Figure 10.14 *Structure of a boundary lubricant layer subjected to different levels of compression. At light pressures of 10–200 MPa, the molecular chains remain upright but the terminal CH$_3$ groups develop gauche defects (top right). A sharp AFM tip or sharp asperity displaces the lubricant layer and penetrates the film when the contact exceeds about 1 GPa (bottom left). For blunt contacts, the molecules remain trapped in the contact zone and the solid surfaces plastically deform around them (bottom right). Reproduced with permission from Carpick and Salmeron (1997). Copyright 1997, American Chemical Society.*

by linear chain boundary lubricants (Carpick and Salmeron 1997, Perry et al. 2001, Salmeron 2001). For relatively light contact pressures of 10–200 MPa, the terminal $-CH_3$ groups are bent over in a gauche distortion that slightly reduces the film thickness while maintaining packing density (see also Figure 10.12). With increasing pressure under the contacting asperity, the molecular layer compresses by the linear chains collectively adopting discrete tilt angles away from normal (not shown in Figure 10.14). For sharp asperity contacts with $R < 100$ nm, the asperity breaks through the boundary lubricant layer, displacing all the lubricant molecules from the contact zone to make solid–solid contact when the contact pressure exceeds a threshold on the order of 1 GPa. For blunt contacts (typically when $R > 100$ nm), the lubricant molecules remain trapped in the contact zone while the solid surfaces undergo plastic deformation.

10.2.3 Example: growth of a ZDDP anti-wear boundary layer

One shortcoming of using the organic films described above as boundary lubricants is that they typically lose their effectiveness at temperatures between 80°C to 150°C depending on the situation. Fortunately, numerous additives have been designed to provide friction reduction and anti-wear protection at temperatures above those where organic films are effective. As mentioned above, anti-wear and extreme pressure additives are either molecules or nanoparticles which form a beneficial surface-bound tribofilm at the sliding interface through the combined action of compressive and shear contact stresses. Such films can be very effective in reducing wear in the boundary lubrication regime, particularly at high temperatures and contact pressures. However, the tribochemical processes responsible for the growth and eventual chemical structure of most tribofilms remain poorly understood.

Much of the work on understanding the growth mechanisms and their eventual impact on the tribological performance of tribofilms has focused on ZDDP, which is the most common anti-wear additive used in engine oils. Figure 10.15 illustrates schematically the structure of a ZDDP tribofilm.

Even though recent studies have helped determine the structure and composition of these films (Grossiord et al. 1999, Aktary et al. 2002, Spikes 2004, Pereira et al. 2006,

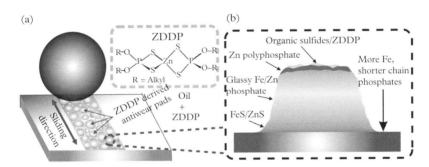

Figure 10.15 *The formation and structure of anti-wear tribofilms. (left) In a sliding contact (idealized here as a ball on a flat) where sliding occurs immersed in oil containing the anti-wear additive ZDDP, molecules and molecular fragments of ZDDP trapped at the sliding interface are subjected to compressive and shear stresses. The stresses and heat combine to cause the molecules to react with the surface and with each other to form tribofilms along the contact path. ZDPP tribofilms often appear patchy, with pad-like structures. Inset: Schematic of a ZDDP molecule, where R represents an alkyl ligand that is often varied. (right) Cross-section of a typical ZDDP tribofilm pad, showing the compositionally-graded structure which is typically 50–150 nm in thickness (for clarity, the z dimension is greatly exaggerated). Reprinted with permission from Gosvami et al. (2015), Copyright 2015, American Association for the Advancement of Science.*

Martin et al. 2012), the mechanisms that drive the nucleation and growth of ZDDP films remain poorly understood, despite more than 50 years in commercial use. Experimental work has shown that contact stresses, both in compression (Mosey et al. 2005) and in shear (Zhang and Spikes 2016), are essential for forming the film. Some work indicates that nanoparticle debris from the steel substrate can be incorporated into these tribofilms and can catalyze the reactions that form the film (Martin 1999). However, several key questions still remain: for example, why do the ZDDP tribofilms develop with a graded composition and correspondingly graded mechanical properties, and why do they exhibit a self-limiting thickness in the range 50–150 nm (Bec et al. 1999)? Next we discuss some results from AFM experiments by Gosvami et al. (2015) that help to elucidate the nanoscale mechanisms of how these macroscopic ZDDP tribofilms form with these properties.

In these experiments, an AFM operates in a lubricant oil formulation with 0.8 wt. % ZDDP, which is heated to typical engine temperatures ~80–150°C (these temperatures are known to be required to form robust ZDDP-derived tribofilms). When the tip slides on the substrate, some ZDDP molecules (or their thermal decomposition products that are present in the oil) become confined at the interface, and undergo stress-assisted chemical reactions that build up a film only where sliding contact occurs.

Gosvami et al. (2015) observed that the measured growth rate $\Gamma_{\text{growth-rate}}$ (nm^3/s) fits well to a stress-activated Arrhenius model, as shown in Figure 10.16; that is, the tribofilm growth rate follows Arrhenius rate law kinetics from transition state theory:

$$\Gamma_{\text{growth-rate}} = \Gamma_0 \exp\left(-\frac{\Delta G_{\text{act}}}{k_B T}\right), \tag{10.5}$$

where the pre-factor Γ_0 depends on the effective attempt frequency of the reaction, ΔG_{act} is the free activation energy barrier of the rate-limiting reaction in the growth process, k_B is the Boltzmann's constant, and T is the absolute temperature. The activation barrier ΔG_{act} depends on the applied stresses through the relationship

$$\Delta G_{\text{act}} = \Delta U_{\text{act}} - \sigma \Delta V, \tag{10.6}$$

where ΔU_{act} is the energy barrier in the absence of stress, σ is mean value of the stress component affecting the activation barrier, and ΔV is the activation volume (the characteristic volume involved in transforming from the reactants to the transition state of the reaction). The stress component σ can be either the contact pressure or the shear stress, or some combination of these. This stress dependence of an energy barrier comes from reaction rate theory (Hanggi 1990), and is a well-known effect in many thermally-activated processes including the motion of dislocations (Gibbs 1965) and the process of wear via bond breaking (Jacobs et al. 2010, Jacobs and Carpick 2013). Reaction rate theory is further discussed in Chapter 12 in connection with wear rates.

The good fit of the experimental wear rate-vs.-temperature in Figure 10.16(a) to an exponential function indicates that this tribofilm growth is a *thermally activated* process described by eq. (10.5), and the good fit to wear rate-vs.-contact pressure in

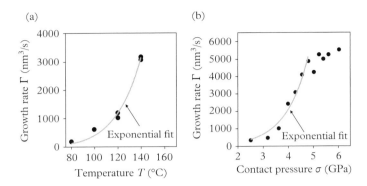

Figure 10.16 *The circles are ZDDP tribofilm growth rates within the contact area of an AFM tip sliding over a steel surface in a lubricant oil (a polyalphaolefin or PAO) with 0.8 wt. % ZDDP anti-wear additive mixed in. Growth experiments are done as a function of temperature (a) and contact pressure (b). The grey lines show the fit to the exponential dependence predicted by Arrhenius kinetics for the stress-assisted thermally activated process expressed by eqs. (10.5) and (10.6). Adapted with permission from Gosvami et al. (2015), Copyright 2015, American Association for the Advancement of Science.*

Figure 10.16(b) indicates that it is a *stress-assisted* thermally activated process described by the combination of eqs. (10.5) and (10.6) (Jacobs and Carpick 2013, Spikes and Tysoe 2015). While multiple reaction steps are surely involved, the fits to the data indicate that one step dominates, with $\Delta U_{act} = 0.8\pm0.2$ eV and $\Delta V = 3.8\pm1.2$ Å^3; these values are consistent with the energy and size of molecular-scale bond breaking or formation processes.

The strong dependence of the growth rate on contact pressure helps to explain the graded structure of the ZDDP film and why growth levels off after ~100 nm thickness. The elastic modulus of the fully formed ZDDP tribofilm is rather low (10–30 GPa, far less than that of the steel substrate (Bec et al. 1999)). Consequently, as the film grows, the effective modulus of the film and substrate combination gradually reduces below that of the steel substrate. Essentially, the film is acting as a cushion between the tip and substrate. This leads to a progressive reduction in contact pressure as the film grows, since the thicker the film, the less the influence of the substrate on the elastic properties. This lower contact pressure reduces the growth rate and thus the amount of crosslinking and other tribofilm-forming reactions, resulting in a graded structure and graded composition as the film grows and with the film having a weaker structure the thicker it grows. Eventually, the film thickness reaches a value where the stress and thermal energy are no longer sufficient to overcome the intrinsic energy barrier to the reactions and growth tapers off. Once this thickness is reached, the topmost layer is readily removed and modified, though it can be quickly replaced by new ZDDP species that get trapped in the contact.

10.2.4 Molecularly thin liquid boundary lubricant layers

The strong cohesion between the close packed chains of the boundary lubricants depicted in Figures 10.9, 10.12, 10.13, and 10.14 results in these lubricant layers behaving like a soft-solid film. While this strong cohesion is good for resisting penetration of contacting asperities, the soft-solid character means that these types of films are slow to recover from damage. So, while these types of films may protect the surface the first time a sliding contact passes over it, the film may be too damaged to protect the surface on subsequent passes unless a mechanism exists for quickly replenishing the displaced lubricant. Within bearings that are immersed in a lubricating oil, this replenishment is accomplished by having a sufficiently high concentration of the boundary lubricant in the base oil (typically 0.1–1%) to ensure rapid adsorption onto the exposed surface.

For surfaces that cannot be immersed in oil or covered with a grease containing a boundary layer additive, a closely packed molecular layer may not provide sufficient boundary lubrication for those interfaces that experience frequent repeated sliding contacts. In these cases, a more appropriate choice is to have a very thin liquid film (often only a monolayer in thickness) of lubricating molecules, as depicted in Figures 10.17 and 10.18. Then, when contact displaces lubricant from an asperity summit, the liquid lubricant molecules from the surrounding area flow or diffuse to cover the exposed surface. Since the molecules within these films maintain their liquid character, they tend not to form the ordered, upright structures, but instead have the much more disordered liquid molecular structure.

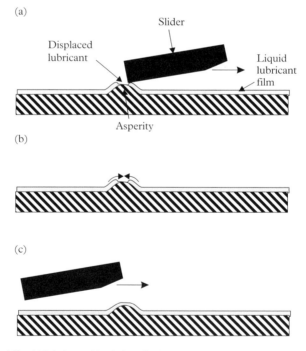

Figure 10.17 *(a) A liquid lubricant film being displaced from an asperity by a contacting slider. (b) Lubricant from the surrounding area then flows or diffuses to replenish the lost lubricant before (c) the slider recontacts the asperity on a subsequent pass.*

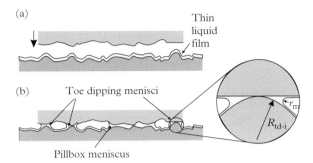

Figure 10.18 *(a) Prior to contact, the bottom surface is coated with a thin film of liquid lubricant. (b) After contact the lubricant redistributes to form "toe dipping" and "pillbox" menisci around the contact points.*

As discussed in the example below, reactive or polar end-groups help improve the lubricating properties, as the bonding of the end-group helps resist displacement during contact. These reactive end-groups also act as "anchors" impeding the diffusion of these molecules across surfaces, greatly slowing down the replenishment process (Min et al. 1995, Ma et al. 1999, Scarpulla et al. 2003, Mate 2013). Consequently, in technological applications (such as for the lubricant film deposited on a disk in a disk drive) close attention is paid to controlling the degree to which the liquid lubricant molecules are "bonded" and the degree to which the remaining film is "free" to flow or to diffuse over a surface as this is an important factor for the lubricant mobility on the surface, which in turn determines the speed with which the lubricant can replenish that loss through contact (as illustrated in Figure 10.17) and the speed with which the lubricant can wick up around contact points to form menisci (as illustrated in Figure 10.18(b)).

For those lubricant films exposed to ambient air or vacuum conditions, the lubricants need to have very low volatility. For tribological systems in a sealed environment, however, more volatile lubricants can be used, and this can provide the additional replenishment mechanism of vapor redeposition. While used infrequently, vapor lubrication, where a reservoir of volatile boundary lubricant provides for a continual replenishment source, remains a potentially attractive means of providing boundary lubrication in sealed environments (Hanyaloglu and Graham 1994, Henck 1997, Gellman 2004, Barthel and Kim 2014).

10.2.4.1   *Example of the importance of end-groups in a liquid lubricant film*

The AFM provides a convenient way to study how molecular end-groups alter the surface forces and lubrication properties of a single asperity contact at the molecular level. Figure 10.19 shows the force vs. distance curves as an AFM tip is brought into contact with silicon wafers covered with thin liquid films of perfluoropolyether polymer lubricants, which differ slightly in that one has a reactive alcohol (–OH) end-group while the other has an unreactive and relatively neutral –CF$_3$ end-group.

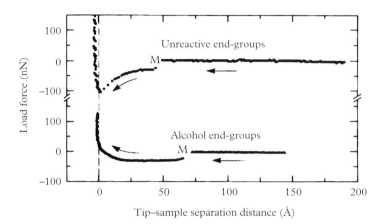

Figure 10.19 *Force versus separation distance as a tungsten AFM tip with ~100 nm radius approaches a silicon wafer covered with 30 Å thick perfluoropolyether polymer lubricant films that have either (top) unreactive −CF₃ end-groups or (bottom) reactive alcohol end-groups. These perfluoropolyether lubricants (with the chemical structure R–CF₂(OC₂F₄)ₘ(OCF₂)ₙO–R, where R=−CF₃ or −CH₂OH) are linear chain polymers, about 100 Å in length and 7 Å in diameter, and are commonly used as lubricants in disk drive, aerospace, and MEMS applications. The zero separation distance is defined to occur at the onset of hard wall repulsion. A negative load force corresponds to an attractive force acting on the tip. Reprinted with permission from Mate (1992b). Copyright 1992, by the American Physical Society.*

When the tip first contacts the lubricant films at point M on the curve, a sudden attractive force is observed due to the liquid lubricant's forming a meniscus around the tip, as discussed in Section 6.3.1.1. For the lubricant with unreactive end-groups, the normal force becomes more attractive with decreasing separation distance. This attractive force is the sum of the capillary force from the meniscus and the solid–solid attraction, mostly likely the van der Waals force mediated by the intervening liquid lubricant. The force quickly turns repulsive once the hard wall of solid–solid contact is reached.

When the lubricant has alcohol end-groups, a very different force behavior is observed as the tip penetrates the lubricant film: the normal force becomes increasingly more repulsive with decreasing separation distance, indicating that the attractive van der Waals force is counteracted by a repulsive force associated with the alcohol end-groups. This repulsive force comes first from the compression of the polymer molecules beneath the tip, followed by the force needed to overcome the hydrogen bonding between the alcohol end-groups and the silicon oxide surface. Figure 10.20 illustrates how the AFM tip interacts with these two polymer lubricants with different termination.

A major way in which reactive end-groups on lubricant molecules facilitate boundary lubrication is to dramatically increase in the contact pressure that a thin layer of lubricant molecules can support before the lubricant is squeezed out and solid–solid contact occurs. For the two lubricants shown in Figure 10.19, an extra 100 nN load force has to

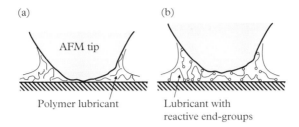

Figure 10.20 *Schematic diagram of a polymer lubricant forming a liquid meniscus around an AFM tip or asperity contact. (a) If the lubricant has neutral end-groups, it is easily squeezed out. (b) If the lubricant has reactive or polar end-groups, it resists being squeezed out due to the bonding forces between the end-groups and the tip and sample surfaces.*

be exerted on the tip to make hard wall contact for the polymer lubricant with alcohol end-groups compared to the same polymer with neutral end-groups. As the contact zone underneath this AFM tip is roughly about 10^2 nm^2 in area, this extra 100 nN load corresponds to an average contact pressure of 1 GPa for squeezing out the lubricant with alcohol end-groups.

10.3 Physics of lubricant menisci

As mentioned in previous chapters, for contacting surfaces not under vacuum or immersed in a liquid, menisci can form around the contact points, either from the capillary condensation of vapors or the migration of adsorbates toward the contact points. Rather than relying on this adventitious adsorption of contaminants for lubrication, which can lead to unpredictable results, tribologists often use the following methods in order to achieve better control of the material that ends up in these menisci:

- Control the humidity surrounding the contacting surfaces to control the degree to which water vapor condenses around the contact point to form menisci, and/or
- Apply a thin film of mobile lubricant to the contacting surfaces (as discussed in the previous section) so that a known lubricant ends up forming the menisci around the contact points.

In the next couple of sections, we discuss the physics of how these menisci form and the adhesion and friction forces that result from them. First, we cover physics of liquid lubricant films and the impact of the resulting menisci; then we cover how menisci from the capillary condensation of vapors impact adhesion and friction at sliding surfaces.

10.3.1 Disjoining pressure of a liquid lubricant film

An important design parameter for a liquid lubricant film is determining the optimal amount or film thickness to be deposited, in that enough lubricant needs to be deposited so as to provide adequate lubrication for the life of the device, but not so much that it results in high adhesion and friction forces or that it creates other problems. We start with the physics of menisci formation, then we discuss how these menisci generate the adhesion and friction forces.

As illustrated in Figure 10.18, after contact is made, the liquid lubricant film redistributes to form menisci around the contact points. The two counteracting forces affecting this redistribution are the capillary pressure, which pulls the liquid lubricant into the menisci surrounding the contact points, and the disjoining pressure, which draws the liquid lubricant out of the meniscus and back onto the solid surfaces (Mate 1992a, 2011). The physical origins of capillary pressure were discussed in Section 5.2. The disjoining pressure of a liquid film was briefly discussed in Section 7.3.8 in connection with the van der Waals interaction between the film and substrate; in this section, we present a fuller discussion of this fundamental parameter of a liquid lubricant film.

The disjoining pressure $\pi(h)$ represents the interaction energy per unit volume between a liquid film and the adjacent solid substrate, as measured relative to the interaction energy experienced by the same liquid molecules at the surface of the bulk liquid. Formally, the disjoining pressure is defined as the negative derivative of the Gibbs energy per unit area with respect to film thickness; but, informally, the disjoining pressure can be thought of as the extra force F' experienced by a molecule at the surface of the liquid film due to the presence of the nearby substrate divided by the area A_{mol} of that molecule

$$\pi(h) = F'/A_{\text{mol}}, \tag{10.7}$$

where h is the thickness of the liquid film. This is illustrated in Figure 10.21.

If we adopt the sign convention of F' being positive when the substrate exerts an attractive interaction on the liquid film higher than that of the bulk liquid, then a positive disjoining pressure π corresponds to an attractive interaction that acts to thicken the liquid film. Such an attractive interaction acts to pull liquid out of reservoirs and to thicken the liquid film surrounding this reservoir (Robbins et al. 1991). This is illustrated

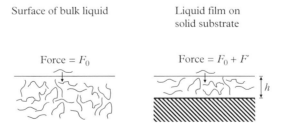

Surface of bulk liquid   Liquid film on
                                solid substrate

Force = F_0 Force = $F_0 + F'$

Figure 10.21 *A simplistic view of disjoining pressure. A polymer molecule on the surface of the bulk liquid experiences a force F_0 pulling that molecule towards the liquid, while the same molecule on the surface of a liquid film experiences a slightly different force $F_0 + F'$. The disjoining pressure corresponds to F'/A_{mol}, where A_{mol} is the molecular area.*

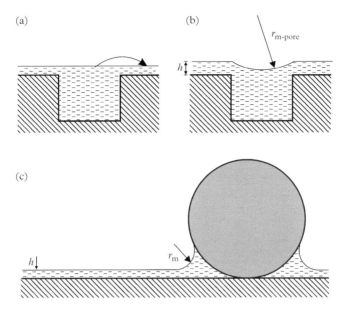

Figure 10.22 *How a liquid redistributes over a pore opening. (a) If the liquid surface is initially flat over the pore, the attraction of the solid for the liquid (positive disjoining pressure) pulls the liquid out of the pore, thickening the liquid film around the pore opening. (b) Equilibrium is reached when the capillary pressure of the liquid in the pore balances the disjoining pressure of the liquid film. (c) Liquid from a thin film with thickness h wicking up around a spherical particle on a solid surface.*

in Figure 10.22, where the reservoir is either the meniscus surrounding a contact point or a surface cavity filled with liquid. This attractive interaction also provides a driving force, in addition to surface tension, that pulls flat a non-uniform liquid film.

If the strength of the interaction between the substrate and the liquid film is less than that of the liquid with itself, the disjoining pressure π is negative, and the interaction is considered repulsive. With such a repulsive interaction, the substrate repels the liquid in the film, causing the film to thin, with the excess liquid dewetting as droplets (Kim et al. 1999).

Following Derjaguin and Churaev (1974), the disjoining pressure can be represented as the sum of three main contributors

$$\pi(h) = \pi_{\text{VDW}} + \pi_{\text{bond}} + \pi_{\text{struct}}, \tag{10.8}$$

where the π_{VDW} is the contribution from the van der Waals forces acting between the film and the substrate, π_{bond} is the contribution from the other bonding interactions between the liquid molecules and substrate (electrostatic, covalent, ionic, hydrogen bonding, etc.), and π_{struct} is the contribution from the liquid molecules having a different structure in a film than in the bulk liquid.

For lubricant films without reactive or polar functional groups, the van der Waals term π_{VDW} usually dominates the disjoining pressure. This is also the easiest term for which to derive a mathematical expression (as discussed in Section 7.3.8) and is expressed as

$$\pi_{vdw}(h) = \frac{A_{SLV}}{6\pi h^3},\tag{10.9}$$

where A_{SLV} is the Hamaker constant for this solid–liquid–vapor system. Equation (10.9) predicts that the van der Waals disjoining pressure diminishes rapidly with a h^{-3} dependence as the film thickness h increases. Also, $\pi \to 0$ as $h \to \infty$, which is to be expected since the film properties must approach those of the bulk liquid as $h \to \infty$. As discussed in Chapter 7, the Hamaker constant A_{SLV} can either be theoretically calculated by several methods or estimated from values in the literature. For example, for thin films of perfluoropolyether lubricants of the type used in disk drives, A_{SLV} has been found by experiment and theory to lie in the range from 0.2 to 1 $\times$ 10^{-19} J (Mate 2011). The dashed line in Figure 10.23 is for the case where $A_{SLV} = 10^{-19}$ J.

As shown for the example in Section 10.2.4.1, using lubricants with reactive functional groups can greatly improve the lubricating properties of liquid lubricant films. However, the addition of these functional groups to the lubricant can lead to films where the π_{bond} and π_{struct} terms in eq. (10.8) are no longer insignificant. Unfortunately, as of 2019 no comprehensive theories exist which quantitatively predict how these $\pi_{bond}(h)$ and $\pi_{struct}(h)$ terms vary as a function of lubricant film thickness h (Mate 2011). However, several functional forms have been developed for qualitatively explaining the behavior of liquid lubricant films. For example, Figure 10.23 shows several functional forms of $\pi(h)$ developed to explain the behavior of lubricants in disk drives:

- The $\pi_{bond}(h)$ term is chosen to have an exponential dependence on film thickness h similar to that for the electrostatic double layer interaction (Section 7.4.2.1) or the hydrophobic interaction (eq. (7.46) in Section 7.4.2.2):

$$\pi_{bond} = -B\exp(-h/\lambda).\tag{10.10}$$

- The $\pi_{struct}(h)$ term is chosen to have the same form as the solvation pressure (eq. (7.43) in Section 7.4.1) for spherical shaped molecules sandwiched between two smooth surfaces separated by a distance h, that is,

$$\pi_{struct}(h) = -C\cos\left(\frac{2\pi h}{h_m} + \alpha\right)\exp\left(-\frac{h}{h_m}\right).\tag{10.11}$$

This expression represents the tendency of the liquid molecules adjacent to a solid surface to form ordered layers with thickness h_m.

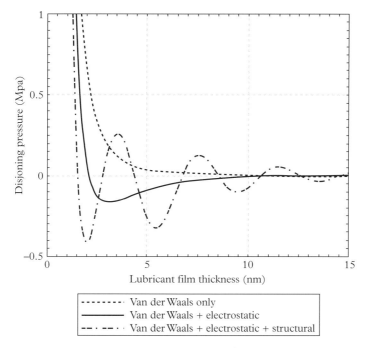

Figure 10.23 *Examples of different functional forms for disjoining pressure π (h) for perfluoropolyether lubricant films, illustrating how the addition of the π$_{bond}$(h) and π$_{struct}$(h) terms to the π$_{VDW}$(h) term can influence the shape of the disjoining pressure curve. The dashed line is for the case where the disjoining pressure is only due to van der Waals interactions: π$_{VDW}$(h) = A$_{SLV}$/6π h³ with A$_{SLV}$ = 10⁻¹⁹ J. The solid line plots π (h) = π$_{VDW}$(h) + π$_{bond}$(h) where π$_{bond}$(h) is from an electrostatic double layer interaction: π$_{bond}$(h) = −Bexp(−h/λ) with B = 1.6 MPa and λ = 2 nm. The dash-dot line plots the total disjoining pressure calculated from eq. (10.12) with C = 1 MPa, h$_m$ = 4 nm, and α = 1.2π. Reprinted with permission from Mate (2011). Copyright 2011, IEEE.*

Adding these expressions for $\pi_{\text{bond}}(h)$ and $\pi_{\text{struct}}(h)$ to that for van der Waals interaction leads to following expression for the total disjoining pressure:

$$\pi(h) = \pi_{\text{vdw}} + \pi_{\text{bond}} + \pi_{\text{struct}}$$

$$= \frac{A_{\text{SLV}}}{6\pi h^3} - B\exp\left(-h/\lambda\right) - C\cos\left(\frac{2\pi h}{h_{\text{m}}} + \alpha\right)\exp\left(-\frac{h}{h_{\text{m}}}\right), \qquad (10.12)$$

which is plotted as the dot-dashed line in Figure 10.23.

From Figure 10.23 we see that, if the magnitude of the B and C parameters describing these electrostatic and structural forces are large enough, the disjoining pressure is negative for some ranges of film thickness h, which leads to dewetting of the lubricant into droplets on the disk surfaces, consistent with experimental observations (Kim et al. 1999, Mate 2011). For disk lubricants in disk drives lubricant dewetting is avoided by using lubricant films with a thickness less than its first crossing of disjoining pressure $= 0$, that is, using a film thickness less than the dewetting thickness. (For modern disk drive lubricants, the dewetting thickness ranges from 1.5 to 3.0 nm (Mate 2011, Guo et al. 2012).)

Also, we see from Figure 10.23 that, as $h \to 0$, the different disjoining pressure curves become dominated by the van der Waals contribution. Because of this and the inherent simplicity of the expression in eq. (10.9), it is quite common to only use the van der Waals contribution to the disjoining pressure when calculating the behavior of lubricant films when their thickness is thinner than the dewetting thickness. This is the approach that we will take in the follow sections.

10.3.2 Equilibrium distribution of a liquid film

Figure 10.22(a)-(b) illustrates how a liquid film redistributes itself around a pore opening on a surface so as to minimize energy. At equilibrium, the differential in energy dE equals zero or

$$dE = \left(\mu_{\text{surface}} - \mu_{\text{pore}}\right) dn + P_{\text{cap}} dV = 0, \tag{10.13}$$

where μ_{surface} and μ_{pore} are the chemical potentials of the liquid on the surface and in the pore and P_{cap} is the capillary pressure of the liquid in the pore. The first term in the middle portion of eq. (10.13) represents how the potential energy of the liquid molecules changes as they move from the pore to the film on the surrounding surface; the second term represents the work done by the capillary pressure during this move. As the disjoining pressure π represents the interaction energy per unit volume between the liquid film and the surface with respect to the bulk liquid in the pore, we can write

$$-\pi \, dV = \left(\mu_{\text{surface}} - \mu_{\text{pore}}\right) dn. \tag{10.14}$$

So, the equilibrium condition eq. (10.13) becomes simply

$$\pi(h) = P_{\text{cap}} \left(r_{\text{m-pore}}\right). \tag{10.15}$$

Therefore, the liquid redistributes itself until the disjoining pressure of the liquid film around the pore opening equals the capillary pressure in the pore opening. Equation (10.15) is a fairly general result, applicable to other situations where a thin liquid film is in equilibrium with a nearby meniscus, such as the meniscus that forms from a liquid wicking up around the spherical contact as illustrated in Figure 10.22(c).

For a pore with a circular opening where the meniscus has a spherically-shaped surface with a radius of curvature $r_{\text{m-pore}}$ over the opening, the capillary pressure is given by eq. (5.5): $P_{\text{cap}} = 2\gamma_{\text{L}}/r_{\text{m-pore}}$, where γ_{L} is the liquid surface tension. If the liquid film's disjoining pressure is from van der Waals attraction (eq. (10.9)), then, when a liquid film with thickness h surrounds the pore opening (Figure 10.22(b)), the equilibrium meniscus radius $r_{\text{m-pore}}$ is given by

$$\frac{A_{\text{SLV}}}{6\pi h^3} = \frac{2\gamma_{\text{L}}}{r_{\text{m-pore}}}$$

$$r_{\text{m-pore}} = \frac{12\pi \gamma_{\text{L}} h^3}{A_{\text{SLV}}}. \tag{10.16}$$

How small can this meniscus radius $r_{\text{m-pore}}$ be? The smallest meniscus radius occurs for the thinnest liquid film that is still continuous, about one layer of molecules lying flat or $h \sim 0.5$ nm. If we also assume that $\gamma_{\text{L}} = 20$ mN/m and $A_{\text{SLV}} = 10^{-19}$ J, then the smallest possible radius would be

$$r_{\text{m-pore}} = \frac{12\pi \, (20 \text{ mN/m}) \, (0.5 \text{ nm})^3}{10^{-19} \text{ J}}$$

$$= 0.94 \text{ nm}. \tag{10.17}$$

Equation (10.16) indicates that $r_{\text{m-pore}} > h$ for any liquid film thickness greater than one molecular layer.

An analysis similar to that done for eq. (10.16) can also be done for a liquid covering a groove on a surface, except that in this case the capillary pressure is given by

$$P_{\text{cap}} = \frac{\gamma_{\text{L}}}{r_{\text{m-groove}}}, \tag{10.18}$$

where $r_{\text{m-groove}}$ is the radius of the cylindrically shaped meniscus over the groove. In Section 7.3.8, Figure 7.13 showed an example where the lubricant distribution over such a groove was used to measure the disjoining pressure as a function of film thickness on the surrounding surface (Fukuzawa et al. 2004).

10.3.3   Lubricant distribution between contacting surfaces

Figure 10.18 shows schematically how a thin liquid film wets the gap between rough contacting surfaces. As discussed in Section 6.5, the different wetting regimes between contacting surfaces are labeled toe dipping, pillbox, flooded, and immersed (Matthewson and Mamin 1988).

As illustrated in the left-hand inset of Figure 10.24, when the equilibrium meniscus radius r_{m} is less than $D/2$ (assuming that $h << D$ and where D is a typical gap separation between the surfaces), the system conforms to the toe dipping regime. If the interaction between the liquid film and the solid surfaces is predominately van der Waals attraction, then equating the capillary pressure in the meniscus around a toe dipping asperity contact to the lubricant film's disjoining pressure yields

$$P_{\text{cap}} = \frac{\gamma_{\text{L}}}{r_{\text{m}}} = \frac{A_{\text{SLV}}}{6\pi h^3}. \tag{10.19}$$

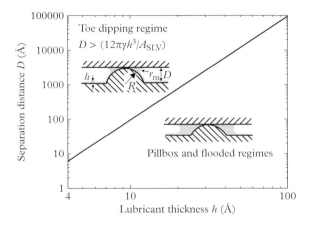

Figure 10.24 *The plot shows the boundary between the toe dipping and pillbox wetting regions for a liquid lubricant film with disjoining pressure solely coming from van der Waals interactions with a Hamaker constant of $A_{SLV} = 10^{-19}$ J. The lubricant surface tension $\gamma_L = 25$ mN/m, and assuming $h << D$. For conditions to the left of this line, a small toe dipping meniscus of the liquid lubricant forms around the contact point. When the lubricant thickness h is increased, the meniscus grows in size and enters the pillbox region to the right of the line when $2r_m > D$. Reproduced from Mate and Homola (1997) with permission from Springer Science and Business Media. Copyright 1997, Kluwer Academic Publishers.*

(This expression for P_{cap} assumes $r_m << R$ (Section 5.2.1).) The equilibrium meniscus radius around the contacting asperity is then

$$r_m = \frac{6\pi \gamma_L h^3}{A_{SLV}}. \tag{10.20}$$

From the discussion of eq. (10.17), we see that for typical values A_{SLV} the meniscus radius r_m is always larger than the lubricant film thickness h and increases rapidly as h increases.

   If the thickness h is large enough to generate a meniscus radius r_m greater than $D/2$ (still assuming that $h << r_m$ and D), the meniscus enters the pillbox regime—conditions to the right of the line plotted in Figure 10.24—and starts to flood the gap between the surfaces. In the pillbox regime, the capillary pressure is set by the separation distance: $P_{cap} = 2\gamma_L/(D - h)$, where we assume that a contact angle $= 0°$ on both surfaces. Initially, P_{cap} is higher than $\pi (h)$ and pulls lubricant into the pillbox shaped menisci. This flooding process continues until either the film thickness is reduced to the point where the disjoining pressure balances the capillary pressure or until the entire gap floods.

10.3.4   Meniscus force

While a thin film of liquid lubricant can be effective for lubricating surfaces that cannot be fully immersed in the lubricant, care must be taken to control the lubricant film thickness

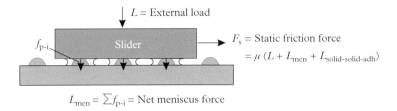

$L_{\mathrm{men}} = \Sigma f_{\mathrm{p\text{-}i}} = $ Net meniscus force

Figure 10.25 *How the meniscus force L_{men} adds to the external loading force L to increase static friction F_s.*

and the roughness of the contacting parts to ensure against excessive meniscus and friction forces. The meniscus force comes from the capillary pressure of the lubricant that wicks up around a contact point while the contacting parts are at rest.

Modeling this meniscus formation process can help guide in deciding what combination of lubricant film thickness and surface roughness best ensures that the meniscus and friction forces generated by the equilibrium distribution of the lubricant are in an acceptable range (Mate 1992a, Gui and Marchon 1995). Usually, the principal concern is that the meniscus force from the lubricant might result in excessive static friction. As illustrated in Figure 10.25, the net meniscus force L_{men} adds to the externally applied load L, which through Amontons' law adds to the friction force. From Amontons' law, the static friction F_s is given by

$$F_s = \mu\,(L + L_{\mathrm{men}} + L_{\mathrm{solid-solid-adh}}),\qquad(10.21)$$

where μ is the coefficient of friction and $L_{\mathrm{solid\text{-}solid\text{-}adh}}$ is the adhesion force from solid–solid contact within the menisci.

10.3.4.1   Example: stiction of a recording head slider

In Figure 10.26, the force needed to initiate sliding of a recording head slider on a disk surface is plotted as a function of the amount of time the slider has been sitting at rest on that disk. This static friction force is often referred to as *stiction*. The disk surface used in these experiments was textured with bumps fabricated by high power laser pulses; these laser-textured bumps have a uniform height of 38 nm and are spaced every 50 μm. (Laser texturing has been previously discussed in Sections 3.4.1.3 and 6.5.1.1.) The use of a laser-textured disk provides a well-defined separation of $D \sim 38$ nm over most of the apparent contact area, as illustrated in Figure 10.27.

From the plot in Figure 10.24, we can estimate that for $D \sim 38$ nm the crossover from the toe dipping and pillbox behavior should occur when the lubricant thickness is increased above 16 Å. The data in Figure 10.26 agree quite well with this crossover estimate: For lubricant thicknesses of 6 and 13 Å, the static friction remains constant at 18 mN out to rest times as long as 6000 minutes (4.2 days), consistent with toe dipping behavior, where the meniscus force is independent of the meniscus size. For a lubricant thickness greater than 16 Å, however, static friction increases with rest time, consistent

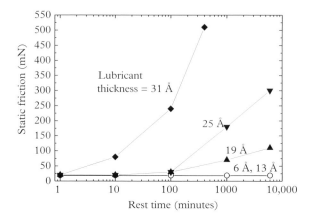

Figure 10.26 *Experimental measurements from author C.M. Mate's lab of the force needed to initiate sliding (static friction or "stiction") of a recording head slider after sitting on a laser-textured disk for different periods of time. The average height of the laser-textured bumps is 38 nm, making the separation distance D also 38 nm. The lubricant is the perfluoropolyether polymer with alcohol end-groups.*

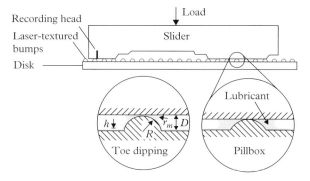

Figure 10.27 *Disk drive recording head slider sitting on a disk surface textured with a laser to have bumps of a uniformed height. Expanded views illustrate toe dipping and pillbox lubricant menisci forming around the bump–slider contacts.*

with pillbox menisci that slowly grow with time, increasing the meniscus loading force and the resulting static friction force. The thicker the lubricant film in this pillbox regime, the faster lubricant flows into the pillbox shaped menisci, and the faster static friction increases.

10.3.4.2   *Calculating the meniscus force*

To minimize the meniscus force and resulting static friction, the lubricant film thickness needs to be sufficiently thin and the roughness sufficiently large to ensure that the system stays in the toe dipping regime to the left of the line plotted in Figure 10.24. In the toe dipping regime, the meniscus force f_{td-i} acting on a single asperity contact with radius R_{td-i} is given by eq. (6.63)

$$f_{td-i} = 4\pi\gamma_L R_{td-i}. \tag{10.22}$$

(Note that this equation for the meniscus force assumes that the lubricant contact angle ~0°; this is reasonable as the lubricant needs to wet the contacting surfaces in order to spread as a thin film.) The total meniscus force L_{men} in the toe dipping regime is the sum over the individual meniscus asperity forces:

$$
\begin{aligned}
L_{men} &= \sum f_{td-i} \\
&= 4\pi\gamma_L \sum R_{td-i} \\
&= 4\pi\gamma_L n_{td} R_{td-ave}
\end{aligned}
\tag{10.23}
$$

where n_{td} is the number of asperities in contact and R_{td-ave} is their average radius. (This is the same result as eq. (6.64)). From eq. (10.23), we see that, when in the toe dipping regime, the total meniscus force depends only on the number of asperities in contact and their average radius and is independent of lubricant film thickness and separation distance. As pointed out in the example in Section 6.5.1.1, L_{men} and n_{td} are coupled: an increase in the number of contacting asperities n_{td} increases the total loading force $(L + L_{men})$, which leads to a further increase in n_{td}. This increasing loading force also reduces the separation distance D, and, at high enough loading force, portions of the contacting interface are pushed into the pillbox regime.

As the lubricant film thickness is increased or the separation between the contacting surfaces is made smaller, those portions of the contacting interface with the smallest separations eventually cross over to the pillbox side of the line plotted in Figure 10.24. In the pillbox regime, the meniscus force f_{p-i} generated by each pillbox shaped meniscus is given by the product of its area A_{p-i} and its capillary pressure $P_{cap-i} = 2\gamma_L/D_i$, where D_i is the separation distance at the edge of that pillbox meniscus and assuming $h << D_i$. In the pillbox regime, the net meniscus force L_{men} is the sum over the toe dipping and pillbox menisci

$$L_{men} = 4\pi\gamma_L \sum R_{td-i} + \sum \frac{2\gamma_L}{D_i} A_{p-i}. \tag{10.24}$$

Estimating the meniscus forces from pillbox menisci is tricky for several reasons:

- The pillbox meniscus areas A_{p-i} continue to grow until the disjoining pressure equilibrates with the menisci's capillary pressure. The meniscus growth rate depends on the flow of a viscous lubricant through nanometer thick films, which can be difficult to estimate. The time scales for these menisci to reach equilibrium can range from fractions of a second to many months. For example, in Figure 10.26, the pillbox menisci that form for lubricant thickness > 16 Å do not reached equilibrium even after 100 hours.

- As the pillbox menisci grow, the lubricant film thins until its disjoining pressure equals the pillbox menisci capillary pressure. Sometimes the entire contacting interface floods before these pressures equilibrate and the system enters the immersed wetting regime (Section 6.5.3). This is actually an easy system for which to calculate the meniscus force, as the meniscus area is the total area immersed and the capillary pressure is equal to the disjoining pressure of the lubricant film surrounding the contacting interface.

- As the pillbox menisci grow, the meniscus force grows, pulling the contacting surfaces closer together. The interface can collapse if the separation reduction results in a meniscus force rising faster than the elastic force from the contacting asperities. This is especially a concern for those surfaces where the loading force is supported by a low density of asperities (Gui and Marchon 1995, Gui et al. 1997).

10.4   Capillary condensation of water vapor

Since humans find relative humidities in the range 30–50% to be most comfortable, most commercial and industrial sites, as well as many homes, actively try to maintain the relative humidity (RH) in this range. As discussed in Section 5.2.2, at these humidities, capillary condensation of this water vapor can occur around contacting asperities if the surfaces are hydrophilic, that is, if the water contact angles are small. For a RH in the middle of the comfortable range (RH = 40%), the Kelvin equation (eq. (5.11)) predicts that the Kelvin radius r_k should be

$$
\begin{aligned}
r_k &= \frac{2\gamma_{\text{water}} V_{\text{molar}}}{RT \ln (P_v/P_{\text{sat}})} \\
&= \frac{2 \, (73 \text{ mN/m}) \left(18 \text{ cm}^3\right)}{(8.31 \text{ J/K} \cdot \text{mol}) \, (295 \text{ K}) \ln(0.4)} \\
&= -1.2 \text{ nm}
\end{aligned}
\tag{10.25}
$$

Since this radius is only on the order of a nanometer, this implies that water menisci at these humidities are very small, sometimes only a few cubic nanometers in volume. Even though their sizes are small, these menisci can still generate a significant meniscus force.

As discussed in the previous section, the total meniscus force L_{men} acting on the contacting surfaces is the sum of the areas wetted by the individual menisci times the capillary pressure within those menisci. Since r_k is only on the order of a nanometer for moderate humidities, it is impossible to flood gaps greater than a few nanometers, and, most likely, all the contacting asperities will be in the toe dipping regime (Figure 10.28). Taking into account the possibility of finite water contact angles (θ_1 and θ_2), it is more appropriate to rewrite eq. (10.22) as

$$f_{\text{td}-i} = 2\pi R_{\text{td}-i}\gamma_{\text{water}}\left(\cos\theta_1 + \cos\theta_2\right),\qquad(10.26)$$

as the expression for the equilibrium meniscus force on an individual asperity in this toe dipping situation when $R_{\text{td}-i} \gg |r_m|$. Then, similar to eq. (10.23) the total equilibrium meniscus force L_{men} is given by

$$L_{\text{men}} = 2\pi\gamma_L n_{\text{td}} R_{\text{td}-\text{ave}}\left(\cos\theta_1 + \cos\theta_2\right).\qquad(10.27)$$

Since the meniscus force L_{men} adds to the external loading force, eq. (10.21) indicates that the presence of water menisci from humidity adds a contribution μL_{men} to the friction force.

As illustrated in Figure 10.28(b), once the contacting surfaces start sliding relative to each other, the menisci become distorted and eventually break, since the contact lines where the individual menisci meet the solid surfaces may be pinned by defects. Also, the faster the sliding speed, the less time that new water capillary bridges have to nucleate and grow around individual contacts. Consequently, for hydrophilic surfaces, the adhesion and friction forces are expected to decrease with increasing sliding speed and decreasing humidity. Indeed, several AFM experiments have found that adhesion and friction decrease linearly with logarithmic increases in sliding speed for hydrophilic surfaces at moderate humidities, examples of which are shown in Figures 10.29 and 10.30 (Riedo et al. 2002, Szoszkiewicz and Riedo 2005, Greiner et al. 2010, 2012, Noel et al. 2012).

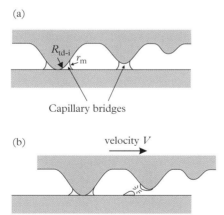

(a)

$R_{\text{td-i}}$

r_m

Capillary bridges

(b)   velocity V

Figure 10.28 *(a) A capillary bridge forming around a contacting asperity (left-hand asperity) and an asperity nearly in contact (middle asperity). For the right-hand asperity, the gap between the asperity summit and the opposing surface is too big for a capillary bridge to nucleate. For asperities in the toe-dipping regime where |asperity radius R_{td-i}| $\gg$ |meniscus radius r_m|, then $r_m = r_k/2$. (b) How the capillary bridges distort and break with a sliding velocity V.*

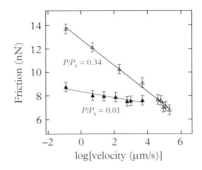

Figure 10.29 *The average friction force acting on an AFM tip as a function of log(velocity V) when sliding over a hydrophilic CrN surface at relative humidities P/P$_s$ of 1% and 34%. For the very dry condition of P/P$_s$ = 1%, friction is only weakly dependent on velocity, consistent with no water meniscus condensing around the tip–surface contact. For a moderate humidity of P/P$_s$ = 34%, friction is much higher at low velocities due to the presence of water condensing around the tip-surface contact, but decreases rapidly with increasing log V since water menisci have less time to nucleate and grow as the tip slides to new regions of the surface. Reprinted with permission from Riedo et al. (2002). Copyright 2002 by the American Physical Society.*

For the example shown in Figure 10.30, the adhesion force is measured by pulling the AFM tip off the surface while it is sliding over a range of velocities V on a hydrophilic gold surface and a hydrophobic graphite surface at an RH of 48% (Noel et al. 2012). For the hydrophobic surface, the adhesion force is constant with sliding speed, consistent with no water meniscus condensing around the tip–surface contact. For the hydrophilic surface, however, once the sliding velocity exceeds a threshold velocity V_{start}, the adhesion force decreases dramatically with increasing velocity. Presumably, on this hydrophilic surface, the capillary menisci break due to sliding and do not have enough time to grow to their equilibrium size when the sliding speed exceeds V_{start}, while for sliding speeds $< V_{start}$ the menisci do have enough time.

The logarithmic dependence of adhesion and friction forces with sliding speed can be modeled as the combination of two thermally activated processes:

1. First, water vapor overcomes the energy barrier ΔE_{nuc} to nucleate a new capillary bridge or meniscus, either around a contacting asperity or bridging the gap between an asperity and a nearby surface, as illustrated in Figure 10.28(a) (Riedo et al. 2002).

2. Once nucleated, the water menisci grows to its equilibrium size, but, during this growth process, the contact line has to overcome the energy barriers for passing over surface defects (Noel et al. 2012). (These are the same irregularities responsible for contact angle hysteresis (Section 5.4.2.1).) The energy barrier for growth ΔE_{growth} is expected to be proportional to the perimeter of the meniscus $2\pi r_2$, where r_2 is the radius of the area wetted by the meniscus.

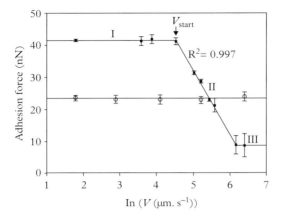

Figure 10.30 *Average adhesion force for an AFM tip sliding over a hydrophilic gold surface (solid squares) and a hydrophobic graphite surface (open circles) at a 48% relative humidity. For the gold hydrophilic surface, the adhesion is characterized by three sliding regimes: regime I, where the adhesion force is constant with sliding velocity V, indicating that the water menisci still have sufficient time to grow to their full equilibrium size; regime II, where the adhesion decreases linearly with the logarithmic increasing velocity, due to menisci not being able to grow to their full size before they break; and regime III, where the constant adhesion indicates that the asperity contact times are too short for water vapor to nucleate around asperities at these highest sliding speeds. Reprinted with permission from Noel et al. (2012). Copyright 2012 by the American Physical Society.*

Let's first consider the thermally activated process to nucleate a new capillary bridge. The energy barrier that needs to be overcome to nucleate a volume of liquid v_d can be expressed as (Restagno et al. 2000)

$$\Delta E_{nuc} = v_d \rho_L \Delta \mu = v_d \rho_L T \log \left(\frac{P_{sat}}{P_v} \right), \qquad (10.28)$$

where ρ_L is the liquid density, $\Delta \mu$ is the change in chemical potential when the vapor condenses to the liquid, and P_v and P_{sat} are the partial and the saturation pressures of the vapor. So the probability Π that condensation occurs around a particular asperity in a time period t_w is

$$\Pi(t_w) = 1 - \exp \left(-\frac{t_w}{\tau} \right)$$

$$\tau = \tau_0 \exp \left(\frac{\Delta E_{nuc}}{k_B T} \right) \qquad (10.29)$$

where τ_0 is the pre-exponential constant.

For a sliding contact with a contact diameter d, the residence time for contact is $t_{res} = V/d$, where V is the sliding velocity. During this time t_{res}, water menisci only nucleate around a fraction $f(t_{res})$ of the asperities in contact and near contact, where $f(t_{res})$ is given by (Riedo et al. 2002)

$$f(t_{\text{res}}) = \frac{\ln(t_{\text{res}}/\tau_0)}{\lambda A \rho_L \ln(P_{\text{sat}}/P_v)}$$

$$= -\frac{\ln(V/V_0)}{\lambda A \rho_L \ln(P_{\text{sat}}/P_v)} \qquad (10.30)$$

where λ is the range of interstitial heights between the sliding surfaces, A is the characteristic area of a meniscus around an asperity, and $V_0 = d/\tau_0$. Combining eq. (10.30) with eq. (10.27) predicts a meniscus force during sliding of

$$L_{\text{men}} = -\left[2\pi n_{\text{td}} R_{\text{td-ave}} \gamma_L \left(\cos\theta_1 + \cos\theta_2\right)\right] \frac{\ln(V/V_0)}{\lambda A \rho_L \ln(P_{\text{sat}}/P_v)}$$

$$\propto -\ln(V/V_0), \qquad (10.31)$$

where n_{td} is the number of asperities and $R_{\text{td-ave}}$ is the average radius of curvature of the asperity summits. From eq. (10.31) we see that the menisci adhesion force [and from eq. (10.21) the friction force] decrease logarithmically with increasing velocity V and goes to zero as $V \to V_0$. So, above the threshold velocity V_0, water menisci do not have time to form, giving rise to regime III in Figure 10.30 where adhesion is constant with velocity.

For the experimental results shown in Figure 10.30, where a single asperity in the form of an AFM tip is sliding across a smooth surface, the decrease in adhesion force with increasing velocity in regime II is most likely from the meniscus around the contacting AFM tip not having sufficient time to grow to an equilibrium size during the residence time of the moving contact (Noel et al. 2012). Since this is also a thermally activated process, the probability that the meniscus will grow to an area A in a time t_w is given by an equation similar to eq. (10.29):

$$\Pi(t_w) = 1 - \exp\left(-\frac{t_w}{\tau}\right)$$

$$\tau = \tau_0' \exp\left(\frac{\Delta E_{\text{grow}}}{k_B T}\right) \qquad (10.32)$$

where τ_0' is the pre-exponential constant for this process. Since this is an Arrhenius law probability, it follows that the average meniscus area, and hence the average meniscus force, scales the same way as with nucleated menisci (eq. (10.31)), that is, $L_{\text{men}} \propto -\ln(V/V_0)$.

10.5   Example of cold welding: what happens in the absence of lubrication

Before leaving this chapter on lubrication, it is useful to discuss what can happen in the absence of lubrication. As discussed in Chapters 5 and 6, clean surfaces of metals and other high surface energy materials experience high adhesion forces when brought into contact. By *clean surface* we mean a surface where all the contamination layers that would lower the surface energy have been removed.

Adhesion measurements on clean surfaces are often performed in high vacuum environments, as once a clean surface is prepared and exposed to ordinary air, it is likely to adsorb adventitious contamination from the air, and can also oxidize. For example, most clean metal surfaces will typically form an oxide layer when exposed to air, and on top of the oxide layer, a thin layer water and hydrocarbon molecules will adsorb from any vapors in the air. Doing experiments in high vacuum then enables the surfaces to stay clean for the extended periods of time needed to conduct experiments. The most extensive series of these types of measurements were done in the 1970s by Donald H. Buckley, who found that the adhesion and friction forces rise dramatically as the contamination is progressively removed (Buckley 1981).

Results from a more recent high vacuum experiment by Gellman and Ko (2001) are shown in Figure 10.31, where the friction of clean copper surfaces is measured as a function of exposure to trifluoroethanol. For the initially bare Cu(111) surfaces, the

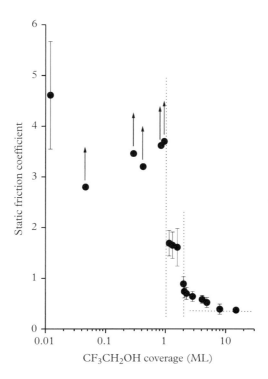

Figure 10.31 *Static coefficient of friction measured at 120 K in an ultra-high vacuum chamber between two copper single crystals with (111) surfaces. The static friction decreases as the coverage of trifluoroethanol, CF_3CH_2OH, increases; ML = monolayer coverage. Reprinted from Gellman and Ko (2001) with permission from Springer Science and Business Media. Copyright 2001, Plenum Publishing Corporation.*

static friction is very high, with $\mu_s > 4$. This result is consistent with what Buckley and others have found for clean metal surfaces in vacuum. Indeed, total seizure frequently occurs between the contacting metal surfaces, which is usually attributed to *cold welding*, a process where two metal surfaces fuse together as if they were welded, even though no heat is applied to liquefy or soften the metals. If such cold welded interfaces are forced to slide over each other, the high adhesive and friction forces between them usually leads to severe damage occurring in the form of *galling* or *fretting*.

Figure 10.31 also shows that, as the small prototype lubricant molecule, trifluoroethanol, is absorbed on the Cu(111) surfaces, the friction decreases with sudden drops occurring when the first and second monolayers complete, a limiting value of $\mu_s \approx 0.3$–0.4 being reached at high coverages. This dramatic drop in friction illustrates the major impact that a molecularly thin layer of the right sort of atom or molecule can have on lowering the adhesion and shear strength of a contacting junction.

Very high friction and adhesion sometimes also occurs in practical contacts operating outside of high vacuum conditions. For example, plain bearings and gear teeth can be susceptible to scuffing and seizure (variations of the cold welding phenomenon) if the lubricant film breaks down or is absent. When this happens, considerable damage occurs to the bearing or gear, often resulting in rapid and catastrophic failure due to adhesive or abrasive wear.

....

10.6 PROBLEMS

1. Consider a hard sphere sliding over a hard plate with a very thin coating of low shear strength material (the situation depicted in Figure 10.8(c)).

    (a) Derive an expression for the frictional coefficient μ as a function of the loading force L, the sphere radius R, the composite elastic modulus E_c of the sphere and plate, and the shear strength s of the thin coating. Neglect the influence of the coating itself on the contact area.
    (b) MoS$_2$ coatings can exhibit coefficients of friction as low as $\mu = 0.02$. If this value of μ is observed for a 6 mm diameter steel bearing sliding at a load of 80 N across a steel plate coated with a thin coating of MoS$_2$, what value of shear strength would this imply for the coating? Assume that composite elastic modulus for the steel bearing and plate is $E_c = 200$ GPa.

2. Section 10.2.3 discusses how ZDDP tribofilms tend to stop growing after reaching a thickness of approximately 100 nm, and how this may be due to the reduction in the applied stress due to the low modulus of the tribofilm cushioning the applied load. Let us consider this effect for a steel asperity of radius $R = 20$ nm and a steel flat.

    (a) Using the Hertz model, find the applied load L if the mean pressure in the contact is initially 3 GPa when there is no film.

(b) Consider the case where a tribofilm of thickness $h = 100$ nm has grown on the steel surface. Assuming that the film has a Young's modulus of $E = 20$ GPa and a Poisson's ratio of $\nu = 0.4$ and that the load L stays the same as in (a), find the mean pressure in the contact zone. As the geometry is now a thin film on substrate, the Hertz model is no longer valid. Instead, use the following contact mechanics equation for a sphere indenting a thin film on a substrate:

$$L = \frac{E_u A^2}{4\pi R h},$$
(10.33)

where A is the area of contact and E_u is the uniaxial strain modulus of the tribofilm, which is given by

$$E_u = \frac{(1 - \nu) E}{(1 + \nu)(1 - 2\nu)}.$$
(10.34)

This equation, developed by Reedy (2011), assumes that the film modulus is much lower than that of the sphere or the substrate, and thus treats them as rigid.

(c) Using eqs. (10.5) and (10.6), by what factor has the tribofilm growth rate reduced from when it first started to when it reached the thickness of 100 nm? Assume the activation volume is 3.8 $\mathring{A}^3$ and that the tribofilm is grown at 100 °C.

10.7 REFERENCES

Aktary, M., M. T. McDermott and G. A. McAlpine (2002). "Morphology and nanomechanical properties of ZDDP antiwear films as a function of tribological contact time." *Tribology Letters* **12**(3): 155–62.

Alba-Simionesco, C., B. Coasne, G. Dosseh, G. Dudziak, K. E. Gubbins, R. Radhakrishnan and M. Sliwinska-Bartkowiak (2006). "Effects of confinement on freezing and melting." *Journal of Physics: Condensed Matter* **18**(6): R15–R68.

Barrena, E., S. Kopta, D. F. Ogletree, D. H. Charych and M. Salmeron (1999). "Relationship between friction and molecular structure: alkylsilane lubricant films under pressure." *Physical Review Letters* **82**(14): 2880–3.

Barthel, A. J. and S. H. Kim (2014). "Lubrication by physisorbed molecules in equilibrium with vapor at ambient condition: effects of molecular structure and substrate chemistry." *Langmuir* **30**(22): 6469–6478.

Bec, S., A. Tonck, J. M. Georges, R. C. Coy, J. C. Bell and G. W. Roper (1999). "Relationship between mechanical properties and structures of zinc dithiophosphate anti-wear films." *Proceedings of the Royal Society of London, Series A: Mathematical, Physical and Engineering Sciences* **455**(1992): 4181–203.

Bowden, F. P. and D. Tabor (1950). *The friction and lubrication of solids*. Oxford: Clarendon Press.

Braun, O. M. and M. Peyrard (2003). "Dynamics and melting of a thin confined film." *Physical Review E* **68**(1): 11506.

Buckley, D. H. (1981). *Surface effects in adhesion, friction, wear, and lubrication.* Elsevier.

Carpick, R. W. and M. Salmeron (1997). "Scratching the surface: fundamental investigations of tribology with atomic force microscopy." *Chemical Reviews* **97**(4): 1163–94.

Chandross, M., C. D. Lorenz, M. J. Stevens and G. S. Grest (2008). "Simulations of nanotribology with realistic probe tip models." *Langmuir* **24**(4): 1240–6.

Chandross, M., E. B. Webb, M. J. Stevens, G. S. Grest and S. H. Garofalini (2004). "Systematic study of the effect of disorder on nanotribology of self-assembled monolayers." *Physical Review Letters* **93**(16): 166103.

Cui, S. T., C. McCabe, P. T. Cummings and H. D. Cochran (2003). "Molecular dynamics study of the nano-rheology of *n*-dodecane confined between planar surfaces." *The Journal of Chemical Physics* **118**(19): 8941–4.

Derjaguin, B. V. and N. V. Churaev (1974). "Structural component of disjoining pressure." *Journal of Colloid and Interface Science* **49**(2): 249–55.

Drummond, C. and J. Israelachvili (2000). "Dynamic behavior of confined branched hydrocarbon lubricant fluids under shear." *Macromolecules* **33**(13): 4910–20.

Friedenberg, M. C. and C. M. Mate (1996). "Dynamic viscoelastic properties of liquid polymer films studied by atomic force microscopy." *Langmuir* **12**(25): 6138–42.

Fukuzawa, K., J. Kawamura, T. Deguchi, H. Zhang and Y. Mitsuya (2004). "Measurement of disjoining pressure of a molecularly thin lubricant film by using a microfabricated groove." *IEEE Transactions on Magnetics* **40**(4): 3183–5.

Gao, J. P., W. D. Luedtke, D. Gourdon, M. Ruths, J. N. Israelachvili and U. Landman (2004). "Frictional forces and Amontons' law: from the molecular to the macroscopic scale." *Journal of Physical Chemistry B* **108**(11): 3410–25.

Gao, J. P., W. D. Luedtke and U. Landman (2000). "Structures, solvation forces and shear of molecular films in a rough nano-confinement." *Tribology Letters* **9**(1–2): 3–13.

Gee, M. L., P. M. McGuiggan, J. N. Israelachvili and A. M. Homola (1990). "Liquid to solid-like transitions of molecularly thin-films under shear." *Journal of Chemical Physics* **93**(3): 1895–906.

Gellman, A. J. (2004). "Vapor lubricant transport in MEMS devices." *Tribology Letters* **17**(3): 455–61.

Gellman, A. J. and J. S. Ko (2001). "The current status of tribological surface science." *Tribology Letters* **10**(1–2): 39–44.

Gibbs, G. B. (1965). "The thermodynamics of thermally-activated dislocation glide." *Physica Status Solidi(b)* **10**(2): 507–12.

Gosvami, N. N., J. A. Bares, F. Mangolini, A. R. Konicek, D. G. Yablon and R. W. Carpick (2015). "Mechanisms of antiwear tribofilm growth revealed in situ by single-asperity sliding contacts." *Science* **348**(6230): 102–6.

Granick, S. (1991). "Motions and relaxations of confined liquids." *Science* **253**(5026): 1374–9.

Granick, S. (1999). "Soft matter in a tight spot." *Physics Today* **52**(7): 26–31.

Greiner, C., J. R. Felts, Z. Dai, W. P. King and R. W. Carpick (2010). "Local nanoscale heating modulates single-asperity friction." *Nano Letters* **10**(11): 4640–5.

Greiner, C., J. R. Felts, Z. Dai, W. P. King and R. W. Carpick (2012). "Controlling nanoscale friction through the competition between capillary adsorption and thermally activated sliding." *ACS Nano* **6**(5): 4305–13.

Grossiord, C., J. M. Martin, T. Le Mogne and T. Palermo (1999). "UHV friction of tribofilms derived from metal dithiophosphates." *Tribology Letters* **6**(3–4): 171–9.

Gui, J. and B. Marchon (1995). "A stiction model for a head–disk interface of a rigid disk-drive." *Journal of Applied Physics* **78**(6): 4206–17.

Gui, J., D. Kuo, B. Marchon and G. C. Rauch (1997). "Stiction model for a head–disc interface: experimental." *IEEE Transactions on Magnetics* **33**(1): 932–7.

Guo, X. C., B. Marchon, R. H. Wang, C. M. Mate, Q. Dai, R. J. Waltman, H. Deng, D. Pocker, Q. F. Xiao, Y. Saito and T. Ohtani (2012). "A multidentate lubricant for use in hard disk drives at sub-nanometer thickness." *Journal of Applied Physics* **111**: 0240503.

Hanggi, P., Talkner, P. & Borkovec, M. (1990). "Reaction-rate theory: fifty years after Kramers." *Reviews of Modern Physics* **62**(2): 251–341.

Hanyaloglu, B. and E. E. Graham (1994). "Vapor-phase lubrication of ceramics." *Lubrication Engineering* **50**(10): 814–20.

Hardy, W. B. and I. Bircumshaw (1925). "Boundary lubrication – plane surfaces and limitations of Amontons' law." *Proceedings of the Royal Society of London, Series A: Mathematical and Physical Sciences* 108: 1–27.

Henck, S. A. (1997). "Lubrication of digital micromirror devices." *Tribology Letters* **3**(3): 239–47.

Hu, H. W., G. A. Carson and S. Granick (1991). "Relaxation-time of confined liquids under shear." *Physical Review Letters* **66**(21): 2758–61.

Israelachvili, J. N. (2005). "Importance of pico-scale topography of surfaces for adhesion, friction, and failure." *MRS Bulletin* **30**(7): 533–9.

Jabbarzadeh, A. and R. I. Tanner (2006). "Molecular dynamics simulation and its application to nano-rheology." *Rheology Reviews* 2006: 165–216.

Jabbarzadeh, A., P. Harrowell and R. I. Tanner (2006a). "Crystal bridge formation marks the transition to rigidity in a thin lubrication film." *Physical Review Letters* **96**(20): 206102.

Jabbarzadeh, A., P. Harrowell and R. I. Tanner (2006b). "Low friction lubrication between amorphous walls: unraveling the contributions of surface roughness and in-plane disorder." *The Journal of Chemical Physics* **125**(3): 034703.

Jabbarzadeh, A., P. Harrowell and R. I. Tanner (2007). "The structural origin of the complex rheology in thin dodecane films: three routes to low friction." *Tribology International* **40**(10): 1574–86.

Jacobs, T. D. B. and R. W. Carpick (2013). "Nanoscale wear as a stress-assisted chemical reaction." *Nature Nanotechnology* **8**(2): 108–12.

Jacobs, T. D., B. Gotsmann, M. A. Lantz and R. W. Carpick (2010). "On the application of transition state theory to atomic-scale wear." *Tribology Letters* **39**(3): 257–71.

Kim, H. I. and J. E. Houston (2000). "Separating mechanical and chemical contributions to molecular-level friction." *Journal of the American Chemical Society* **122**(48): 12045–6.

Kim, H. I., C. M. Mate, K. A. Hannibal and S. S. Perry (1999). "How disjoining pressure drives the dewetting of a polymer film on a silicon surface." *Physical Review Letters* **82**(17): 3496–9.

Klein, J. and E. Kumacheva (1998). "Simple liquids confined to molecularly thin layers. I. Confinement-induced liquid-to-solid phase transitions." *Journal of Chemical Physics* **108**(16): 6996–7009.

Krass, M.-D., N. N. Gosvami, R. W. Carpick, M. H. Müser and R. Bennewitz (2016). "Dynamic shear force microscopy of viscosity in nanometer-confined hexadecane layers." *Journal of Physics: Condensed Matter* **28**(13): 134004.

Kumacheva, E. and J. Klein (1998). "Simple liquids confined to molecularly thin layers. II. Shear and frictional behavior of solidified films." *Journal of Chemical Physics* **108**(16): 7010–22.

Lio, A., D. H. Charych and M. Salmeron (1997). "Comparative atomic force microscopy study of the chain length dependence of frictional properties of alkanethiols on gold and alkylsilanes on mica." *Journal of Physical Chemistry B* **101**(19): 3800–5.

Ma, X., J. Gui, B. Marchon, M. S. Jhon, C. L. Bauer and G. C. Rauch (1999). "Lubricant replenishment on carbon coated discs." *IEEE Transactions on Magnetics* **35**(5): 2454–6.

Major, R. C., H. I. Kim, J. E. Houston and X.-Y. Zhu (2003). "Tribological properties of alkoxyl monolayers on oxide terminated silicon." *Tribology Letters* **14**(4): 237–44.

Martin, J. M. (1999). "Antiwear mechanisms of zinc dithiophosphate: a chemical hardness approach." *Tribology Letters* **6**(1): 1–8.

Martin, J. M., T. Onodera, C. Minfray, F. Dassenoy and A. Miyamoto (2012). "The origin of anti-wear chemistry of ZDDP." *Faraday Discussions* **156**(1): 311–23.

Mate, C. M. (1992a). "Application of disjoining and capillary-pressure to liquid lubricant films in magnetic recording." *Journal of Applied Physics* **72**(7): 3084–90.

Mate, C. M. (1992b). "Atomic-force-microscope study of polymer lubricants on silicon surfaces." *Physical Review Letters* **68**(22): 3323–6.

Mate, C. M. (2011). "Taking a fresh look at disjoining pressure of lubricants at slider–disk interfaces." *IEEE Transactions on Magnetics* **47**(1): 124–30.

Mate, C. M. (2013). "Spreading kinetics of lubricant droplets on magnetic recording disks." *Tribology Letters* **51**(3): 385–95.

Mate, C. M. and A. M. Homola (1997). "Molecular tribology of disk drives." In: *Micro/nanotribology and its applications*, B. Bhushan, Ed. Dordrecht: Kluwer Academic Publishers, pp. 647–61.

Matthewson, M. J. and H. J. Mamin (1988). "Liquid-mediated adhesion of ultra-flat solid surfaces." *Proceedings of Materials Research Society Symposia* **119**: 87–92.

Mikulski, P. T., L. A. Herman and J. A. Harrison (2005). "Odd and even model self-assembled monolayers: links between friction and structure." *Langmuir* **21**(26): 12197–206.

Min, B. G., J. W. Choi, H. R. Brown and D. Y. Yoon (1995). "Spreading characteristics of thin liquid films of perfluoropolyalkylethers on solid surfaces. Effects of chain-end functionality and humidity." *Tribology Letters* **1**(2–3): 225–32.

Mosey, N. J., M. H. Müser and T. K. Woo (2005). "Molecular mechanisms for the functionality of lubricant additives." *Science* **307**(5715): 1612–15.

Mukhopadhyay, A., J. Zhao, S. C. Bae and S. Granick (2002). "Contrasting friction and diffusion in molecularly thin confined films." *Physical Review Letters* **89**(13): 136103.

Noel, O., P.-E. Mazeran and H. Nasrallah (2012). "Sliding velocity dependence of adhesion in a nanometer-sized contact." *Physical Review Letters* **108**(1): 015503.

O'Shea, S. J. and M. E. Welland (1998). "Atomic force microscopy at solid-liquid interfaces." *Langmuir* **14**(15): 4186–97.

O'Shea, S. J., N. N. Gosvami, L. T. W. Lim and W. Hofbauer (2010). "Liquid atomic force microscopy: solvation forces, molecular order, and squeeze-out." *Japanese Journal of Applied Physics* **49**(8S3): 08LA01.

Peachey, J., J. Vanalsten and S. Granick (1991). "Design of an apparatus to measure the shear response of ultrathin liquid-films." *Review of Scientific Instruments* **62**(2): 463–73.

Pereira, G., A. Lachenwitzer, D. Munoz-Paniagua, M. Kasrai, P. R. Norton, M. Abrecht and P. U. P. A. Gilbert (2006). "The role of the cation in antiwear films formed from ZDDP on 52100 steel." *Tribology Letters* **23**(2): 109–19.

Perry, S. S., S. Lee, Y. S. Shon, R. Colorado and T. R. Lee (2001). "The relationships between interfacial friction and the conformational order of organic thin films." *Tribology Letters* **10**(1–2): 81–7.

Ramin, L. and A. Jabbarzadeh (2012a). "Effect of load on structural and frictional properties of alkanethiol self-assembled monolayers on gold: some odd–even effects." *Langmuir* **28**(9): 4102–12.

Ramin, L. and A. Jabbarzadeh (2012b). "Frictional properties of two alkanethiol self-assembled monolayers in sliding contact: odd–even effects." *The Journal of Chemical Physics* **137**(17): 174706.

Reedy, E. D. (2011). "Thin-coating contact mechanics with adhesion." *Journal of Materials Research* **21**(10): 2660–8.

Restagno, F., L. Bocquet and T. Biben (2000). "Metastability and nucleation in capillary condensation." *Physical Review Letters* **84**(11): 2433–6.

Riedo, E., F. Lévy and H. Brune (2002). "Kinetics of capillary condensation in nanoscopic sliding friction." *Physical Review Letters* **88**(18): 185505.

Robbins, M. O., D. Andelman and J. F. Joanny (1991). "Thin liquid-films on rough or heterogeneous solids." *Physical Review A* **43**(8): 4344–54.

Salmeron, M. (2001). "Generation of defects in model lubricant monolayers and their contribution to energy dissipation in friction." *Tribology Letters* **10**(1–2): 69–79.

Scarpulla, M. A., C. M. Mate and M. D. Carter (2003). "Air shear driven flow of thin perfluoropolyether polymer films." *Journal of Chemical Physics* **118**(7): 3368–75.

Sivebaek, I. M. and B. N. J. Persson (2016). "The effect of surface nano-corrugation on the squeeze-out of molecular thin hydrocarbon films between curved surfaces with long range elasticity." *Nanotechnology* **27**(44): 445401.

Sivebaek, I. M., V. N. Samoilov and B. N. J. Persson (2004). "Squeezing molecularly thin alkane lubrication films: layering transitions and wear." *Tribology Letters* **16**(3): 195–200.

Spikes, H. (2004). "The history and mechanisms of ZDDP." *Tribology letters* **17**(3): 469–89.

Spikes, H. and W. Tysoe (2015). "On the commonality between theoretical models for fluid and solid friction, wear and tribochemistry." *Tribology Letters* **59**(1): 21.

Szoszkiewicz, R. and E. Riedo (2005). "Nucleation time of nanoscale water bridges." *Physical Review Letters* **95**(13): 135502.

Thompson, P. A. and M. O. Robbins (1990). "Origin of stick-slip motion in boundary lubrication." *Science* **250**(4982): 792–4.

Thompson, P. A., G. S. Grest and M. O. Robbins (1992). "Phase transitions and universal dynamics in confined films." *Physical Review Letters* **68**(23): 3448–51.

Tutein, A. B., S. J. Stuart and J. A. Harrison (2000). "Role of defects in compression and friction of anchored hydrocarbon chains on diamond." *Langmuir* **16**(2): 291–6.

Weinstein, A. and S. A. Safran (1998). "Surface and bulk ordering in thin films." *Europhysics Letters* **42**(1): 61–6.

Wong, S.-S., H. Takano and M. D. Porter (1998). "Mapping orientation differences of terminal functional groups by friction force microscopy." *Analytical Chemistry* **70**(24): 5209–12.

Yang, Y., A. C. Jamison, D. Barriet, T. R. Lee and M. Ruths (2010). "Odd–even effects in the friction of self-assembled monolayers of phenyl-terminated alkanethiols in contacts of different adhesion strengths." *Journal of Adhesion Science and Technology* **24**(15–16): 2511–29.

Zhang, J. and H. Spikes (2016). "On the mechanism of ZDDP antiwear film formation." *Tribology Letters* **63**(2): 24.

11

Atomistic Origins of Friction

In Chapter 4, we introduced the macroscopic laws of friction and began to discuss some of the possible microscopic origins of friction. In this chapter we delve deeper into how friction originates at the atomic scale, specifically for "dry" or unlubricated surfaces.

When Amontons originally proposed his laws of friction in 1699, many, including Amontons, thought that friction originated from the interlocking of the roughness on opposing surfaces (see Figure 4.3). However, by the middle of the twentieth century this roughness mechanism had been ruled out and replaced by the notion that the two main contributors to friction of dry surfaces are:

1. *Adhesion*—resistance to sliding that comes from the atomic level interaction forces between the atoms on the opposing contacting surfaces;
2. *Plastic deformation and material damage*—resistance during sliding that comes from the work of the asperities of the harder surface plastically deforming the softer surface and the work needed to remove material during wear processes, either with or without plastic deformation.

The friction due to plastic deformation and plowing was discussed in Section 4.2.2, and tends to be a fairly small contributor to friction. Material damage during wear processes and the resulting friction will be discussed in the next chapter. In this chapter we focus on the intricacies of *adhesive friction* (also called *interfacial friction*), or how the interactions between atoms generate a lateral force opposing the motion of one smooth surface sliding over another without plastic deformation and wear.

11.1 Concept of adhesive friction

The notion that friction should be related to adhesion seems fairly natural: the same bonding mechanism that makes it difficult to pull surfaces apart should also make it difficult to slide them over each other. The experimental evidence that friction and adhesion are related goes as far back as Desaguliers who in 1734 noticed that friction and adhesion forces between metals greatly increased after their surfaces were polished to a smooth finish (Desaguliers 1734).

Tribology on the Small Scale: A Modern Textbook on Friction, Lubrication and Wear. Second edition. C. Mathew Mate and Robert W. Carpick. © Oxford University Press 2019. Published in 2019 by Oxford University Press.
DOI: 10.1093/oso/ 9780199609802.001.0001

One obvious way that adhesion contributes to friction is by increasing the total loading force that pushes the surfaces into contact. If L_{ext} is the externally applied loading force and L_{adh} is the adhesive contribution to the loading force, then Amontons' law for the friction force F becomes

$$F (L_{ext} + L_{adh}) . \tag{11.1}$$

So, one consequence of adhesion is that friction can occur even when no external load is applied.

Based on experiments they conducted from the 1930s through the 1960s, Bowden and Tabor (1950) promoted the concept that adhesion also contributes to the coefficient of friction μ in eq. (11.1) (also discussed Section 4.2.1). They argued that this adhesive component of friction F_{adh} is proportional to the real area of contact A_r:

$$F_{adh} = A_r s, \tag{11.2}$$

where s is the stress needed to initiate or to maintain sliding—that is, s is the *shear strength* of the contact junctions. That friction is proportional to normal force—as expressed by eq. (11.1)—then comes about because the real area of contact is proportional to load, which, as explained in Section 3.4, is generally true for contacting rough surfaces when the asperities are undergoing either plastic or elastic deformation.

A fundamental feature of Bowden and Tabor's concept of adhesive friction is that the shear strength s is determined by the interaction forces between the sliding atoms. That adhesion on its own is insufficient to cause friction can be seen from the following simple argument: If we consider two solids with ideally flat surfaces, without any atomic structure, and where an attractive interaction pulls them into contact, it then takes no energy to slide one flat surface over another as no work is done against the intermolecular forces. However, once the surfaces have atomic structure, we have to consider the work to somehow displace the atoms of one surface vertically relative to those on the opposing surface.

11.1.1   Cobblestone model

We start with a very simple atomic level model for adhesive friction referred to as the cobblestone model, which makes the analogy that the frictional process at the atomic level resembles pushing a cart over a cobblestone road (Tabor 1982, Homola et al. 1989, 1990, Berman and Israelachvili 1998, Israelachvili 2005). As illustrated in Figure 11.1, in the cobblestone model, the cartwheels represent lubricant molecules or atoms on an upper surface that is being moved over the cobblestones, which represent the atomic roughness of the lower surface.

In order to initiate sliding, a "cartwheel" must first be raised a small distance ΔD against the attractive surface interactions so that the atoms on the opposing surfaces have enough room to slide over each other. If the adhesion force is L_{adh}, the energy required to separate the two surfaces this distance is $\Delta D \times L_{adh}$, which is a small fraction of the total

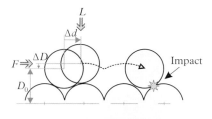

Figure 11.1 *Schematic of the cobblestone model, where the atomic level friction process is assumed to be analogous to pushing a cartwheel across a road of cobblestones. In this model, each atom on the upper surface needs to be raised a distance ΔD in order to be slid a distance Δd to the top of the next bottom atom, while being pushed by a lateral force F. The top atom then slides down the other side of the bottom atom, and a fraction ε of the kinetic energy is dissipated on impact. Reproduced with permission from Israelachvili (2005). Copyright 2005, Materials Research Society.*

adhesion energy $2\gamma A_r$, where γ is the interfacial surface energy. Since $\gamma \propto D^{-2}$ where $D(t)$ is the vertical separation distance between atom centers as a function of time (eq. (7.25)), we can approximate this change in energy as

$$2\gamma A_r \left(1 - \frac{D_0^2}{(D_0 + \Delta D)^2}\right) \simeq 4\gamma A_r \frac{\Delta D}{D_0}. \tag{11.3}$$

This energy must come from applying enough lateral force F over the distance Δd to equal the work required to overcome the force of adhesion:

$$\Delta d \times F = \Delta D \times L_{adh} \simeq 4\gamma A_r \frac{\Delta D}{D_0}. \tag{11.4}$$

Not all this work is lost to frictional processes, as part of the work expended on separating the surfaces is recouped when they approach again. Instead, the work done against the adhesive friction force F_{adh} should correspond to energy loss during the cycle of separation and approach:

$$\Delta d \times F_{adh} = 4\Delta\gamma A_r \frac{\Delta D}{D_0}, \tag{11.5}$$

where $\Delta\gamma = (\gamma_s - \gamma_a)$ is the difference between the separation and approach interfacial energies. (The adhesion hysteresis discussed in Section 5.6.) This leads to the following expression for the shear strength of the interface:

$$s = \frac{F_{adh}}{A_r} = \frac{4\Delta\gamma \cdot \Delta D}{\Delta d \cdot D_0} = \frac{4\gamma\varepsilon \cdot \Delta D}{\Delta d \cdot D_0}, \tag{11.6}$$

where $\varepsilon = \Delta\gamma/\gamma$ is the coefficient of energy dissipation and corresponds to the fraction of energy lost in each separation/approach cycle.

As an example, a typical measured value of the shear strength for shearing a hydrocarbon film between two mica surfaces in the surface force apparatus (SFA) is 2×10^7 Pa (Homola et al. 1990). This value can also be estimated by eq. (11.6), if we assume that $\gamma = 25$ mN/m (a typical value for a hydrocarbon surface) and that 10% of the surface energy is lost each time the surfaces slide an atomic dimension (i.e., $\Delta d \sim 1$ Å):

$$s = \frac{4\left(25 \text{ mJ/m}^2\right)\left(0.5 \text{ Å}\right)}{\left(2 \text{ Å}\right)\left(1 \text{ Å}\right)}$$

$$= 2.5 \times 10^7 \text{ Pa.} \tag{11.7}$$

This is in reasonable agreement with the experimentally measured value.

An important outcome of the cobblestone model is that eq. (11.6) predicts that friction should be proportional to the adhesion energy hysteresis $\Delta\gamma$. Suda (2001) has derived a similar correlation between friction and adhesion hysteresis from rupture dynamics. This hypothesis is also consistent with the SFA experiments of Israelachvili and coworkers who found friction to be correlated with adhesion hysteresis (Chen et al. 1991, Yoshizawa et al. 1993). Israelachvili and coworkers have also shown that adhesion hysteresis can be associated with structural hysteresis at the interface, such as interdigitation of chain-like molecules at an interface, which occurs more easily for soft, amorphous molecular films and less easily for well-aligned, well-packed crystalline molecular films (Israelachvili et al. 1994).

Even though friction is often found experimentally to be correlated with adhesion energy hysteresis and sometimes with surface energy, the range of measured shear strengths is many times greater than that expected from the plausible range of surface energies or adhesion hysteresis. Indeed, the shear strengths for wear-free sliding of crystalline interfaces are experimentally found to vary over a tremendous range—from 10^{-2} to 10^{10} Pa (Krim 1995, 1996). This suggests that much more is going on than indicated by the simple cobblestone model. The rest of this chapter is devoted to further unraveling the atomic level origins of shear strength and interfacial friction.

11.1.2   Pressure dependence of the shear strength

One aspect of the shear strength that should be mentioned here is that the shear strength s is sometimes found to increase linearly with the contact pressure P (Briscoe and Tabor 1978, Briscoe 1981, Briscoe and Smith 1982, Gee et al. 1990, Berman et al. 1998, Piétrement and Troyon 2001a, 2001b, Israelachvili 2005):

$$s = s_0 + \alpha P. \tag{11.8}$$

A pressure-dependent shear strength is typically observed in those experiments where the interactions across the sliding interface are relatively weak, particularly for soft organic films and polymers. In these situations, the contact pressure from the externally applied load is comparable to the pressure generated by the adhesive force.

In special cases where the shear strength has a strong linear dependence on contact pressure *and the real area of contact is constant*, Amontons' law of friction being proportional to load will continue to appear to be valid for macroscopic contacts (Gao et al. 2004, Israelachvili 2005). In most typical situations, however, the shear strength is independent of the applied loading pressure P, and adhesive friction scales with the real area of contact (e.g., Carpick and Salmeron 1997). In this case, Amontons' law is applicable to the contact of rough surfaces, where the contact area is proportional to the load L, as discussed in Chapter 3.

11.1.3   Maximum possible shear strength

Just as a solid with no defects or dislocations should reach a theoretical maximum yield strength $Y_{\text{theoretical}} \sim E/10$ as discussed in Section 3.3.1, similarly, simple calculations show that a commensurate contact should reach a theoretical maximum shear strength $s \sim G/30$ (MacMillan 1972, Hurtado and Kim 1999a), where G is the reduced shear modulus for the two materials in contact, given by:

$$G^* = \left(\frac{2 - v_1}{G_1} + \frac{2 - v_2}{G_2} \right)^{-1}, \tag{11.9}$$

with G_1, G_2 and v_1, v_2 the shear moduli and Poisson ratios of the two materials, respectively. (Note that this equation is equivalent to the composite elastic modulus given in eq. (3.8), but for shear.) Indeed, experimental values approaching this limit have been observed by Carpick et al. (1996) for clean, strongly adhering contacts in the atomic force microscope (AFM), such as a platinum tip sliding on mica in ultra-high vacuum (UHV).

However, most measurements report shear strengths several times or even decades lower than the theoretical maximum (but still much higher than found for cases involving superlubricity discussed in Section 11.2.2). An explanation for this has been suggested by Hurtado and Kim (1999a, 1999b), who considered how a commensurate single asperity interface behaves as a function of size using concepts from fracture mechanics and dislocation theory. They find that, for small contact areas, the entire contact slips all at once (*concurrent slip*) with a shear strength equal to the theoretical limit, while for larger contact sizes the energy required to nucleate a dislocation around the edge of the contact zone becomes favorable compared to the energy required to cause concurrent slip. For these larger contact areas, the interface slides by a dislocation sweeping across the contact interface, allowing slip by one atomic lattice site; this *single dislocation-assisted slip* reduces the shear strength for these larger contact areas below the theoretical limit. At even larger contact sizes, multiple dislocations sequentially nucleate, pile up, and eventually slip cooperatively. This *multiple dislocation-cooperated* shear strength produces a lower limit known as the Peierls stress, which can be hundreds or thousands of times lower than the shear modulus. These modes of slip at the interface are analogous to plastic slip across a slip plane in a ductile material.

The inherent roughness typically present on most practical surfaces also reduces the shear strengths of contacting interfaces below the theoretical maximum. In particular, simulations of Prof. Mark Robbins and coworkers have shown that the atomic roughness present for amorphous surfaces, as well as the atomic structural features that exist for nominally curved asperities, like surface steps or just local surface curvature, can lead to strong deviations from the predictions of continuum contact mechanics for contact area, contact stresses, contact stiffness, friction, and shear strength (Luan and Robbins 2005, 2006). Consequently, it is crucial to consider the atomic structure of the materials to understand nanoscale contact behavior.

11.2 Atomistic models for static friction

For these atomistic models we will only be considering dry friction (i.e., no lubricant present) and that no plastic deformation or wear occur, so that the atoms are only elastically disturbed from their equilibrium positions. For *static friction*, we can divide the process for energy dissipation into three steps:

1. The interfacial atoms are elastically displaced from their equilibrium positions.
2. When displaced far enough, the atoms reach an unstable configuration and suddenly slip to a new equilibrium position.
3. During this slip process, the elastic strain energy is converted into atomic vibrational energy that manifests itself as frictional heating.

We examine how the first two steps impact static friction first by considering the simple Frenkel–Kontorova and Prandtl–Tomlinson models and then by considering the more sophisticated modeling done using molecular dynamic (MD) simulations.

Later, in Section 11.3, we discuss how energy is dissipated in the third step when the interfacial atoms slip to the next stable configuration by examining the energy dissipation mechanisms of phononic and electronic excitations. The discussion of these mechanisms provides insights into the atomic origins of *kinetic friction*.

11.2.1 Frenkel–Kontorova model

First introduced to describe the physics of dislocations in solids, the *Frenkel–Kontorova model* (Frenkel and Kontorova 1939, Kontorova and Frenkel 1939) is now widely used to describe many physical phenomena, including the friction between crystalline interfaces (Braun and Kivshar 2003, Hirano 2006, Vanossi et al. 2013). As illustrated in Figures 11.2 and 11.3, the Frenkel–Kontorova model consists of a one-dimensional chain of atoms placed in a periodic potential. The atoms on the chain interact with their neighbors on the chain via harmonic potentials represented by springs, and a lateral

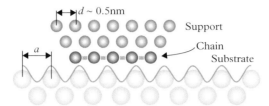

Figure 11.2 *Schematic illustration of how the atoms at the summit of a nanoscale asperity contacting a substrate resemble the atoms in a one-dimensional Frenkel–Kontorova chain. Reproduced with permission from Bylinskii et al. (2016). Copyright 2016, Springer Nature.*

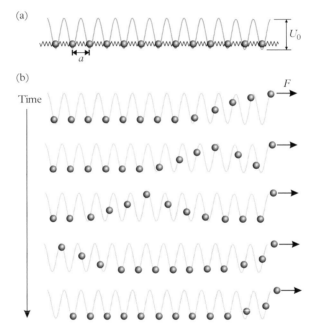

Figure 11.3 *Schematic diagram of the one-dimensional Frenkel–Kontorova model. (a) A chain of atoms interact with each other via overdamped springs while experiencing a periodic potential from the substrate. (b) When the chain (without the springs being shown) is pulled to the right by a lateral force F, a kink develops at the right edge that then moves leftward along the chain. Once the kink reaches the leftmost atom, the chain has moved one lattice spacing. The atoms not in the kink reside at minima of the interfacial potential, while those in the kink are spread out over the depth of interfacial potential. Figure (b) is reproduced with permission from Ward et al. (2015). Copyright 2015, Springer Nature.*

force F pulls the chain along the interfacial periodic potential, either by being applied at one end of the chain, or equally distributed at each atom. As illustrated in Figure 11.2, the atoms on the chain could represent the atoms of one of the contacting surfaces; as these atoms slide, the atoms feel a lateral force generated from the gradient of the periodic potential created by the atoms on the substrate surface. Here we discuss only the one-dimensional case, but the Frenkel–Kontorova model can be straightforwardly extended to a more realistic two-dimensional simulation of actual friction situations (McClelland 1989, Hirano 2006).

Figure 11.3(a) illustrates the special case where the equilibrium spacing of atoms on the chain equals the spacing of the periodic potential. When no lateral force is applied, the equilibrium configuration is for all the atoms to reside at the bottoms of each potential well. If the springs were infinitely stiff, all the atoms would move in unison up the sides of the potential wells when the chain end is pulled to one side; when they come up to the top of the potential peak, they all slide in unison down into the next potential well. In this special case of equal density of atoms and minima, the maximum lateral force needed to initiate sliding (which occurs with infinite chain stiffness) equals the force to overcome each individual potential well multiplied by the number of atoms on the chain. This type of friction is called "atomistic locking" (Hirano 2003).

For chains where the springs are not infinitely stiff, the atoms no longer move in unison, and it becomes easier to slide the chain across the surface. When the chain spring stiffness becomes comparable to the gradients of the interfacial potential wells, the lateral force needed to initiate sliding is greatly reduced by the formation of kinks. For example, if we pull on one end of the chain with enough force to move the first atom a lattice spacing a, as illustrated in Figure 11.3(b), we will introduce a "kink" in the chain, as the tension in the connecting springs has caused the neighboring atoms to adopt a new minimum energy configuration to accommodate the displacement of the edge atom, while most of the remaining atoms stay near the potential well minima. The importance of kinks is that the activation barrier for moving a kink is always much smaller than the amplitude of the interfacial potential U_0; this mean that kinks can be moved along the chain much more easily than moving all the atoms in unison. The higher the kink concentration is, the higher the mobility of the system. For finite chain lengths, kinks are generated at one end, propagate down the length of the chain, and disappear at the other end. The onset of sliding for a finite contact area is then initiated by creating interface dislocations at the edge of the contact zone that then propagate across the contact zone, in a manner similar to the dislocations responsible the ductility of materials (Sections 3.1 and 12.6.2.1).

An important outcome of the Frenkel–Kontorova model is the possibility of frictionless sliding if the chain is incommensurate with the interfacial potential. (The system is said to be incommensurate if the ratio of the intrinsic period of the chain to the substrate period a is an irrational number.) An incommensurate chain becomes "free" or "unpinned" if the stiffness of the interfacial potential is small enough compared to the chain stiffness, leading to the chain atoms staying near their intrinsic locations along the chain (i.e., higher chain stiffnesses resist pinning). For an infinitely long, incommensurate chain of atoms in a interfacial potential with $U_0 < U_c$ where U_c is the threshold potential amplitude for pinning, the total sum of the forces acting on all the atoms is strictly zero, as the loss and gain of each atom's interaction energy cancel each other out; in this situation, any applied lateral force moves the chain. The simple explanation for this is that for every atom going up a side of a potential barrier, somewhere along the infinite chain another atom is going down at just the right point on the potential to cancel the other one. Even for commensurate systems, the force needed to initiate sliding can be greatly reduced by having the ratio of chain atoms to substrate atoms much different than one, as the atoms within the repeating structure are out of phase with each other as they move along the potential curve.

The condition of frictionless sliding $U_0 < U_c$ occurs when the interactions across the sliding interface are much weaker than the interactions between atoms within the solids. This frictionless behavior disappears when $U_0 > U_c$, and the chain becomes locked to the substrate; in these situations, the chain's elastic constant g is less than a threshold value called g_{Aubry}, after S. Aubry who first observed this transition within the Frenkel–Kontorova model (Aubry and Le Daeron 1983). When this locking occurs, local regions are created with a common periodicity (with the atoms close to the minima of the interfacial potential as illustrated in Figure 11.3(b)) that are separated by kinks that require a finite force to move, but one which is smaller than moving the locked atoms en masse.

11.2.2 Superlubricity

The frictionless sliding discussed in the previous section is one of the more exciting predictions of the Frenkel–Kontorova analysis, as it implies that static friction should vanish for an infinitely long, incommensurate chain. That this prediction of ultra-low friction should also extend to realistic two-dimensional sliding systems was initially demonstrated theoretically by Sokoloff (1984) and McClelland (1989), and later confirmed experimentally by Hirano et al. (1991), who named this phenomenon of ultra-low friction *superlubricity*. In the experiments of Hirano et al., incommensurability was achieved by rotating away from commensurability contacting crystals of muscovite mica in an SFA. Since then, ultra-low friction has also been observed experimentally in numerous other incommensurate sliding systems consisting of atomically clean crystalline surfaces, as reviewed by Meyer and Gnecco (2014), Zheng and Liu (2014), and Martin and Erdemir (2018).

A necessary condition for superlubricity is that no wear occurs, particularly from chemical reactions or cold welding. Consequently, the bonding across the sliding interface needs to be much less than the bulk bonding or hardness in order to achieve superlubricity. This means that superlubricity is generally achieved at low loads and between chemically inert materials.

By its very nature superlubricity means that there are no atomic scale stick-slip instabilities. Since atomic scale stick-slip is often the dominant mode for frictional energy dissipation, ultra-low friction results when this channel is shut off. However, other energy dissipation channels remain open, such as phonon emission and electronic excitations. So, even though the static friction vanishes for those incommensurate interfaces undergoing superlubricious sliding, the viscous or kinetic friction does not vanish during sliding. Consequently, the term *superlubricity* is something of a misnomer, as the resistance to motion is never completely absent as it is for other "super" effects in physics, such as superconductivity or superfluidity. The name superlubricity, however, has now stuck for describing all types of ultra-low friction situations, even though Müser

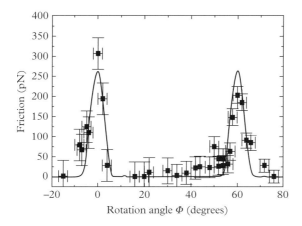

Figure 11.4 *Average friction force measured as a function of sample rotation for an AFM tip sliding over the basal plane surface of a graphite crystal, with a small graphite flake sandwiched between the tip and the graphite surface, as depicted in Figure 11.5. Reprinted with permission from Dienwiebel et al. (2004). Copyright 2004, the American Physical Society.*

(2004) has pointed out that the superlubricity phenomenon in incommensurate sliding systems would be better denoted as *structural lubricity* since it is purely a structural effect.

11.2.2.1 *Superlubricity of a graphite flake on graphite*

A particularly striking example of superlubricity is shown in Figure 11.4. In these experiments by Dienwiebel et al. (2004), an AFM tip is scanned over the basal plane of a graphite crystal. It is well known that friction is very low for AFM tips rubbing against graphite (Mate et al. 1987), and this is now attributed to a small graphite flake being picked up at the end of the AFM tip, as illustrated in Figure 11.5. The small flake leads to the low friction for two reasons:

1. Weak adhesion occurs between the contacting planes of graphite atoms, while the atom lattice within the graphite planes is fairly stiff (Section 10.2.1).
2. The presence of a finite angle between the lattice vectors of graphite flake and graphite substrate generally results in the surfaces being effectively incommensurate.

In the experiment by Dienwiebel et al. shown in Figure 11.4, the graphite sample was rotated underneath the AFM tip and the attached flake. Every 60°, the misfit angle between the lattices of the graphite planes of the flake and the sample becomes zero, and the friction rises sharply as these two surface structures become commensurate; that is, all the graphite atoms on the tip flake become locked with the potential minima of the (0001) surface of the graphite substrate. Rotating the sample a few degrees away from a zero misfit angle is all that is needed to make the flake's atomic structure incommensurate

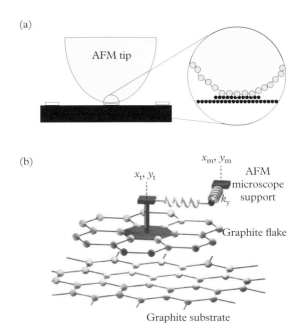

(a)

AFM tip

(b)

x_m, y_m

x_t, y_t

AFM microscope support

k_y

Graphite flake

Graphite substrate

Figure 11.5 *(a) Schematic diagram of an AFM tip dragging a small graphite flake across a graphite sample. (b) Atomic structure of the graphite flake and the (0001) surface of the graphite sample. Figure (b) is reprinted from Verhoeven et al. (2004) with permission from the American Physical Society, copyright 2004.*

with the graphite substrate; this drastically reduces the friction down to near the detection limit of the AFM.

11.2.2.2   Other examples of superlubricity

Since 1991 when structural superlubricity was first observed for mica, almost all subsequent observations have been limited to nano-sized contacts occurring in either high vacuum conditions or in inert atmospheres. The three main challenges to achieving superlubricity over larger length scales are:

1. The surfaces need to be atomically smooth and crystalline over a large area.
2. The surfaces need to be kept completely clean over the areas where contact is expected to occur. As a practical matter, achieving and maintaining surfaces with this degree of cleanliness is only possible under high vacuum conditions or with surfaces with a suitably low surface energy maintained under an inert atmosphere; though Cihan et al. (2016) observed superlubricity in ambient air for gold islands on graphite material, attributed to the ability of the gold-graphene interface to resist contaminant infiltration.

3. Once the area becomes large enough, the interplay between the interface interaction energies and the elastic compliance of the 3-D material surrounding the
   contact interface causes dislocations to form within the contact zone, and these
   dislocations require a finite lateral force to move (Sharp et al. 2016, Dietzel et al.
   2017).

Since a prerequisite for obtaining superlubricity is that the corrugation potential acting
across the sliding interface be weak enough ($U_0 < U_c$), most experimental observations
of this phenomenon have been for lamellar materials—mica, MoS_2, graphite, and
graphene—where the weak bonding between the crystal basal planes makes it easy for a
small flake to slide over a basal plane surface.

However, the phenomenon of superlubricity is not limited to lamellar materials. For
example, it has also been observed in AFM friction experiments on Si(111) (Hirano
et al. 1997) and Ti_3SiC_2 (Crossley et al. 1999). Superlubricity has also been observed
for isolated islands of one type of atom deposited onto a crystalline surface of another
material (Dietzel et al. 2013, Cihan et al. 2016).

Even for clean metal surfaces, it is theoretically possible for superlubricity to occur
between the (111) surfaces of face-centered cubic (fcc) metals, due to their lower surface
energies compared to other metal crystal faces (Singh-Miller and Marzari 2009). To
illustrate this, Figure 11.6 shows the friction force predicted from a molecular dynamics
simulation for a Pt asperity terminated with the (111) crystal face sliding against an
Au(111) surface (Li et al. 2011, Dong et al. 2013). As with the graphite flake experiment
described in the previous section, when the two crystalline surfaces are aligned, they
become somewhat commensurate with each other as the interatomic spacings are similar
($a_{Pt} = 0.392$ nm, $a_{Au} = 0.408$ nm); this results in high friction dominated by atomic
scale stick-slip motion. Once the asperity is rotated away from alignment, the mean
friction quickly drops to near zero, indicating that the surfaces are becoming sufficiently
incommensurate for superlubricity to occur.

11.2.2.3 *Impact of finite contact area on superlubricity*

So far, the discussion has focused on incommensurate interfaces, where both materials
at the contact are crystalline; for these interfaces low friction occurs because the lateral
forces between two incommensurate, infinite, and rigid solids *systematically cancel*. However, ultra-low friction can also occur if one or both of the materials is *amorphous* instead
of crystalline: in this case the lateral forces *randomly cancel* rather than systematically
cancel over the large contact area.

Moreover, only incommensurate or amorphous systems of *infinite extent* can be truly
superlubricious. For incommensurate or amorphous systems with $U_0 < U_c$ in contact
over a *finite extent*, the static friction force can be shown to vary as follows:

- First we take advantage of the statistical result that the net average lateral force
  scales with $\sqrt{N}$, when we sum over the random forces acting on each individual
  atom for amorphous interfaces or the nominally random forces for incommensurate

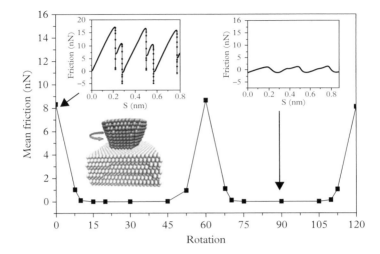

Figure 11.6 *The mean friction predicted from a molecular dynamics simulation for Pt asperity terminated with the (111) crystal face sliding against an Au(111) surface plotted against the rotation between the two surfaces. The two inset plots show representative traces of the friction when aligned at 0° rotation (and effectively commensurate) and misaligned at 90° rotation (and effectively incommensurate); S is the lateral position of the support structure. High friction and stick-slip motion occurs when the surfaces are aligned, and low friction occurs when misaligned due to the superlubricity effect. Simulation parameters: load = 0 nN, temperature = 10 K, speed = 1 m/s, and contact area = 7.3 nm². Reproduced with permission from Dong et al. (2013). Copyright 2013, American Vacuum Society.*

interfaces, where N is the number of atoms in contact (Müser 2004). Since the real area of contact A_r is proportional to N, the net lateral force should scale as $\sqrt{N}$.

- Next we assume that the corrugation amplitude scales with the loading pressure L/A_r.

- Then we can write

$$F_{static} \propto (force\ corrugation\ amplitude) \left[\sqrt{N} \right]$$

$$\propto \frac{L}{A_r} \sqrt{A_r}$$

$$\propto \frac{L}{\sqrt{A_r}}$$

$$\mu_{static} = \frac{F_{static}}{L} \propto \frac{1}{\sqrt{A_r}}. \tag{11.10}$$

From eq. (11.10) we can see that, as the contact area goes to infinity ($A_r \to \infty$), the static friction coefficient goes to zero ($\mu_{static} = F_{static}/L \to 0$), as expected for a superlubricious system. We also see that, for superlubricious systems of finite extent, the static friction force still follows Amontons' law of friction being proportional to load even though the real of contact area is fixed, when the corrugation pressure is proportional to the loading pressure.

A useful way to experimentally examine how contact area impacts the static friction is to use an AFM to measure the lateral force needed to slide an island of one material across the smooth surface of another material, as illustrated in Figure 11.7(a). In such experiments, the interface between the underside of the island and the surface of the substrate onto which it is deposited provides a well-defined interface where the contact area can be measured from the AFM image of the island.

Figure 11.7(b) shows how the friction force varies with the island area as determined in experiments by Dietzel et al. (2013) for amorphous antimony and crystalline gold islands deposited onto a graphite substrate:

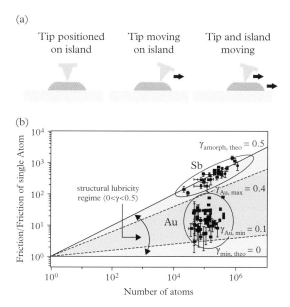

Figure 11.7 *(a) Schematic diagram showing how an AFM tip is used to slide a particle or a small island of material over a smooth substrate. (b) How the friction force needed to slide an island varies with the island area as determined in experiments by Dietzel et al. (2013) for amorphous antimony (circles) and crystalline gold islands (squares) deposited onto a graphite substrate under ultra-high vacuum conditions. The friction force F is normalized by an estimated value of the lateral force acting on an individual atom $F_0 = \Delta E/a$, where ΔE is the estimated energy barrier for sliding a single atom and a is the lattice spacing of the substrate. The island area is normalized by the atomic density to produce the number of interface atoms N. For superlubricity one expects that $F/F_0 = N^\gamma$ with $\gamma = 0.5$; this value of the exponent γ is observed for the amorphous antimony islands but not for the crystalline gold islands. (a) Reproduced from Dietzel et al. (2009) with permission from AIP Publishing; (b) reproduced from Dietzel et al. (2013) with permission from the American Physical Society.*

- For the amorphous antimony islands, the friction scales as $A_r^{0.53\pm0.05}$. This agrees well with the theoretical prediction that, at constant loading pressure, $F_{static} \propto \sqrt{A_r}$.
- For the crystalline gold islands, the friction scales as $A_r^{0.33\pm015}$, a much lower exponent than for the amorphous antimony interface. Dietzel et al. (2013) attribute this lower value and the wider range of exponents to variations in the shapes and the crystal orientations of the islands, a feature that has also been predicted by de Wijn (2012).

11.2.3   Example of extreme atomistic locking: cold welding

A phenomenon somewhat opposite to superlubricity is *cold welding*, which was also previously discussed in Section 10.5. Cold welding occurs when clean metal surfaces are brought together in high vacuum after they have been clean of any absorbed contaminants or oxide layers. Since these metal surfaces typically have very high surface energies (100–1000 mN/m), fairly strong forces are generated on any atoms that come into contact with them. Only for a few metal surfaces are the interaction forces between contacting interfaces weak enough to allow for stable stick-slip sliding without wear; these are typically the low surface energy (111) surfaces for the noble metals, such as for the Pt(111)/Au(111) example discussed in the previous section. For other metal contacts, high adhesive forces drive the interfacial metal atoms to rearrange so as to maximize the metal–metal contact and to minimize the metal area exposed to vacuum, which minimizes the overall surface and interfacial energy. Even if the atoms at the interface are initially incommensurate or amorphous, this rearrangement of atoms results in patches of the interface becoming commensurate. Once commensurate, high friction and adhesion forces result that are on the order of the bulk yield shear strength, resulting in extremely high friction coefficients of $\mu > 4$ (Bowden and Rowe 1956, Buckley 1981, Pashley et al. 1984, Gellman and Ko 2001). In cold welding these friction and adhesion forces are high enough that fracture occurs within the bulk during sliding and microscopic bits of metals are pulled out of one surface and adhere to the other, a process known as *adhesive wear*. In vacuum and other environments where the surfaces stay clean, the friction and adhesion are often high enough for seizure to occur, since the two metal surfaces fuse together as if they were welded. Since this "welding" takes with no heat being applied, this phenomenon is called *cold welding*.

Figure 11.8 shows what can happen at the atomic scale when the adhesion forces are strong enough for cold welding to occur during sliding. In this molecular dynamics simulation, the lattices of the two copper (100) crystal surfaces are aligned so that atoms on the opposing surfaces are commensurate, facilitating atomistic locking. So, when the tip moves across the surface, slippage occurs along the (111) crystal planes *within* the tip rather than at the (100) crystal plane of the original interface; this is because the (100) planes have a higher critical shear yield stress than the (111) planes. Eventually, a pyramid-shaped pile of atoms detaches from the tip to form a potential wear particle.

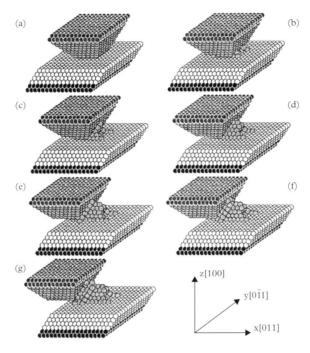

Figure 11.8 *Simulation of an Cu(100) tip sliding against a Cu(100) surface with a commensurate alignment of the two. During sliding, the atoms at the tip's apex adhere to the opposing surface and become detached from the tip due to slip along planes within the tip. Reprinted with permission from Sorensen et al. (1996). Copyright (1996), American Physical Society.*

11.2.4   Prandtl–Tomlinson model

A somewhat simpler and earlier model for friction than the Frenkel–Kontorova model was developed by Prandtl (1928) and Tomlinson (1929) and is usually referred to as the *Prandtl–Tomlinson model* or the independent oscillator model. A one-dimensional Prandtl–Tomlinson model is illustrated in Figure 11.9. As with the Frenkel–Kontorova model, a slider atom feels the periodic potential of the substrate surface atoms as it slides over them; but rather than being connected to each other along a chain as in the Frenkel–Kontorova model, the slider atom is independently connected by a spring to the rigid slider.

In the Prandtl–Tomlinson model, each individual asperity contact is modeled as a single entity contacting a corrugated potential landscape, and the elasticity of this asperity contact is represented by a spring system with an *effective stiffness* k_{eff} (also called the *system stiffness*) and an *effective mass* M_{eff}. In the idealized case displayed in Figure 11.9, a single atom at the end of the asperity or AFM tip apex experiences a periodic potential as it scans laterally over the individual atoms of the substrate surface.

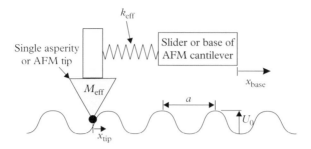

Figure 11.9 *One-dimensional Prandtl–Tomlinson model where a small collection of atoms at the apex of single asperity are connected to a slider by an overdamped spring; these atoms experience a periodic potential from the atomic corrugation of the substrate surface as they slide across this surface. A practical realization of the Prandtl–Tomlinson model is an AFM configured to measure the friction force as an AFM tip is scanned over the surface at a loading force low enough that only a small number of tip atoms are in contact with the sample.*

Initially, as in the Frenkel–Kontorova model, the system is in a configuration of minimum potential energy. When the slider starts to move, the system exhibits static friction if the potential barrier height U_0 is sufficiently high, until sufficient elastic stress is built up in the spring to overcome this potential barrier, at which point the atom jumps rapidly to their next stable state. During these jumps or slips, the stored elastic energy is converted into kinetic energy, which is dissipated as heat; this dissipation process is incorporated into the model as the damping constant for the spring-atom oscillator. If the potential barrier height is lowered to below the threshold for slips to occur, the atoms move smoothly through the potential with negligible net friction.

For the case where an AFM is used to realize Prandtl–Tomlinson behavior, the effective spring constant k_{eff} is usually not just simply the lateral spring constant of the AFM cantilever k_{lever}: rather the lateral stiffness comes from multiple contributions. One way of decomposing these various contributions is

$$\frac{1}{k_{eff}} = \frac{1}{k_{lever}} + \frac{1}{k_{tip}} + \frac{1}{k_{contact}}. \tag{11.11}$$

In addition to the cantilever stiffness k_{lever}, this expression takes into account the other contributions to the overall lateral stiffness within the AFM experiment coming from k_{tip} the lateral stiffness of the shank of the AFM tip (Lantz et al. 1997a) and $k_{contact}$ the lateral contact stiffness of the elastic contact between the tip and the sample (Carpick et al. 1997, Lantz et al. 1997b).

While values of k_{lever} typically range from 8 to 110 N/m, Lantz and coworkers have shown that AFMs with slender tip geometries can have values of k_{tip} in the range 39–84 N/m (Lantz et al. 1997a, 1997b). The lateral contact stiffness can be shown to be given by $8\,G^*a$ for an elastic contact where a is the contact radius and G^* is the reduced shear modulus, given in eq. (11.9) (Carpick et al. 1997, Piétrement and Troyon 2001a).

Values of $k_{contact}$ thus depend on the materials and tip size, but as one example, Lantz and coworkers estimate $k_{contact} = 79$ N/m for a 1.4 nm radius Si tip in contact with a $NbSe_2$ sample at a load of 4.5 nN. That k_{tip}, $k_{contact}$, and k_{lever} fall in overlapping ranges shows that none of these terms can be ignored when considering the lateral force response of an AFM friction measurement.

11.2.4.1 Example: an AFM tip sliding across an NaCl crystal at ultra-low loads

A nice example of using the Prandtl–Tomlinson model to analyze an AFM experiment was done by Socoliuc et al. (2004). In their experiments an AFM tip was scanned over a NaCl surface in UHV at sufficiently small loads that only a few atoms at the end of the tip were in contact. Having only a few atoms in contact is important for obtaining a periodic interfacial potential, since the tip atoms are incoherent with the sample surface atoms and having a large number of atoms in contact would have averaged out the periodic potentials experienced by the individual tip atoms.

Figures 11.10(a)–(c) show the lateral force acting on the tip as it is scanned back and forth across the NaCl surface at different loads. At the highest load of 4.7 nN (Figure 11.10(a)), the trace of the lateral force exhibits a sawtooth structure with the periodicity of the NaCl crystal surface, characteristic of the atomic scale stick-slip motion discussed in Section 4.3.1.3; the presence of atomic scale stick-slip at this load of 4.7 nN indicates that the load is still low enough that only a few atoms at the end of the AFM tip touch the NaCl surface. The slip motion occurs when the gradient of the lateral periodic force exceeds the effective lateral spring constant of the AFM, as illustrated in Figure 4.15; at this point, the tip suddenly slips to the next potential minimum, where it sticks until enough elastic stress is again built up for the next slip. When the tip is scanned in the opposite direction, the slips occur in the other direction, and a hysteresis or friction loop is observed. The area of the friction loop equals the energy dissipated by friction during the back-and-forth-scan; this energy dissipation occurs during the slip portions of the scan when the tip's kinetic energy is dissipated.

When the load is decreased to 3.3 nN (Figure 11.10(b)), the amplitude of the periodic lateral force decreases, leading to a decrease in the magnitude of the static friction prior to the atomic scale slip and a decrease in the energy dissipated in the friction loop. When the loading force is reduced below a threshold value (Figure 11.10(c)), the slips and the hysteresis disappear, and the forward and backward scans overlay each other perfectly, indicating that a negligible amount of energy is now dissipated during the scan. The disappearance of stick-slip motion at low loading force indicates that the amplitude of the lateral periodic friction force becomes sufficiently small at low loads that its gradient no longer exceeds the effective lateral spring constant of the AFM. Since no slips occur, no energy is dissipated during a back and forth scan, and the average friction is zero. This is another form of superlubricity, which occurs for atomic scale contacts at ultra-low loads.

This transition from stick-slip motion to smooth modulation for AFM friction can be explained in straightforward manner within the Prandtl–Tomlinson model. We assume

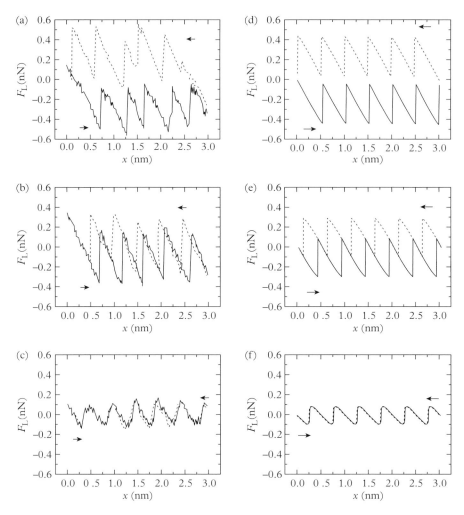

Figure 11.10 *(a–c) Measurement of the lateral force F_L acting on an AFM tip as it scans back and forth over a (001) surface of a NaCl crystal. The applied loading force is (a) $F_N = 4.7$ nN, (b) $F_N = 3.3$ nN, and (c) $F_N = -0.47$ nN. For (c) the negative value of F_N indicates that a slight adhesive force holds the tip in contact as it slider over the NaCl surface. (d–f) Results from the Prandtl–Tomlinson model that correspond well with the experimental results. This model uses an effective lateral spring constant $k_{eff} = 1$ N/m, a lattice periodicity $a = 0.5$ nm, and $\lambda = 5$ for (d), 3 for (e), and 1 for (f). Reprinted with permission from Socoliuc et al. (2004). Copyright 2004, by the American Physical Society.*

that the tip experiences a sinusoidal potential for the lateral force with amplitude U_0 and period a (Figure 11.9) and that x_{tip} and x_{base} are the lateral motion of the tip atom and the base of the AFM cantilever, respectively. Then the tip's energy configuration is described by the combined potentials

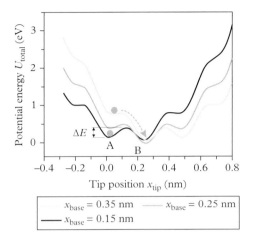

Figure 11.11 *Total potential energy for the lateral interaction acting on an AFM tip as it scans over an atomic surface corrugation, as illustrated in Figure 11.9 and described by eq. (11.12). The parameters for this plot are an effective stiffness $k_{eff} = 1.8$ N/m, a substrate lattice spacing $a = 0.25$ nm, and a corrugation amplitude $U_0 = 0.5$ eV (for this case, the corrugation amplitude is large enough to induced atomic stick-slip, i.e., $\lambda > 1$). Initially, an atom is at the minimum of potential well A, but undergoes a jump to the minimum of B when the energy barrier height ΔE is sufficiently reduced for a thermally activated jump to occur. Adapted from Krylov and Frenken (2014) with permission from Wiley, copyright 2014.*

$$U_{total}\left(x_{tip}, x_{base}\right) = -\frac{U_0}{2}\cos\left(2\pi\frac{x_{tip}}{a}\right) + \frac{1}{2}k_{eff}\left(x_{tip} - x_{base}\right)^2, \tag{11.12}$$

where k_{eff} is the effective spring constant between the tip atom and the AFM base (eq. (11.11)).

Figure 11.11 shows an example of how this total potential evolves as the base of the AFM cantilever moves relative to the sample. For this case the corrugation amplitude U_0 is great enough to result in stick-slip motion, so the potential is a corrugated parabola with a series of minima. During the "stick" phase, the atom at the tip apex initially resides at the minimum of the A well. As cantilever support moves, the energy barrier ΔE between the A and B wells is reduced. When it is reduced to zero, the barrier height vanishes, and the tip "slips" to the lower B well minimum.

Whether the motion is continuous or a series of stick-slips is described by the parameter

$$\lambda = \frac{2\pi^2 U_0}{k_{eff}a^2}. \tag{11.13}$$

When $\lambda > 1$, multiple wells occur on the total potential curve, as shown in Figure 11.11, separated by finite energy barriers; this results in stick-slip motion. When $\lambda < 1$, the energy barriers disappear and the potential curve has only one minimum; this results in continuous motion without stick-slip. Figures 11.10(d)–(f) show the numerical results for the motion for three different values of λ that reproduce well the experimental results of Figures 11.10(a)–(c).

This fitting of the Prandtl–Tomlinson model to the experimental results further indicates that the effective lateral spring constant for this experiment was on the order of 1 N/m. This is a factor of fifty less than the lateral spring constant of the AFM cantilever, indicating that the effective stiffness is dominated by the lateral stiffness of the contact (the combination of k_{tip}, k_{apex}, and $k_{contact}$), which is on the order 1 N/m.

11.2.4.2   *Thermal activation of stick-slip events*

For cases where $\lambda > 1$, an energy barrier ΔE exists between the A and B minima of the total potential, as shown in Figure 11.11. As the sample position x_{base} increases from zero, the barrier height decreases until $\Delta E = 0$; at this point the system is mechanical unstable and a slip occurs from well A to B. At finite temperatures, however, thermally activated jumps are possible over the energy barrier. This jump process is described by transition state theory where the probability rate k for such a thermally activated jump or slip is given by

$$k = f_0 \exp\left(-\frac{\Delta E}{k_B T}\right), \tag{11.14}$$

where k_B is the Boltzmann constant and f_0 is the pre-exponential factor, which is typically associated with the attempt frequency or the characteristic frequency of the tip confined in the potential well (Spikes and Tysoe 2015).

The phenomenon of thermally activated jumps results in several important consequences. The first consequence is that distributions are observed in atomic stick-slip experiments for the slip length and the maximum lateral force F_{max} (Schirmeisen et al. 2006). Also, the F_{max} reached in any particular stick event is less than the maximum possible value F_{max}^* (except at $T = 0$ K, when $F_{max}^* = F_{max}$). For a sinusoidal lateral potential, the maximum possible lateral force is $F_{max}^* = \pi U_0/a$ The second consequence is that the average lateral force needed to induce slip decreases with increasing temperature. This trend of friction decreasing with increasing temperature has been observed experimentally in AFM friction experiments on a variety of materials, including silicon, SiC wafers, ionic crystals, and graphite (Schirmeisen et al. 2006, Brukman et al. 2008, Zhao et al. 2009, Greiner et al. 2010, Jansen et al. 2010, Barel et al. 2011). At extremely low temperatures, friction may reduce again due to a different physical phenomenon from the thermal energy not being enough for strong interfacial bonds to form in the first place, resulting in a smaller interfacial corrugation potential U_0 (Barel et al. 2010).

If one assumes that the energy barrier ΔE increases linearly with lateral force near the critical point for a jump and that jumps in the reverse direction can be neglected, then the exponential dependence of the jump rate on ΔE in eq. (11.14) implies that the average maximum force $\langle F_{max} \rangle$ should decrease linearly with temperature and should also increase logarithmically with velocity v (Gnecco et al. 2000):

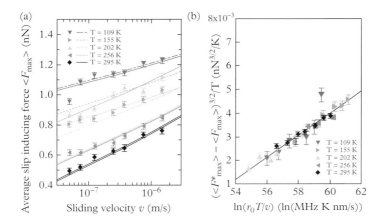

Figure 11.12 *(a) The symbols are the average measured values of the slip inducing force $\langle F_{max} \rangle$ as a function of scanning velocity v, as measured in AFM experiments by Jansen et al. (2010) for an AFM tip scanning over a graphite (0001) surface in ultra-high vacuum. The solid lines show the fit of the measured data to eq. (11.16). (b) This plot shows that all the data measured as a function of velocity and temperature collapse to a single curve predicted by the thermal activation theory described by eq. (11.16). Reprinted with permission from Jansen et al. (2010). Copyright 2010, by the American Physical Society.*

$$\langle F_{max} \rangle = F^*_{max} - \frac{k_B T}{\zeta} \ln\left(\frac{v}{v_1}\right), \tag{11.15}$$

where ζ is of the order π / a and $v_1 = f_0 k_B T / \zeta k_{eff}$. A similar relationship as in eq. (11.15) also holds for the average friction force $\langle F \rangle$, since $\langle F \rangle$ differs from $\langle F_{max} \rangle$ by a constant approximately equal to $1/2 \, a k_{eff}$. The reason for this logarithmic dependence on velocity is that, the faster the AFM tip slides over the surface, the less time it has at each minimum to do a thermally activated jump to the next minimum.

Figure 11.12 shows an example for AFM friction measurements on graphite in UHV that shows both the behavior of that average friction decreasing with increasing temperature and the related phenomenon of the average friction associated with thermal activated stick-slip increasing logarithmically with velocity.

By doing an analysis taking into account the non-linear dependence of $(F^*_{max} - F)$ on ΔE, it can be shown for velocities less than a critical velocity $(v < v_c)$ that the relationship between friction, temperature, and velocity is given in the stick-slip regime by

$$\langle F_{max} \rangle = F^*_{max} - \left[\beta k_B T \ln\left(\frac{v_c}{v}\right) \right]^{2/3}, \tag{11.16}$$

where β is a parameter that depends on the shape of the corrugation potential; for a sinusoidal potential, $\beta = \left(3\pi \sqrt{F^*_{max}}\right) / \left(2\sqrt{2}a\right)$ (Sang et al. 2001, Riedo et al. 2003, Jansen et al. 2010, Krylov and Frenken 2014). Equation (11.16) indicates that $\langle F_{max} \rangle$ increases logarithmically with velocity for low velocities $(v \ll v_c)$; but, as the velocity approaches the critical velocity v_c, $\langle F_{max} \rangle$ deviates from logarithmic behavior of eq. (11.16), and above v_c, it plateaus at $\langle F^*_{max} \rangle$, as the thermal energy is no longer able to assist the slip process. The critical velocity v_c is proportional to the temperature T and prefactor f_0 of the jump rate:

$$v_c = \frac{2\beta f_0 k_B T}{3 k_{eff} \sqrt{F^*_{max}}}. \tag{11.17}$$

Figure 11.12(b) shows how well the experimental data of friction versus temperature and velocity in Figure 11.12(a) fit to the thermal activation theory described by the relation given by eq. (11.16) for this AFM tip sliding over a graphite surface in UHV.

The plateauing of friction at higher scanning velocity is increased towards a critical velocity has been observed in experiments by Liu et al. (2015) for a silicon AFM tip sliding over an Au(111) surface in UHV. This can be understood as follows: at sufficiently high speeds, thermal energy no longer has enough time to assist the tip in jumping over the potential barrier. Thus, one expects a plateau in this essentially athermal regime. A modified version of eq. (11.16) predicts this plateau (Riedo et al. 2003).

This *logarithmic increase* in friction with increasing velocity for AFM friction experiments done in dry environments should be contrasted with the *logarithmic decrease* in friction with increasing velocity for similar AFM experiments done in humid environments that were discussed in Section 10.4. While these opposite trends in friction as function of velocity both originate from thermal activated processes (which accounts for them both having a logarithmic dependence on velocity), the opposite direction of the trends comes from the underlying physical mechanisms being entirely different. In humid environments, the decreasing friction with velocity comes from the capillary menisci of water having less time to fully form around the AFM tip as the velocity increases; while, for atomic scale stick-slip, the increasing friction with velocity comes from atoms having less time to do a thermally activated jump over an energy barrier as the velocity increases.

11.2.5 Role of stiffness in the Frenkel–Kontorova and Prandtl–Tomlinson models

Within both the Frenkel–Kontorova and Prandtl–Tomlinson models, we can intuitively see that the average friction force should increase as the ratio of the interfacial potential amplitude to the stiffness of the spring system increases. (Detailed analyses indicate that the average friction increases linearly with this ratio.) For the Frenkel–Kontorova model, the spring stiffness corresponds to the bonding stiffness of the surface atoms. For Prandtl–Tomlinson model, however, this stiffness corresponds to the stiffness of the measurement system. As expressed by eq. (11.11), this stiffness in AFM experiments comes from the combination of the lateral spring constant of the AFM cantilever, the internal stiffness of the AFM tip to lateral deflections, and the contact stiffness. While it is unsurprising that friction should depend on an interface property corresponding to the potential corrugation, it is somewhat counterintuitive that friction should equally depend on a property of the system that is driving the sliding, that is, the effective spring constant connecting the contact interface to the rest of the driving system.

The dependence of friction on the driving system comes from the sequence of events that occurs for energy to be dissipated during friction:

1. energy is first stored as elastic energy within the stiffness of the driving system;
2. then this energy is transferred to the interfacial atoms in the form of kinetic energy;
3. and finally, this kinetic energy is dissipated into heat via phononic and electronic excitations.

That the average friction within the Prandtl–Tomlinson model is inversely proportional to the effective stiffness indicates that the amount of energy stored in the spring is more important than the efficiency of the transfer of this mechanical energy into heat: at the same force, a stiffer spring stores less energy, resulting in a lower average friction force (Krylov and Frenken 2014).

11.2.6   Molecular dynamic (MD) simulations

While the Prandtl–Tomlinson model has been successfully applied to many situations (for a list, see Szlufarska et al. (2008), Müser (2011), or Liu et al. (2015)), it can be unclear whether the Frenkel–Kontorova or Prandtl–Tomlinson model is the better choice for describing static friction at the atomic scale for a particular situation. For three-dimensional solids, both the Frenkel–Kontorova and Prandtl–Tomlinson models are likely to be equally inadequate for describing the friction: the Frenkel–Kontorova model neglects the interactions of the interfacial layer (represented by the chain) with the bulk of the slider body; the Prandtl–Tomlinson model includes this interaction in a simplified way, but neglects the interactions between the atoms at the slider surface.

A reasonable question to ask is: How well do the general conclusions of the one-dimensional Frenkel–Kontorova and Prandtl–Tomlinson models carry over to more realistic three-dimensional models? MD simulations are one of the main ways how the dependence of friction on atomic structure and on interaction potentials has been systematically studied (for reviews, see Robbins and Müser (2000), Dong et al. (2013), and Vanossi et al. (2013)). For bare surfaces sliding against each other, the basic concepts of the Frenkel–Kontorova and Prandtl–Tomlinson models are found to remain valid in MD simulations (Müser and Robbins 2000, Hirano 2006). In particular, commensurate sliding surfaces are always found to be "pinned," meaning that a threshold lateral force has to be applied to overcome the static friction associated with the pinned atoms. Incommensurate sliding surfaces in these simulations are unpinned and exhibit frictionless sliding, unless one of the surfaces becomes so deformable that the atoms rearrange themselves to accommodate the atoms on the opposing surface; within the Frenkel–Kontorova model, this rearrangement is equivalent to the system being unpinned when the elastic constant $g > g_{Aubry}$ and pinned when $g < g_{Aubry}$.

11.2.7   Why static friction occurs in real-life situations

Since both the Frenkel–Kontorova and Prandtl–Tomlinson models predict vanishingly small friction (superlubricity) to occur when the atomic surface corrugation is low enough, one might suspect superlubricity to be a fairly widespread phenomenon for macroscopic objects contacting at sufficiently low loads. In reality, however, this friction-less behavior turns out to be fairly exotic form of friction that has only been observed in experiments carefully constructed for this purpose. So why, in practice, is static friction is so universal, even though the atoms at most sliding interfaces are incommensurate or amorphous, and the surface forces are usually not strong enough to pull the atoms on opposing surfaces into commensurability?

To resolve this dilemma, Robbins, Müser, and coworkers pointed out that realistic contacts always contain third bodies or "dirt" in the form of adsorbed molecules or atoms. As discussed several times in this book, surfaces invariably adsorb some contamination from the surrounding environment, typically in the form of small hydrocarbon or water molecules that adhere to the surfaces. (This adsorption of contamination can only be prevented by placing the surfaces in ultra-high vacuum.) MD simulations (He et al. 1999, Müser and Robbins 2000, He and Robbins 2001, Müser et al. 2001) indicate that this sandwiching of mobile molecules between the sliding surfaces provides another mechanism for static friction: The key feature these molecules bring to the sliding interface is enough mobility to diffuse within the interface until they find the local energy minima between the opposing surface atoms. Once in these local minima positions, the molecules become locked in and resist lateral motion, as illustrated in Figure 11.13. In the simulations, this locking of the adsorbates generates static and kinetic friction between an otherwise incommensurate interface with a coefficient of friction that is independent of contact area.

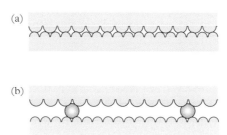

Figure 11.13 *How the introduction of contamination molecules or atoms between two incommensurate surfaces suppresses structural lubricity and generates static friction. (a) Two "clean" incommensurate surfaces in contact, which can experience frictionless sliding at low loads. (b) The same incommensurate surfaces, but now with mobile contamination or "dirt" molecules trapped at the contacting interface and separating the two incommensurate surface potentials from each other. Since the equilibrium configuration is for these contamination molecules to be trapped near potential minima, a lateral force needs to be applied to initiate sliding.*

Another effect that contributes to the suppression of superlubricity and to the creation of static friction is the rearrangement of atoms and molecules at a contacting interface. Besides the adsorbed "dirt" layer between the contacting surfaces discussed above, this rearrangement can also come from the surface atoms and nearby atoms in the bulk that elastically or plastically deform under the influence of strong adhesion and shear forces at the contacting interface, such as discussed in Section 11.2.3. This rearrangement also produces the phenomenon of *adhesion hysteresis* discussed in Section 5.6, where the energy gained when the two surfaces are brought into contact is less than the energy required to separate them. As discussed in Section 11.1.1, higher friction often correlates to higher values of adhesion hysteresis. Consequently, to obtain superlubricity at a contacting interface, the energies required to strain or displace atoms at the interface should be large compared to magnitude of the interaction potentials.

11.3   Energy dissipation mechanisms for friction

Once sliding commences, a kinetic friction force F_k acts on the sliding body opposing its motion across the surface. This kinetic friction force originates from atomic level mechanisms that convert the mechanical work, supplied to make the body slide, into other forms of energy. Figure 11.14 shows schematically the different types of energy into which this frictional work or energy is eventually dissipated. Heat is usually the dominant channel into which frictional energy is finally dissipated. The next major dissipation channels are plastic deformation, generation of new surfaces, and mechanical vibration/sound; the remaining possible dissipation channels, while interesting in their own right, tend to receive insignificant fractions of frictional energy.

One might have suspected that plastic deformation and wear to be a more dominant channel into which frictional energy is finally dissipated, since high friction is often accompanied by high wear rates. This is only true in special situations like for the case of fretting and plowing friction, where the friction force originates from the work

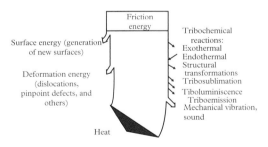

Figure 11.14 *Typical distribution of energy loss through friction during steady sliding. Reproduced from Markov (2013) with permission from Springer Nature, copyright 2013.*

needed to plastically deform the plowed material. However, even in situations where plowing and other forms of plastic deformation and wear are particularly severe, such as for the friction of annealed metals and alloys, experiments have shown that only 10–16% of the energy consumed in friction is typically lost through deformation and wear (Maksimkin and Gusev 2001). Indeed, when surfaces are either oxidized, separated by a lubricant film, or otherwise contaminated, plastic deformation is often minimal, and friction originates mostly from interfacial effects.

This conversion of mechanical energy into heat should be thought of as a multi-stage process, with the friction force being determined only by the particular mechanism that causes the sliding body to irretrievably to lose momentum. The subsequent processes that dissipate this energy into heat do not necessarily contribute to the friction force and may not occur on the same temporal or spatial scales. However, even after many decades of extensive study, for the most part the details of the atomic level mechanisms that generate kinetic friction for unlubricated surfaces still remain hidden.

In the following sections we discuss several of the mechanisms other than deformation and wear that generate interfacial kinetic friction at the atomic level. We begin by outlining what is known about how friction converts mechanical energy into heat via phononic and electronic excitations. Another mechanism for interfacial friction, adhesion hysteresis, was discussed briefly within the cobblestone model of friction (Section 11.1.1) and will not be covered further.

11.3.1 Friction of atoms, molecules, and monolayers sliding over surfaces

A good starting point for considering how kinetic friction originates at the atomic level is to discuss what happens when individual atomic or molecular adsorbates slide over a surface. Figure 11.15 illustrates the idealized situation where an individual atom or molecule slides across a solid surface. In this situation, an average force F is applied to the adsorbate in order to promote its diffusion in the direction of the applied force. This diffusion process occurs by the adsorbate performing thermally activated jumps (or slips, if it is attached to an opposing sliding body that provides the effective spring constant as discussed in Section 11.2.4.2) from one equilibrium site to a neighboring one. During this jump, mechanical work is converted into kinetic energy as the adsorbate picks up speed; then the adsorbate's momentum is irretrievably lost when it essentially "collides" with the surface atoms surrounding the new equilibrium site (Figure 11.1), and the mechanical energy is transferred into exciting the atomic vibrations of the surface atoms.

This process where motion is impeded by atomic level collisions can be described by the Einstein model for diffusion of small particles through a viscous medium. In this case, the average velocity v for diffusion due to an applied force F is given by

$$v = \chi F, \tag{11.18}$$

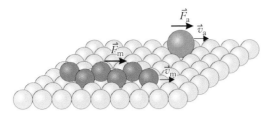

Figure 11.15 *Schematic diagram of an atom and a molecule diffusing across a surface in presence of a driving force. An adsorbate moves across the surface at an average velocity v in response to an average driving force F.*

where χ is the adsorbate's mobility, which is related to the diffusion coefficient D by the Einstein equation

$$D = \chi k_B T$$

$$\chi = \frac{D}{k_B T}.$$ (11.19)

An alternative way of describing this motion is by the relation

$$F = m \eta_m v$$ (11.20)

where m is the molecular mass of the diffusing entity, v is its average velocity, and η_m represents the viscosity it experiences.

The physics of the friction of diffusing adsorbed molecules has been thoroughly reviewed by Krim (2012).

11.3.1.1   *Quartz crystal microbalance*

Since the pioneering experiment of Krim and Widom (1988), where a krypton monolayer was slid over a gold substrate, the *quartz crystal microbalance* (QCM) has been the main technique for measuring friction dissipation for adsorbed layers sliding over surfaces. As illustrated in Figure 11.16, a QCM consists of a thin single crystal of quartz with metal electrodes deposited in its top and bottom surfaces. Quartz crystals are used as they have very little internal dissipation or friction, enabling them to oscillate with a very sharp resonance frequency, typically between 5 and 10 MHz, driven by a voltage applied to the electrodes. When atoms or molecules are adsorbed on the electrode surfaces, the added mass causes the resonance frequency of the crystal to shift to a lower frequency. Due to the very sharp resonance, very small shifts in the resonance frequency can be readily detected, enabling the coverage of a minute fraction of a monolayer to be measured.

Since the resonance motion is a transverse shear oscillation of the quartz crystal, the electrode surfaces move laterally, as illustrated in Figure 11.16. If the adsorbed layer cannot keep up with this motion, it slides over the surface with a relative velocity v_{rel} and

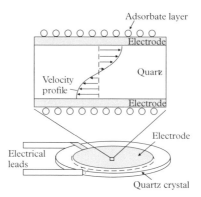

Figure 11.16 *Schematic diagram of the quartz crystal microbalance (QCM) used for measuring the friction dissipation of adsorbed layers sliding across surfaces (bottom). An enlarged cross-sectional view of the center of the QCM showing the motion of the resonance oscillation of the quartz crystal as well as a submonolayer coverage of adsorbed noble gas atoms on the metal electrode surfaces (top).*

dissipates its kinetic energy via friction, which broadens the resonance. This can also be measured with great sensitivity. QCM measurements of interfacial friction are limited to sliding systems with very low friction such as absorbed rare gases or small, physisorbed molecules that can slide easily over the electrode surface, while those adsorbed layers with higher friction or chemisorbed layers that are more strongly bonded to the substrate result very little slippage. Typically, noble metals are used for the electrode surface, as adsorbates generally bond to them with a lower binding energy, promoting slippage. These metals are deposited in such a manner that the exposed surface is predominately a smooth, close packed crystalline surface, which further promotes ease of sliding.

Generally, for these weakly bound atoms and molecules that slide easily on a QCM surface, no threshold force is needed to induce sliding (i.e., no static friction); the adsorbed layers experience only viscous friction during sliding described by eq. (11.20), which can also be described by a slip time,

$$t_{slip} = 1/\eta_m, \tag{11.21}$$

that is inversely proportional to the frictional damping. This slip time t_{slip} corresponds to the time for the velocity of the adsorbed layer to decay by $1/e$ after the driving force F is removed.

11.3.1.2 *Example: Xe on Ag(111)*

Figure 11.17 shows the slip time for xenon on a Ag(111) surface as a function of coverage (Daly and Krim 1996). In these QCM experiments, the xenon films slide a distance of 2 nm, which is about one tenth the peak-to-peak amplitude of the lateral motion of the crystal surface, and have a peak sliding velocity on the order of 1 cm/s. The data in Figure 11.17 have several interesting features:

- As the xenon coverage increases from zero to the completion of an incommensurate monolayer at 0.9 monolayer, the slip time drops from 2.1 ns to 0.8 ns, indicating increasing interfacial friction as the atoms become more closely packed.

- Maxima occur in the slip times when a sparse two-dimensional xenon gas phase exists on either the silver surface or the complete xenon monolayer or bilayer surfaces (i.e., the slip time maxima occur at coverages of <0.3, 1.2, and 2.05 monolayers).

- At one monolayer coverage, the xenon atoms have the same lattice spacing as in bulk xenon and slide over the Ag(111) surface as a single solid layer with a shear strength given by

$$s = \frac{\rho v_{\mathrm{rel}}}{t_{\mathrm{slip}}}, \tag{11.22}$$

where ρ is the mass per unit area of the xenon film.
From the results in Figure 11.17, Daly and Krim deduce that the shear strength for sliding a one-atom-thick monolayer is 11.9 Pa, while for a two-atom-thick bilayer it is 15.1 Pa or about 25% higher. As discussed in the following section, MD simulations by Tomassone et al. (1997) indicate that the friction in this system comes from predominantly phononic excitations.

The shear strengths 11.9 Pa and 15.1 Pa are at the low end of the range 10^{-2}–10^{10} Pa found for crystalline surfaces sliding over each other (Krim 1995, 1996). Indeed, low end

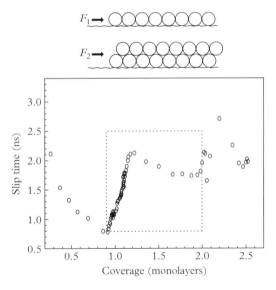

Figure 11.17 *Schematic diagram of the force F_1 needed to slide a one-atom-thick layer and the force F_2 needed to slide a two-atom-thick bilayer of xenon atoms across a Ag(111) surface (top). The slip time τ as function of coverage for xenon on a QCM with Ag(111) electrode surfaces (bottom). Reprinted with permission from Daly and Krim (1996). Copyright 1996, by the American Physical Society.*

of this range mainly comes from QCM experiments for noble gases sliding on surfaces, while the upper portion—such as the 2.5×10^7 Pa calculated by eq. (11.7)—are those measured observed for solid–solid contacts. Even though the shear strengths measured by QCM for gas monolayers on solid surfaces are orders of magnitude less than for solid–solid interfaces, they should still provide a good model system for determining physical origins of kinetic friction dissipation. In the following sections, we consider how phononic and electronic excitations can lead to the observed frictional dissipation in QCM experiments.

11.3.1.3 *Phononic friction*

For most cases, this viscous damping of the atoms and molecules as they move across a surface comes primarily from sliding-induced atomic vibrations (phononic excitations) that are ultimately transformed into heat. The phononic contribution to η_m has been studied analytically and with MD simulations (Cieplak et al. 1994, Tomassone et al. 1997, Persson 1998, Liebsch et al. 1999, Robbins and Müser 2000, Torres et al. 2006). In analytical models for adsorbed noble gases sliding over surfaces, a sinusoidal potential with a corrugation U_0 is typically used to model the adatom–substrate interaction. Then η_m can be expressed as

$$\eta_m = \eta_{\text{subs}} + cU_0^2. \tag{11.23}$$

The first term η_{subs} in eq. (11.23) is the energy dissipation within the substrate from both phonon and non-phonon excitations (such as electronic excitations discussed in the next section), while the second term is the dissipation from phonon excitations occurring within the adsorbate layer. In eq. (11.23) c is a constant that depends on the temperature, the substrate's lattice spacing, and the lifetime of the phonons generated. Good agreement has been found between numerical simulations and experiments for phonon dissipation of noble gas atoms sliding on metal surfaces (Tomassone et al. 1997, Coffey and Krim 2005).

11.3.1.4 *Electronic friction*

In addition to phononic friction, adsorbates can also experience friction from electronic excitations as they move over a conductive material (Persson and Nitzan 1996, Persson 1998, Liebsch et al. 1999, Bruch 2000). The basic mechanism is as follows: when an adsorbate collides with the substrate atoms, in addition to directly exciting atomic vibrations, the collisions also excite conduction electrons near the surface, generating electron-hole pairs. When the electron–hole pairs recombine, the excess energy is converted into phonons and dissipated as heat.

The electronic contribution to friction is more difficult to model than phononic friction and difficult to distinguish in experiments from phononic friction. The clearest indication that it can be a significant contributor to the damping of adsorbate motion on conductors comes from QCM experiments of Prof. Krim and coworkers (Dayo et al. 1998, Highland and Krim 2006). In these experiments the coefficient of damping η_m for several adsorbates on lead was observed to drop abruptly when the lead substrate enters the superconducting state. Since the phonon corrugation potentials and coupling mechanisms are unaffected by the superconductivity transition, this change in adsorbate damping is attributed to electronic mechanisms.

As to the possible mechanism for electronic friction, Persson (1993) has developed a theory for how electronic excitations contribute to viscous damping of adsorbates and how this increases the resistivity in the near surface region of a conductor. As a consequence, measurements of the increase in electrical resistivity when molecules are adsorbed on a surface can be used to estimate the amount of this damping. Several refinements and alternatives to Persson's theory have also been developed and used to estimate the electronic friction contribution. (For a review of these theories and comparison to QCM results, see the review by Krim (2012).) While estimates of the electronic contribution are often comparable the total friction measured in QCM experiments, the accuracy of these theories remains an open question, as it is often possible for the same adsorbate/substrate system to explain the total friction from models or simulations based solely on phononic friction.

11.3.1.5   *Pinning of an absorbed layer*

Previously, when discussing adsorbates sliding across a surface, as illustrated in Figure 11.15, we model the system as adsorbates experiencing only viscous friction as they slide over a smooth surface. In reality, the atomic nature of surfaces means that they can never be perfectly smooth, as atomic corrugation always exists on surfaces. Figure 11.18(a) illustrates the variation in an adsorbed atom's binding energy as it travels across a surface. For physisorbed atoms, the maximum gradient of the energy barrier E_B between adsorption sites typically corresponds to a lateral force on the order of 10^{-3} eV/Å; in a typical QCM experiment, however, the force of inertia F_i is on the order 10^{-9} eV/Å, many orders of magnitude less than what is needed to move a physisorbed atom directly over these barriers. Instead, the external force F lowers the height of the diffusion barrier slightly making, it more likely for adsorbates to diffuse in the direction of F. The sliding velocity v then corresponds to the average drift velocity of the adsorbate layer, which is proportional to F. Since this diffusion is a thermally activated process, the drift velocity at low adsorbates coverages can be shown to be

$$ v \approx F \frac{a^2 f_0}{k_B T} \exp\left(\frac{-E_B}{k_B T}\right), \qquad (11.24) $$

where f_0 is the pre-exponential factor (Persson 1998, Section 8.3.1). So, for a sufficiently large energy barrier E_B or a sufficiently low temperature T, $v \sim 0$, and the adsorbates are effectively pinned to the surface.

Let's consider the example of a QCM experiment performed by Bruschi et al. (2002) with krypton on gold. At a sufficiently low coverage, all the adsorbed krypton atoms are pinned at high binding energy defect sites. When this surface was vibrated in the QCM experiment with a low vibration amplitude, the inertia force was insufficient to overcome the energy barriers around these defect sites. When the vibration amplitude was increased to above 3.4 Å, this generated a sufficient inertia force to eject krypton

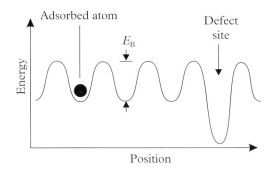

Figure 11.18 *Schematic of how the binding energy of adsorbed atoms varies across a surface, both from the periodic potential of surface atoms and the higher binding energy of a defect site. The adsorbate can move across the surface by thermal excitation over the energy barriers of height E_B.*

atoms from a few of these defect sites, and these loose atoms knocked the remaining pinned krypton atoms free. This resulted in the slip time jumping suddenly from 0 to 7 ns when the adsorbed layer became unpinned. Once unpinned, the adsorbed layer displayed hysteresis in the slip time when the amplitude was decreased. This was attributed to the adsorbed layer staying in the unpinned state down to a smaller vibration amplitude than when the amplitude was increased. In essence, a pinned adsorbed layer exhibits static friction, where a threshold shear stress must be applied to initiate sliding, and since the static friction is higher than the kinetic friction, hysteresis is observed.

11.3.2 Frictional dissipation mechanisms in solid–solid sliding

As discussed previously, when a lateral force is applied so as to slide one solid against another, initially the atoms on the opposing surfaces are pinned in various equilibrium positions if the amplitude of the corrugation potential is high enough to prevent superlubricity. Once enough elastic strain has built up in the solid material, the system becomes mechanical unstable and either slides to the next stable equilibrium position or continues to slide while experiencing an average kinetic friction force. During sliding, mechanical energy is converted into other forms of energy, most predominately into heat. The kinetic frictional force is generated during the initial step of the conversion process when the mechanical energy becomes irretrievably lost.

As discussed in the previous sections, QCM experiments indicate that the dominant mechanisms for generating kinetic friction are phononic and electronic damping. However, the magnitude of the shear stress in a QCM experiment is orders of magnitude less than the shear strength for solid–solid shearing, as measured by AFM and SFA experiments. Still, the handful of AFM experiments designed to investigate mechanisms for kinetic friction have also found evidence of phononic and electronic dissipation during solid–solid sliding.

Evidence that phononic excitations are a major source of dissipation of frictional energy comes from experiments comparing the friction measured on two samples that are identical except that different isotopes are used for one of elemental constituents. Since lattice vibrations mediate the transfer of energy to phononic excitations during the dissipation of frictional energy, changing the vibrational frequencies by changing of the mass of the atoms at one of the contacting surfaces should result in a change in friction if phononic damping is contributing to it. Some examples of experiments that confirm phononic excitations as a mechanism for solid–solid friction:

- Cannara et al. (2007) measured friction for an AFM tip sliding on hydrogen-terminated diamond and silicon surfaces, and on the same surfaces terminated with deuterium. Over a range of loads, the D-terminated surfaces showed ~30% lower friction than the H-terminated surfaces.
- Kajita et al. (2015) observed slightly lower friction for an AFM tip sliding on a diamond crystal isotopically-enriched with^{13}C atoms compared with a mostly^{12}C diamond crystal with natural-abundance of^{13}C.
- de Mello et al. (2017) measured the friction for a microsized diamond probe sliding over amorphous carbon films and found that friction was reduced when deuterium was substituted for hydrogen in the films.

In all three of these experiments, it was found that friction is reduced for materials with higher mass isotopes. The simple idea for why higher mass isotopes lower the amount of energy dissipated through friction is that the dynamic viscosity η the atoms experience when sliding over the interface is reduced when the lattice vibration frequencies responsible for converting the energy into phonon excitations are reduced. The lower viscous damping occurs because the rate of energy dissipation is lowered when the atoms vibrate more slowly, as the lower atomic vibration frequency means fewer opportunities per unit time to transfer energy from the interface into the solids. A similar mechanism also happens during the sliding of adsorbed monolayers over surfaces in QCM experiments (Persson and Ryberg 1985).

For adsorbates sliding on surfaces, Persson (1998) has modeled the dissipation of kinetic energy into vibrational energy of the substrate lattice and predicts that the frictional shear stress on a sliding adsorbate is

$$\tau_{o,\text{vib}} = m\eta\upsilon\sigma, \tag{11.25}$$

where m is the dynamical effective mass of the adsorbate–substrate vibration, η is the damping constant of the interaction, υ is the sliding velocity, and σ is the areal density of the adsorbates. The viscosity η is related to the density ρ and the transverse sound velocity c_T by

$$\eta = \frac{3m\omega^4}{8\pi\rho c_T^3}, \tag{11.26}$$

where m and ω refer to the adsorbate's mass and vibrational frequency, respectively (Persson and Ryberg 1985).

Considering the adsorbate vibration frequency depends on the adsorbate mass as $1/\sqrt{m}$, the viscosity thus varies as $\sim 1/m$. One might think that eq. (11.26) would predict that the friction of the deuterium terminated surface in the experiments by Cannara et al. (2007) would be 50% of that of hydrogen terminated surface. However, the mass m actually represents the reduced mass of the vibrating adsorbate–substrate system, lowering the expected ratio to $\sim 1/1.72$ for H vs. D on a diamond surface. The observed ratio ($\sim 1/1.26$) in Cannara et al. (2007)'s experiments is lower than this; this indicates that other non-phononic (and adsorbate mass-independent) also contribute to friction for both the H- and D-terminated substrates, lowering the observed friction ratio.

Several AFM experiments have also found evidence that electronic excitations can contribute significantly to dissipating frictional energy. For example:

- Park et al. (2006) reported that for doped silicon samples the friction depends on the concentration of conduction electrons near the surface. In these experiments, patterned regions of the silicon were made with alternating p- and n-doping which exhibited different friction forces. The lattice vibrations do not differ significantly between p- and n-doping; rather, the difference in friction is attributed to the difference in the number of electron–hole pairs that can be excited by sliding. Also, the friction force increased by up to a factor of two in the p-doped regions when a sufficiently high positive bias was applied to the sample with respect to the tip; this high positive bias bends the electronic bands in the p-doped regions, causing charge carriers to accumulate at the surface where they can contribute to friction through the excitation of electron–hole pairs.

- A similar result was also seen in AFM experiments with doped GaAs samples by Qi et al. (2008), where they also found that electron accumulation from an applied bias voltage leads to a significant increase in friction. Estimates by Qi et al. (2008) using current theories for dissipation of frictional energy through the creation of electron–hole pair production were too low to explain this increase. This may be because these theories were designed to explain the electronic frictional dissipation for adsorbed monolayers on surfaces as discussed in Section 11.3.1.4, and they do not consider how the high strains generated by the high shear stresses present during solid–solid sliding affect the electronic band structure. Alternately, Qi et al. (2008) propose that the friction may be enhanced due to forces exerted by charges trapped within the thin insulating oxide present on the samples. The trapped charges are produced by the tip–sample field as the tip passes over a given region and act to electrostatically restrain the tip from moving forward, which is observed as increased friction. The authors conclude that while this is undoubtedly an electronic effect, it is not the result of the damping resulting from electron–hole pair production, the standard form of electronic excitation and dissipation.

11.4　PROBLEMS

1. Consider the case where an AFM tip is slid with an applied load of 3.0 nN along the [100] direction of a (001) surface of a hydrogen-terminated silicon single-crystal wafer. As it slides, the tip exhibits atomic lattice stick-slip friction. Assume that:

   - The interfacial shear strength is 500 MPa.
   - The work of adhesion is 60 mJ/m^2.
   - The tip radius is 20 nm.
   - The rectangular AFM cantilever was fabricated from a (001)-oriented single crystal silicon to be 150 μm long, 40 μm wide, and 0.5 μm thick, and that the tip is 10 μm tall. The long axis of the cantilever is aligned with the [100] direction of the silicon crystal from which it is made.
   - The elastic Young's modulus of silicon along the <100> directions is 130 GPa and the Poisson's ratio is 0.28. (Note: you will need this information to calculate the normal and torsional force constants of the cantilever and to apply the DMT model to the contact. You can assume that the shank of the tip is rigid, except for the round end itself which makes contact with the silicon sample.)
   - The DMT model appropriately describes this elastic adhesive contact.

   (a) How big is the contact radius and contact area at this load?
   (b) Create a quantitative, labeled plot of the lateral force as a function of lateral displacement when the sample is displaced laterally 3.0 nm with respect to the cantilever. Label the force and positions at which the tip undergoes atomic stick-slip. (You may wish to refer to the paper by Carpick et al. (1997).)

2. Consider the experimental data shown in Figure 11.10.

   (a) By analyzing the data showing in Figures 11.10(a) and (b), determine your own value of k_{eff} from this data.
   (b) Assuming the transition to smooth sliding occurs at the load given in Figure 11.10(c), find the value of F^*_{max} and from this, calculate U_0.
   (c) Assuming a lattice constant of $a = 0.5$ nm, show that the value of λ is close to 1.

11.5　REFERENCES

Aubry, S. and P. Y. Le Daeron (1983). "The discrete Frenkel-Kontorova model and its extensions. I. Exact results for the ground-states." *Physica D: Nonlinear Phenomena* 8(3): 381–422.

Barel, I., M. Urbakh, L. Jansen and A. Schirmeisen (2010). "Multibond dynamics of nanoscale friction: the role of temperature." *Physical Review Letters* 104(6): 066104.

Barel, I., M. Urbakh, L. Jansen and A. Schirmeisen (2011). "Unexpected temperature and velocity dependencies of atomic-scale stick-slip friction." *Physical Review B* **84**(11): 115417.

Berman, A., C. Drummond and J. Israelachvili (1998). "Amontons' law at the molecular level." *Tribology Letters* **4**(2): 95–101.

Berman, A. D. and J. N. Israelachvili (1998). "Surface forces and microrheology of molecularly thin liquid films." In: *Handbook of micro/nano tribology* (2nd ed.), B. Bhushan, Ed. Boca Raton, FL: CRC Press, pp. 387–449.

Bowden, F. P. and G. W. Rowe (1956). "The adhesion of clean metals." *Proceedings of the Royal Society of London, Series A: Mathematical and Physical Sciences* **233**(1195): 429–42.

Bowden, F. P. and D. Tabor (1950). *The friction and lubrication of solids.* Oxford: Clarendon Press.

Braun, O. M. and Y. S. Kivshar (2003). *The Frenkel–Kontorova model: concepts, methods, and applications.* Berlin: Springer.

Briscoe, B. J. (1981). "Friction and wear of organic solids and the adhesion model of friction." *Philosophical Magazine A: Physics of Condensed Matter, Structure Defects and Mechanical Properties* **43**(3): 511–27.

Briscoe, B. J. and A. C. Smith (1982). "The interfacial shear strength of molybdenum disulfide and graphite films." *ASLE Transactions* **25**(3): 349–54.

Briscoe, B. J. and D. Tabor (1978). "Shear properties of thin polymeric films." *Journal of Adhesion* **9**(2): 145–55.

Bruch, L. W. (2000). "Ohmic damping of center-of-mass oscillations of a molecular monolayer." *Physical Review* B **61**(23): 16201–6.

Brukman, M. J., G. Gao, R. J. Nemanich and J. A. Harrison (2008). "Temperature dependence of single-asperity diamond–diamond friction elucidated using AFM and MD simulations." *The Journal of Physical Chemistry C* **112**(25): 9358–69.

Bruschi, L., A. Carlin and G. Mistura (2002). "Depinning of atomically thin Kr films on gold." *Physical Review Letters* **88**(4): 046105.

Buckley, D. H. (1981). *Surface effects in adhesion, friction, wear, and lubrication.* Elsevier.

Bylinskii, A., D. Gangloff, I. Counts and V. Vuletic (2016). "Observation of Aubry-type transition in finite atom chains via friction." *Nature Materials* **15**(7): 717–21.

Cannara, R. J., M. J. Brukman, K. Cimatu, A. V. Sumant, S. Baldelli and R. W. Carpick (2007). "Nanoscale friction varied by isotopic shifting of surface vibrational frequencies." *Science* **318**(5851): 780–3.

Carpick, R. W. and M. Salmeron (1997). "Scratching the surface: fundamental investigations of tribology with atomic force microscopy." *Chemical Reviews* 97(4): 1163–94.

Carpick, R. W., N. Agraït, D. F. Ogletree and M. Salmeron (1996). "Measurement of interfacial shear (friction) with an ultrahigh vacuum atomic force microscope." *Journal of Vacuum Science & Technology B* **14**(2): 1289–95.

Carpick, R. W., D. F. Ogletree and M. Salmeron (1997). "Lateral stiffness: a new nanomechanical measurement for the determination of shear strengths with friction force microscopy." *Applied Physics Letters* **70**(12): 1548–50.

Chen, Y. L., C. A. Helm and J. N. Israelachvili (1991). "Molecular mechanisms associated with adhesion and contact-angle hysteresis of monolayer surfaces." *Journal of Physical Chemistry* **95**(26): 10736–47.

Cieplak, M., E. D. Smith and M. O. Robbins (1994). "Molecular origins of friction: the force on adsorbed layers." *Science* **265**(5176): 1209–12.

Cihan, E., S. Ipek, E. Durgun and M. Z. Baykara (2016). "Structural lubricity under ambient conditions." *Nature Communications* 7: 12055.

Coffey, T. and J. Krim (2005). "Impact of substrate corrugation on the sliding friction levels of adsorbed films." *Physical Review Letters* **95**(7): 076101.

Crossley, A., E. Kisi, H, J. W. B. Summers and S. Myhra (1999). "Ultra-low friction for a layered carbide-derived ceramic, Ti_3SiC_2, investigated by lateral force microscopy (LFM)." *Journal of Physics D: Applied Physics* **32**(6): 632.

Daly, C. and J. Krim (1996). "Sliding friction of solid xenon monolayers and bilayers on Ag(111)." *Physical Review Letters* **76**(5): 803–06.

Dayo, A., W. Alnasrallah and J. Krim (1998). "Superconductivity-dependent sliding friction." *Physical Review Letters* **80**(8): 1690–3.

de Mello, S. R. S., M. E. H. M. da Costa, C. M. Menezes, C. D. Boeira, F. L. Freire Jr, F. Alvarez and C. A. Figueroa (2017). "On the phonon dissipation contribution to nanoscale friction by direct contact." *Scientific Reports* **7**(1): 3242.

de Wijn, A. S. (2012). "(In)commensurability, scaling, and multiplicity of friction in nanocrystals and application to gold nanocrystals on graphite." *Physical Review B* **86**(8): 085429.

Desaguliers, J. T. (1734). *A course of experimental philosophy.* London: p.182.

Dienwiebel, M., G. S. Verhoeven, N. Pradeep, J. W. M. Frenken, J. A. Heimberg and H. W. Zandbergen (2004). "Superlubricity of graphite." *Physical Review Letters* **92**(12): 126101.

Dietzel, D., J. Brndiar, I. Štich and A. Schirmeisen (2017). "Limitations of structural superlubricity: chemical bonds versus contact size." *ACS Nano* **11**(8): 7642–7.

Dietzel, D., M. Feldmann, H. Fuchs, U. D. Schwarz and A. Schirmeisen (2009). "Transition from static to kinetic friction of metallic nanoparticles." *Applied Physics Letters* **95**(5): 053104.

Dietzel, D., M. Feldmann, U. D. Schwarz, H. Fuchs and A. Schirmeisen (2013). "Scaling laws of structural lubricity." *Physical Review Letters* **111**(23): 235502.

Dong, Y., Q. Li and A. Martini (2013). "Molecular dynamics simulation of atomic friction: a review and guide." *Journal of Vacuum Science & Technology A: Vacuum, Surfaces, and Films* **31**(3): 030801.

Frenkel, Y. I. and T. Kontorova (1939). *Phys. Z. Sowjetunion* **13**: 1 [Sov. Phys. USSR 1, 137 (1939)].

Gao, J. P., W. D. Luedtke, D. Gourdon, M. Ruths, J. N. Israelachvili and U. Landman (2004). "Frictional forces and Amontons' law: from the molecular to the macroscopic scale." *Journal of Physical Chemistry B* **108**(11): 3410–25.

Gee, M. L., P. M. McGuiggan, J. N. Israelachvili and A. M. Homola (1990). "Liquid to solid-like transitions of molecularly thin-films under shear." *Journal of Chemical Physics* **93**(3): 1895–1906.

Gellman, A. J. and J. S. Ko (2001). "The current status of tribological surface science." *Tribology Letters* **10**(1–2): 39–44.

Gnecco, E., R. Bennewitz, T. Gyalog, C. Loppacher, M. Bammerlin, E. Meyer and H.-J. Güntherodt (2000). "Velocity dependence of atomic friction." *Physical Review Letters* **84**(6): 1172.

Greiner, C., J. R. Felts, Z. Dai, W. P. King and R. W. Carpick (2010). "Local nanoscale heating modulates single-asperity friction." *Nano Letters* **10**(11): 4640–5.

He, G. and M. O. Robbins (2001). "Simulations of the kinetic friction due to adsorbed surface layers." *Tribology Letters* **10**(1–2): 7–14.

He, G., M. H. Müser and M. O. Robbins (1999). "Adsorbed layers and the origin of static friction." *Science* **284**(5420): 1650–2.

Highland, M. and J. Krim (2006). "Superconductivity dependent friction of water, nitrogen, and superheated he films adsorbed on Pb(111)." *Physical Review Letters* **96**(22): 226107.

Hirano, M. (2003). "Superlubricity: a state of vanishing friction." *Wear* **254**(10): 932–40.

Hirano, M. (2006). "Atomistics of friction." *Surface Science Reports* **60**(8): 159–201.

Hirano, M., K. Shinjo, R. Kaneko and Y. Murata (1991). "Anisotropy of frictional forces in muscovite mica." *Physical Review Letters* **67**(19): 2642–5.

Hirano, M., K. Shinjo, R. Kaneko and Y. Murata (1997). "Observation of superlubricity by scanning tunneling microscopy." *Physical Review Letters* **78**(8): 1448–51.

Homola, A. M., J. N. Israelachvili, M. L. Gee and P. M. McGuiggan (1989). "Measurements of and relation between the adhesion and friction of two surfaces separated by molecularly thin liquid-films." *Journal of Tribology, Transactions of the ASME* **111**(4): 675–82.

Homola, A. M., J. N. Israelachvili, P. M. McGuiggan and M. L. Gee (1990). "Fundamental experimental studies in tribology: the transition from interfacial friction of undamaged molecularly smooth surfaces to normal friction with wear." *Wear* **136**(1): 65–83.

Hurtado, J. A. and K. S. Kim (1999a). "Scale effects in friction of single–asperity contacts. I. From concurrent slip to single–dislocation–assisted slip." *Proceedings of the Royal Society of London, Series A: Mathematical, Physical and Engineering Sciences* **455**(1989): 3363–84.

Hurtado, J. A. and K. S. Kim (1999b). "Scale effects in friction of single–asperity contacts. II. Multiple-dislocation-cooperated slip." *Proceedings of the Royal Society of London, Series A: Mathematical, Physical and Engineering Sciences* 455(1989): 3385–400.

Israelachvili, J. N. (2005). "Importance of pico-scale topography of surfaces for adhesion, friction, and failure." *MRS Bulletin* **30**(7): 533–9.

Israelachvili, J. N., Y.-L. Chen and H. Yoshizawa (1994). "Relationship between adhesion and friction forces." *Journal of Adhesion Science and Technology* 8(11): 1231–49.

Jansen, L., H. Hölscher, H. Fuchs and A. Schirmeisen (2010). "Temperature dependence of atomic-scale stick-slip friction." *Physical Review Letters* **104**(25): 256101.

Kajita, S., M. Tohyama, H. Washizu, T. Ohmori, H. Watanabe and S. Shikata (2015). "Friction modification by shifting of phonon energy dissipation in solid atoms." *Tribology Online* **10**(2): 156–61.

Kontorova, T. A. and Y. I. Frenkel (1939). *Zh. Eksp. Teor. Fiz.* 8: 89.

Krim, J. (1995). "Progress in nanotribology: experimental probes of atomic scale friction." *Comments on Condensed Matter Physics* 17: 263–80.

Krim, J. (1996). "Friction at the atomic scale." *Scientific American* **275**(4): 74–80.

Krim, J. (2012). "Friction and energy dissipation mechanisms in adsorbed molecules and molecularly thin films." *Advances in Physics* **61**(3): 155–323.

Krim, J. and A. Widom (1988). "Damping of a crystal-oscillator by an adsorbed monolayer and its relation to interfacial viscosity." *Physical Review B* **38**(17): 12184–9.

Krylov, S. Y. and J. W. M. Frenken (2014). "The physics of atomic-scale friction: basic considerations and open questions." *Physica Status Solidi (b)* **251**(4): 711–36.

Lantz, M. A., S. J. O'Shea, A. C. F. Hoole and M. E. Welland (1997a). "Lateral stiffness of the tip and tip-sample contact in frictional force microscopy." *Applied Physics Letters* **70**(8): 970–2.

Lantz, M. A., S. J. O'Shea, M. E. Welland and K. L. Johnson (1997b). "Atomic-force-microscope study of contact area and friction on $NbSe_2$." *Physical Review B* **55**(16): 10776.

Li, Q. Y., Y. L. Dong, D. Perez, A. Martini and R. W. Carpick (2011). "Speed dependence of atomic stick-slip friction in optimally matched experiments and molecular dynamics simulations." *Physical Review Letters* **106**(12): 1261101.

Liebsch, A., S. Goncalves and M. Kiwi (1999). "Electronic versus phononic friction of xenon on silver." *Physical Review B* **60**(7): 5034–43.

Liu, X.-Z., Z. Ye, Y. Dong, P. Egberts, R. W. Carpick and A. Martini (2015). "Dynamics of atomic stick-slip friction examined with atomic force microscopy and atomistic simulations at overlapping speeds." *Physical Review Letters* **114**(14): 146102.

Luan, B. and M. O. Robbins (2005). "The breakdown of continuum models for mechanical contacts." *Nature* **435**(7044): 929–32.

Luan, B. and M. O. Robbins (2006). "Contact of single asperities with varying adhesion: comparing continuum mechanics to atomistic simulations." *Physical Review E: Statistical, Nonlinear, and Soft Matter Physics* **74**(2): 26111.

MacMillan, N. H. (1972). "The theoretical strength of solids." *Journal of Materials Science* **7**(2): 239–54.

Maksimkin, O. P. and M. N. Gusev (2001). "Some features of the energy dissipation in the course of plastic deformation of iron and niobium." *Technical Physics Letters* **27**(12): 1065–6.

Markov, D. P. (2013). "Development of ideas on friction mechanisms." *Journal of Friction and Wear* **34**(1): 70–82.

Martin, J. M. and A. Erdemir (2018). "Superlubricity: Friction's vanishing act." *Physics Today* **71**(4): 40–6.

Mate, C. M., G. M. McClelland, R. Erlandsson and S. Chiang (1987). "Atomic-scale friction of a tungsten tip on a graphite surface." *Physical Review Letters* **59**(17): 1942–5.

McClelland, G. M. (1989). "Friction at weakly interacting interfaces." In: *Adhesion and friction*, M. Grunze and H. J. Kreuzer, Eds. Berlin: Springer-Verlag, pp. 1–16.

Meyer, E. and E. Gnecco (2014). "Superlubricity on the nanometer scale." *Friction* **2**(2): 106–13.

Müser, M. H. (2004). "Structural lubricity: role of dimension and symmetry." *Europhysics Letters* **66**(1): 97.

Müser, M. H. (2011). "Velocity dependence of kinetic friction in the Prandtl–Tomlinson model." *Physical Review B* **84**(12): 125419.

Müser, M. H. and M. O. Robbins (2000). "Conditions for static friction between flat crystalline surfaces." *Physical Review B* **61**(3): 2335–42.

Müser, M. H., L. Wenning and M. O. Robbins (2001). "Simple microscopic theory of Amontons's laws for static friction." *Physical Review Letters* **86**(7): 1295–8.

Park, J. Y., D. F. Ogletree, P. A. Thiel and M. Salmeron (2006). "Electronic control of friction in silicon pn junctions." *Science* **313**(5784): 186–186.

Pashley, M. D., J. B. Pethica and D. Tabor (1984). "Adhesion and micromechanical properties of metal surfaces." *Wear* **100**(1–3): 7–31.

Persson, B. N. J. (1993). "Applications of surface resistivity to atomic scale friction, to the migration of 'hot' adatoms, and to electrochemistry." *The Journal of Chemical Physics* **98**(2): 1659–72.

Persson, B. N. J. (1998). *Sliding friction: physical principles and applications*. Berlin: Springer.

Persson, B. N. J. and A. Nitzan (1996). "Linear sliding friction: on the origin of the microscopic friction for Xe on silver." *Surface Science* **367**(3): 261–75.

Persson, B. N. J. and R. Ryberg (1985). "Brownian motion and vibrational phase relaxation at surfaces: CO on Ni(111)." *Physical Review B* **32**(6): 3586–96.

Piétrement, O. and M. Troyon (2001a). "Study of the interfacial shear strength on carbon fibers surface at the nanometer scale." *Surface Science* **490**(1): L592–6.

Piétrement, O. and M. Troyon (2001b). "Study of the interfacial shear strength pressure dependence by modulated lateral force microscopy." *Langmuir* **17**(21): 6540–6.

Prandtl, L. (1928). "Ein Gedankenmodell zur kinetischen Theorie der festen Körper." *Zeitschrift für Angewandte Mathematik und Mechanik* **8**: 85–106.

Qi, Y., J. Y. Park, B. L. M. Hendriksen, D. F. Ogletree and M. Salmeron (2008). "Electronic contribution to friction on GaAs: an atomic force microscope study." *Physical Review B* **77**(18): 184105.

Riedo, E., E. Gnecco, R. Bennewitz, E. Meyer and H. Brune (2003). "Interaction potential and hopping dynamics governing sliding friction." *Physical Review Letters* **91**(8): 084502.

Robbins, M. O. and M. H. Müser (2000). "Computer simulations of friction, lubrication and wear." In: *Modern tribology handbook*, B. Bhushan, Ed. Boca Raton, FL: CRC Press, pp. 747–96.

Sang, Y., M. Dubé and M. Grant (2001). "Thermal effects on atomic friction." *Physical Review Letters* **87**(17): 174301.

Schirmeisen, A., L. Jansen, H. Hölscher and H. Fuchs (2006). "Temperature dependence of point contact friction on silicon." *Applied Physics Letters* **88**(12): 123108.

Sharp, T. A., L. Pastewka and M. O. Robbins (2016). "Elasticity limits structural superlubricity in large contacts." *Physical Review B* **93**(12): 121402.

Singh-Miller, N. E. and N. Marzari (2009). "Surface energies, work functions, and surface relaxations of low-index metallic surfaces from first principles." *Physical Review B* **80**(23): 235407.

Socoliuc, A., R. Bennewitz, E. Gnecco and E. Meyer (2004). "Transition from stick-slip to continuous sliding in atomic friction: entering a new regime of ultralow friction." *Physical Review Letters* **92**(13): 134301.

Sokoloff, J. B. (1984). "Theory of dynamical friction between idealized sliding surfaces." *Surface Science* **144**(1): 267–72.

Sorensen, M. R., K. W. Jacobsen and P. Stoltze (1996). "Simulations of atomic-scale sliding friction." *Physical Review B* **53**(4): 2101–13.

Spikes, H. and W. Tysoe (2015). "On the commonality between theoretical models for fluid and solid friction, wear and tribochemistry." *Tribology Letters* **59**(1): 21.

Suda, H. (2001). "Origin of friction derived from rupture dynamics." *Langmuir* **17**(20): 6045–7.

Szlufarska, I., M. Chandross and R. W. Carpick (2008). "Recent advances in single-asperity nanotribology." *Journal of Physics D: Applied Physics* **41**(12): 123001.

Tabor, D. (1982). "The role of surface and intermolecular forces in thin film lubrication." In: *Microscopic aspects of adhesion and lubrication*. J. M. Georges, Ed. New York: Elsevier, pp. 651–82.

Tomassone, M. S., J. B. Sokoloff, A. Widom and J. Krim (1997). "Dominance of phonon friction for a xenon film a silver (111) surface." *Physical Review Letters* **79**(24): 4798–801.

Tomlinson, G. A. (1929). "A molecular theory of friction." *Philosophical Magazine* **7**(46): 905–39.

Torres, E. S., S. Goncalves, C. Scherer and M. Kiwi (2006). "Nanoscale sliding friction versus commensuration ratio: molecular dynamics simulations." *Physical Review B* **73**(3): 035434.

Vanossi, A., N. Manini, M. Urbakh, S. Zapperi and E. Tosatti (2013). "Modeling friction: From nanoscale to mesoscale." *Reviews of Modern Physics* **85**: 529–52.

Verhoeven, G. S., M. Dienwiebel and J. W. M. Frenken (2004). "Model calculations of superlubricity of graphite." *Physical Review B* **70**(16): 165418.

Ward, A., F. Hilitski, W. Schwenger, D. Welch, A. W. C. Lau, V. Vitelli, L. Mahadevan and Z. Dogic (2015). "Solid friction between soft filaments." *Nature Materials* **14**(6): 583–8.

Yoshizawa, H., Y. L. Chen and J. Israelachvili (1993). "Fundamental mechanisms of interfacial friction. 1. Relation between adhesion and friction." *Journal of Physical Chemistry* **97**(16): 4128–40.

Zhao, X., S. R. Phillpot, W. G. Sawyer, S. B. Sinnott and S. S. Perry (2009). "Transition from thermal to athermal friction under cryogenic conditions." *Physical Review Letters* **102**(18): 186102.

Zheng, Q. and Z. Liu (2014). "Experimental advances in superlubricity." *Friction* **2**(2): 182–92.

12

Wear

The most common definition of wear is the removal of material when one solid surface rubs against another, though this definition is often broadened to include any form of damage occurring to a sliding surface.

Wear has been used as a manufacturing process going back to the beginning of the Stone Age 3.4 million years ago when the antecedent species of human beings began to fashion the first tools from stones by cutting, grinding, and chipping them into useful shapes. Over the millennia, humans developed ever more sophisticated grinding, cutting, and polishing processes for machining materials into highly controlled shapes and surface finishes. The machining of parts is still one of the most prevalent manufacturing operations, with the expenditure on machining in developed countries estimated to comprise 5% of their GDP (Ivester et al. 2000).

Impressive advances continue to be made in reliably and reproducibly machining and fabricating parts with precise dimensions and finishes (Arrazola et al. 2013). With these advances it is now possible to machine parts with nanometer dimensional tolerances and to achieve control of surface roughness to within a few angstroms. For example, in semiconductor fabrication it is often necessary to remove the irregularities in surface topography generated from earlier processing steps so as to flatten or "planarize" the surface before adding additional circuit elements. This is accomplished through a process called *chemical-mechanical polishing* (*CMP*) or *planarization*, where a polishing pad rubs an abrasive and corrosive chemical slurry against the wafer to remove the uneven material. With advanced CMP processes, heterogeneous surfaces can be polished to a roughness of only a few angstroms. As discussed in Section 2.1, such smooth surfaces are often not the best tribological surfaces, so, in addition to those developed for smoothing surfaces, a wide variety of other machining processes have been developed for roughening surfaces in a controlled manner.

Not all wear is desirable, and wear can lead to catastrophic consequences. For example, on January 31, 2000, Alaska Airlines flight 261 crashed into the Pacific Ocean killing all eighty-eight people on board because improper lubrication of a jackscrew had led to excessive wear and eventually to the loss of control of the plane's horizontal stabilizer.

Another example of a catastrophic wear event occurred on October 17, 2000 when a train derailed in Hatfield, UK, killing four people and injuring over 100. The crash was caused by cracking of a portion of the rail due to rolling contact fatigue that was induced by the tribological stresses over years of train wheels rolling over the rail (Jost 2005,

Tribology on the Small Scale: A Modern Textbook on Friction, Lubrication and Wear. Second edition. C. Mathew Mate and Robert W. Carpick. © Oxford University Press 2019. Published in 2019 by Oxford University Press.
DOI: 10.1093/oso/ 9780199609802.001.0001

Independent Investigation Board 2006). Proper training and inspection by railway staff would have alerted them to this problem, but procedures were not followed partially due to a lack of engineering tribology knowledge in the rail company. Multiple criminal and civil proceedings were filed against the company responsible for the rail infrastructure that led to fines and settlements: this company eventually went bankrupt due to these fines and settlements in combination with the large losses from emergency maintenance and service slowdowns.

12.1   Three stages of component wear

As material deterioration from wear often leads to component failure, mechanical failures from wear are some of the main reasons for shutdowns and unavailability of machinery (Holmberg 2001). When designing components and mechanical systems, one of the main challenges typically faced by a tribology engineer is ensuring that wear is sufficiently minimized so that they can be safely operated over their expected lifetime.

For sliding contacts within machinery, wear typically occurs in three stages as illustrated in Figure 12.1(a):

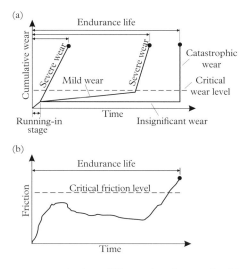

Figure 12.1 *(a) How wear progresses in three different situations, each with a different endurance life (time until the component can no longer perform its intended function). Also indicated is a critical wear level, above which the component still works but the amount of wear is above a safe level. In the first situation, severe wear occurs shortly after a component begins operation: this is usually caused by poor tribological design. The second and third situations respectively correspond to mild and insignificant wear during the operational life of a component, and exhibit three stages of wear: (i) a* running-in stage, *which has a higher rate of wear than the subsequent stages in this particular example; (ii) an* operational stage *with a low rate of wear; and (iii) a* final stage *with a high rate of wear leading to failure. (b) An example where the friction varies as wear changes the contact conditions. In this example, a critical friction level is specified for safe operation. Shortly after exceeding this level, the endurance life is reached. Reproduced with permission from Holmberg (2001). Copyright 2001, Elsevier.*

1. An initial stage where two contacting surfaces adapt to each other through the deformation and wear of asperity contacts. During this *running-in* or *breaking-in stage*, the wear rate is usually higher than the "steady-state" wear rate to which the system eventually evolves.

2. Once a steady-state wear rate is achieved, the wear process enters the mid-age or *steady-state* or *operational stage*. For sliding components, the operational life of the device occurs during this stage.

3. The old-age stage is entered when the wear rate begins to rise rapidly, typically leading to failure of the device. This final *wearing-out stage* has widespread economic consequences ranging from the cost of replacement parts, machine downtime, and lost business.

While underlying wear mechanisms for many specific situations have been determined in detail, in this chapter we will not attempt a comprehensive review of these previous studies of wear (such a review would require many volumes). Instead, we provide only a brief outline of the most common mechanisms contributing to wear and discuss how the modern tribologist analyzes wear processes in terms of these mechanisms.

12.2   Modeling wear

The degree of wear is typically determined by how multiple factors—contact stresses, sliding speeds, environment, material properties, lubricant properties, etc.—combine in a complex manner within the tribological system. At the nanoscale, a wide variety of atomic and molecular phenomena also contribute to the overall wear process, further adding to the difficulty of developing a general understanding of wear processes. Consequently, despite the technological importance of wear, no simple and universal model has been developed to describe it.

Over the decades, many researchers have studied wear and have developed numerous models for wear. A literature review by Meng and Ludema (1995) identified 182 equations for wear published between 1957 and 1992. They pointed out that many of these equations or models, however, are empirically-based, relying on data obtained for specific systems; as a consequence, they only work for specific material pairs, contact geometries, lubrication conditions, and other operating conditions.

Often though, some fundamental understanding exists for the particular wear behavior occurring within a specific system, allowing for an empirical model to be developed for such a system where some degree of scientific guidance is used. These types of models can be divided into two categories (Meng and Ludema 1995):

- *Contact mechanics-based models*—These models start by assuming that the factors important for the mechanics of the contacting interface are also important for determining the wear rate at the interface. The developers of these models then choose those material properties to incorporate into the model that they believe are important to the wear process, usually the elastic modulus E or the hardness H. The Archard wear equation discussed in the following section is an example of this type of model. Often, the impact of topography on the real area of contact is also taken into account.

- *Material failure mechanism-based models*—These models start from the assumption that wear rates should be directly related to the underlying failure mechanisms that initiate surface damage. Consequently, these models emphasize incorporating material quantities related to material flow, fracture toughness, fatigue properties, etc. The delamination wear theory discussed in Section 12.2.2 is an example of this type of model.

12.2.1 Archard wear equation

We start with a very simple model created by Archard (1953) and which is found to have widespread utility in many wear situations. This model usually referred to as the *Archard wear equation.*

This simple model derives an expression for the wear rate W, which is defined as the volume V of material removed per unit sliding distance s:

$$W = \frac{V}{s}.$$ (12.1)

From this definition we see that a starting assumption for the Archard model is that the wear volume is proportional to the sliding distance. While this is frequently borne out in experiments, it should be kept in mind that it is not universally true that the wear volume increases linearly with sliding distance. In particular, the wear rate can change dramatically with sliding distance if the resulting damage changes the nature of sliding surfaces to such an extent that the wear mechanism changes. This is typically what happens when the wear transitions between the different wear stages: from the running-in stage to the steady-state stage, or from the steady-state stage to the wearing-out stage.

Another main assumption of the Archard model is that, due to surface roughness, contact only occurs where the asperities touch. In this situation, the true area of contact is the sum of all the individual asperity contact areas. This assumption implies that the wear occurs by the formation of individual wear particles at individual asperity contacts, as illustrated in Figure 12.2. For an individual asperity contact, the characteristic contact radius is a, with the maximum contact area

$$\Delta A \sim \pi a^2.$$ (12.2)

As illustrated in Figure 12.2, wear occurs when a fragment of material detaches from an asperity. In the Archard wear model, this fragment volume is presumed to be proportional to the cube of the contact dimension a. If we approximate the shape of this wear fragment as a hemisphere with radius a, the volume of this wear fragment is then approximated as

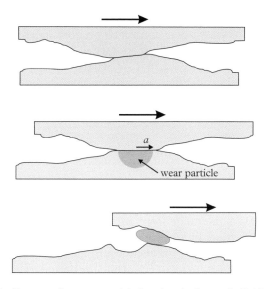

Figure 12.2 *Schematic diagram of a wear particle forming during an individual asperity contact.*

$$\Delta V \sim \frac{2}{3}\pi a^3. \tag{12.3}$$

While in contact, the asperity slides the short distance

$$\Delta s \sim 2a. \tag{12.4}$$

Most asperity contact events, however, do not generate a wear fragment. Instead, detachment only occurs for a small fraction κ of the contact events occurring in a sliding distance s. Then, the average wear rate when a generating wear particle from an individual asperity contact is

$$\frac{\Delta V}{s} = \kappa \frac{\Delta A}{3}. \tag{12.5}$$

The overall wear rate is the sum over the individual asperity contacts:

$$\begin{aligned} W &= \frac{\kappa}{3}\sum \Delta A \\ &= \frac{\kappa}{3}A_{\mathrm{r}}, \end{aligned} \tag{12.6}$$

where A_{r} is the real area of contact between the two surfaces.

In Archard's original model, he considered the wear at metal surfaces where it is safe to assume that deformation of the asperity contacts is predominately plastic. As discussed in Section 3.4.1.2, this leads to the real area of contact A_r being proportional to the loading force L and described by eq. (3.43): $A_r = L/H$, where H is the hardness of the surface region. To simplify the final equation, we will combine the factor of 1/3 in eq. (12.6) with the constant of proportionality by setting $K = \kappa/3$; this leads to the Archard wear equation for the wear rate W:

$$W = \frac{KL}{H}. \tag{12.7}$$

In this equation the dimensionless quantity K is referred to as the *coefficient of wear* or the *wear coefficient*. It is always less than unity and typically much less: for unlubricated materials, K typically lies in the range from 10^{-2} to 10^{-7}.

One of the primary purposes of lubrication is to lower wear rates, with the lowest wear rates being obtained for full-film, hydrodynamic lubrication; for these situations, the wear coefficient is usually insignificant ($<10^{-13}$), while for boundary lubricated systems, the wear coefficients generally range from 10^{-6} to 10^{-10} (as expected this is less than for unlubricated systems.)

For those situations where the deformation of the contacting asperities is primarily elastic, the real area of contact is given by eq. (3.45):

$$A_r \simeq 3\left(\frac{R}{\sigma_s}\right)^{1/2}\frac{L}{E_c}, \tag{12.8}$$

where E_c is the composite elastic modulus of the two surfaces, R the mean radius of curvature of the surface asperities, and σ_s the standard deviation of the asperity heights. In this elastic situation, our simple model indicates that the wear rate should be given by

$$W = \frac{K_e L}{E_c}, \tag{12.9}$$

where K_e is the coefficient of wear given by

$$K_e = \kappa\left(\frac{R}{\sigma_s}\right)^{1/2}. \tag{12.10}$$

As it is often difficult to know whether the asperity deformations are predominantly plastic or elastic, the Archard wear equation is often expressed as

$$W = kL, \tag{12.11}$$

where k is called the *dimensional wear coefficient*. It is usually quoted in units of $\text{mm}^3/(\text{Nm})$. Informally and confusingly, k is often called the *wear rate* or *specific wear rate* for a given material pair, even though the term *wear rate* normally refers to the volume removed per unit sliding distance as given by eq. (12.1).

For dry sliding systems without lubrication, the wear coefficient can vary by many orders of magnitude. This should be contrasted with the friction coefficient, which typically lies in the fairly narrow range between 0.2 and 0.8 for most dry sliding systems (see Figure 4.2). This suggests that the wear rate is much more sensitive than friction to the underlying physical factors impacting wear. While the wear coefficients provide a valuable means for comparing the severity of wear, they tell us nothing about the underlying mechanisms that are responsible for the wear, that is, what causes the wear particle to become detached.

12.2.2 Delamination theory of wear

Another common model for wear is the *delamination theory of wear* proposed by Suh (1973, 1977). This model is often used to explain why wear debris forms with a lamellar shape in wear situations involving metals under dry or boundary lubrication conditions. In these situations, the wear is modeled as a fatigue process, where the repeated loading and unloading during repetitive sliding or rolling contacts causes fatigue crack growth in the subsurface region.

Delamination wear occurs in a series of stages:

1. Initially, some smoothing of the contacting asperities occurs due to the adhesive removal and plastic flattening of protruding asperities. This smoothing initially leads to increasing areas of contact. Alternatively, a series of grooves can be gouged into the softer surface that are aligned with the direction of sliding.

2. Over time, plastic deformation accumulates, from repeated contacts, in the near surface region. For metals, this accumulation of plastic deformation results in work hardening and embrittlement.

3. Next, holes or voids form in the subsurface. These voids are created by the dislocations, generated during plastic deformation, piling up around inclusions and imperfections in the materials, as illustrated in Figure 12.3(a). These dislocations eventually nucleate to form voids, which then grow by trapping subsequent dislocations created during the repeated cycles of plastic deformation.

4. Under the action of repeated sliding, the voids elongate in the direction of the applied shear stress and nucleate cracks that propagate mainly parallel to the surface. For the regions closest to the surface, the higher hydrostatic or compressive stress suppresses crack formation down to a certain depth; so cracks only form below this depth and only in the narrow range where the shear stress is large enough to form a crack, as illustrated in Figure 12.3(b).

5. These cracks tend to propagate parallel to the surface; cracks directed perpendicular tend to be closed by the shear from the asperities sliding over them, and also by the parallel stress component that, except very close the surface, is typically compressive (Suh 1977). Eventually they merge and finally reach the surface,

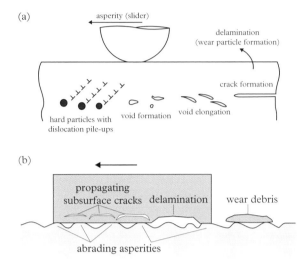

Figure 12.3 *(a) Suh's (1973) model for subsurface crack formation for an individual asperity sliding against a flat surface. During the plastic deformation caused by an asperity sliding in contact, dislocations pile up around particle inclusions and other imperfections. These dislocations nucleate into holes or voids that elongate in the direction of the shear stress. Cracks form at those depths where the shear stress is sufficient to generate a crack, which propagates parallel to the surface. Adapted from Suh (1973) and reprinted from Argibay and Sawyer (2012) with permission. Copyright 2012, Elsevier. (b) A block with a flat surface sliding against a rough surface with abrading asperities. Within the block a network of parallel cracks form below the surface; eventually these cracks merge and some turn towards surface, releasing a long thin lamellar shaped wear debris particle.*

leading to long and thin sheets of material that delaminate from the substrate. One consequence of the cracks occurring at narrow range of depths is that these lamellar debris particles all have a similar thickness. After delamination, fresh surface regions are produced and so the process can then repeat itself.

While originally conceived for metals, further studies have indicated that this type of delamination wear process may also occur for other materials. Also exhibiting wear via flake-like removal are ceramics such as alumina (Deuis et al. 1997), structural composite alloys like Al–Si (Zhang and Alpas 1993), and polymers like Teflon (Blanchet and Kennedy 1992).

12.3   Mechanisms of wear

The Archard wear equation is deceptively simple—it basically says that the wear rate is proportional to the real area of contact, which is proportional to the loading force for rough contacting surfaces. This would tempt one to think that we need only to determine the coefficient of wear for a particular sliding system to adequately describe the wear rate. In practice, however, wear coefficients can change dramatically as the sliding conditions change. We have already mentioned how transient wear rates occur during an initial running-in period while the sliding surfaces evolve toward their steady-state sliding conditions. Wear coefficients can also depend strongly on surface roughness, sliding speed, temperature, and gas or liquid environment; thus, the overall sliding conditions must be considered. Also, wear rates tend to be only proportional to the loading force over the limited range where a particular wear mechanism dominates.

For most sliding systems, no single wear mechanism dominates at all operating conditions. Rather, a variety of wear mechanisms contribute, with their relative importance changing as the sliding conditions change and with abrupt changes in wear rates occurring when one dominant wear mechanism replaces another. That wear can occur in many ways is one reason why many wear models have been developed.

Next, we discuss some of the common wear mechanisms. Here a *wear mechanism* is defined as the process by which material is removed from the contact surfaces. This should be distinguished from a *wear mode*, which is determined by having a specific type of contact geometry, relative motion, or environment. Examples of wear modes are sliding wear, rolling wear, fretting, erosion, and cavitation.

12.3.1   Wear from plastic deformation

For metals sliding at low speeds, mechanical wear processes involving plastic deformation dominate. The general requirement for plastic deformation is that the mechanical stresses generated by the adhesive, loading, and frictional forces exceed the yield stress of one or both of the sliding materials.

For sliding contacts, the magnitude of the friction coefficient is often the key factor in determining whether or not plastic deformation occurs. (For analysis of this for a Hertzian contact see Johnson 1985, Section 7.1). During lateral loading or sliding, the shear stresses generated by static or kinetic friction add to those from the loading and adhesive forces. As discussed in Sections 3.2.3 and 3.3.3, this not only increases the magnitude of the maximum shear stress, but also moves the location of the maximum closer to the surface. For the stationary case ($\mu = 0$), the stress is concentrated a distance ~$0.5a$ below the surface. Using the von Mises criterion for plastic flow, as long as μ stays below 0.3 during sliding, the maximum shear stress and the associated plastic flow remain beneath the surface. This constraining of the plastic flow region beneath the surface limits the possible plastic strain that accumulates with each sliding pass.

As the traction force from friction increases, the location of the maximum stress moves closer to the surface and more towards the trailing edge of the contact zone. When μ is above 0.3, the maximum stress is concentrated at the surface at the trailing edge of the contact zone. When the peak shear stress occurs at the surface, the induced plastic flow is much less constrained than for subsurface yield, resulting in larger shear strains and more surface damage. For coefficients of friction significantly greater than 0.3, wear by surface damage happens immediately, while for coefficients less than 0.3 shear strain accumulates during repeated sliding until a critical value of strain leads to surface damage. One of the ways that lubrication helps to minimize wear from plastic

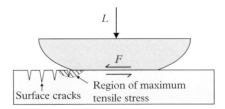

Figure 12.4 *Surface cracks forming in a brittle material behind the trailing edge of a slider, due to the tensile shear stresses generated by the friction F as the slider moves from left to right.*

deformation is by providing for coefficients of friction well below 0.3; the same effect underlies why oxide layers often help to reduce wear on metal surfaces.

Plastic deformation generated by sliding contacts leads to a number of distinct wear mechanisms:

- *Delamination*—As mentioned above in Section 12.2.2, plastic flow from repeated sliding contacts nucleate subsurface cracks that propagate parallel to the surface. These lateral cracks eventually result in delamination when they extend out to the free surface to form platelet-like wear particles.

- *Surface cracks*—Ductile materials can become brittle due to fatigue. For such materials (and also for brittle materials in general), the tensile stresses generated by the traction within the contact region lead to surface fractures and cracks, as illustrated in Figure 12.4. (For example, with a Hertzian contact as discussed in Section 3.2.3, a tensile stress exists at the surface and acts in the radial direction, which is maximal at the edge of the contact.) Wear and fracture for brittle ceramics are discussed further in Sections 12.4.2 and 12.6.3.

- *Fatigue*—Repetitive variations of stresses from repeated sliding or rolling contact results in fatigue failure in the near-surface region. In addition to delamination wear, fatigue is also prevalent in many other wear processes. One particular failure mode arising from fatigue is *pitting* (Totten 2017): often seen in lubricated gears operating in the EHL regime, pitting occurs when embrittlement and subsurface cracks are caused by the repeated contact stresses. These cracks eventually emerge at the surface, resulting in small volumes of material being removed forming "pits" on the surface.

- *Mixing*—With the contact sliding of unlubricated ductile materials, the friction induced plastic flow leads to high plastic shear strains (in the range 10–1000) at the sliding surfaces (Rigney 2000). One of the consequences of these high strains is that the grains near contacting surfaces are often reshaped down to a few nanometers in size. Another consequence of this is mixing of the chemical constituents of the sliding surfaces, resulting in the near surface regions and wear debris being a mixture of the material from the two sliding surfaces (Rigney 2000, Karthikeyan et al. 2005, Rigney and Karthikeyan 2010).

- *Seizure*—During the sliding of metals at extremely high loads, the high contact pressures and shear stresses acting on the contacting asperities cause them to undergo extensive plastic flow and junction growth (Section 4.2.4). Under these circumstances, the real area of contact approaches the nominal area, leading to such high adhesion and shear forces over these enlarged contact zones that momentary seizure of the moving parts takes place followed by severe wear when they move again. The severe damage to the surface caused by this adhesive welding is referred as *scuffing*.

- *Adhesion*—Adhesive forces combined with plastic flow pull out wear particles from the tips of the asperities. Scuffing and seizure are common failure modes caused by excessive adhesive wear. Adhesive wear is elaborated on in the next section.

12.3.2 Adhesive wear

The concept of adhesive wear is based on the notion that the adhesion between asperities when they touch is strong enough to shear portions of these asperities. These portions remain adhered to the opposing surface and can eventually become loose wear debris.

Two modes of adhesive wear have been identified and are illustrated in Figure 12.5:

1. The top row illustrates the mode where adhesion causes gradual smoothing of a sliding asperity by plastic deformation. This type of adhesive wear occurs for ductile materials when a strong adhesive bond forms across the sliding interface, leading to severe plastic shearing at the contacting junction. Repeated contacts

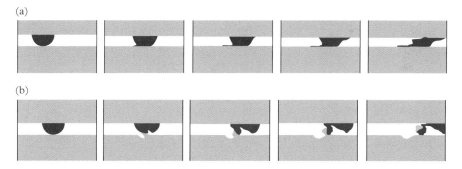

Figure 12.5 *Two possible types of adhesive wear. The panels are snapshots from atomic simulations by Aghababaei et al. (2016) for a single asperity attached to the upper surface as it moves left-to-right while in contact with the flat bottom surface. For clarity the asperity is shown in black even though it is the same material as the upper and bottom surfaces. For a relatively ductile material (a), the sliding asperity experiences severe plastic deformation. Where the material is brittle enough for fracture to occur away from the contacting interface (b), the wear particle is formed, the size of which scales with the junction size, in the manner predicted by Archard's model for wear. Reproduced from the supplementary material of Aghababaei et al. (2016); copyright 2016, Springer Nature; licensed under a Creative Commons Attribution 4.0 International License.*

eventually result in a smoothing of the asperity and in the formation of a transfer film on the opposing surface.

2. The bottom row illustrates the mode where adhesion causes a wear debris particle to be formed by fracture within the material. This type of adhesive wear occurs when the contacting materials are more brittle and have larger contact zones. In this case, the strong adhesion and shear forces induce large elastic strains in the material surrounding the contacting junction due to the lack of ductility. The stored elastic energy builds up and is released when cracks form in the material. These cracks then propagate through the material, eventually reaching the surface and forming a wear debris particle. This particle then rolls between the sliding surfaces and increases in size as it accumulates more material through repetitions of this process.

A model of this second mechanism by Aghababaei et al. (2016) indicates that this fracture mode of adhesive wear only occurs when the dimension of the contact junction exceeds a critical size that scales inversely with the shear modulus G. The fracture mode of adhesive wear is also consistent Archard's simple wear model.

For adhesive wear, the higher the adhesive forces within the contact zone, the more likely a wear fragment will be pulled out. Consequently, higher surface energy typically results in higher wear rates, since adhesive forces scale with surface energy (Chapter 6). Since the surface energy depends strongly on the chemical composition of the surface, wear rates from this adhesive mechanism are sensitive to the presence of contamination layers or lubricant films, since these strongly influence adhesion at the solid–solid contacts. For example, clean metals sliding in vacuum experience very high adhesive and friction forces, as the metals' adsorbed contamination layer, that is present in air, is absent in the vacuum environment (this phenomenon of *cold welding* is also described in Sections 10.5 and 11.2.3). Surfaces do not necessarily have to slide in vacuum for this type of adhesive welding to occur; if the interacting surfaces are insufficiently lubricated, often adhesion occurs at those asperity contacts where the contact pressures are great enough to displace all the lubrication and contamination layers.

Adhesive wear is usually the most common failure mechanism for sliding metal contacts, so needs to be protected against for machinery with metal components to function properly. A fully-formed liquid film is effective in this regard (i.e., operating in the hydrodynamic lubrication regime of the Stribeck curve). In the mixed or boundary lubrication regime, anti-wear additives are used in lubricant oils to reduce wear of all forms including adhesive wear. For example, the most common anti-wear additives in engine oils belong to a family of molecules known as zinc dialkyldithiophosphates, which break down under the combined action of contact stress and temperature to form thin protective films on surfaces that reduce the contact stresses and help prevent adhesion (Spikes 2004).

For dry sliding, oxidation of the metal surface and the adsorption of an adventitious contamination layer can provide some protection against adhesive wear. Wear can be further reduced, often dramatically, by replacing the adventitious contamination layer with a monolayer of a boundary lubricant. Besides lowering the overall surface energy, a good boundary lubricant layer also resists displacement by the contacting asperities, which prevents the underlying metals from coming into direct contact and forming strong metallic bonds.

12.3.3 Abrasive wear

In abrasive wear, the rubbing of hard particles or hard asperities against a surface removes or displaces the material from that surface. For abrasive wear to occur, the particles or one of the contacting surfaces needs to be considerably harder than the surface being abraded (at least 1.2 times harder.)

Figure 12.6 illustrates the two modes of abrasive wear:

1. *Three-body abrasion*—Wear is caused by hard particles that are free to roll or slide between two sliding surfaces. Some common examples of particles that produce abrasive wear: polishing slurries, wear debris generated by other wear mechanisms, and grit contaminants in a lubricated bearing.

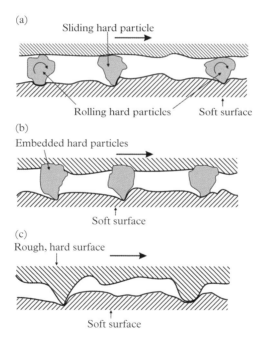

Figure 12.6 *Types of abrasive wear. (a) Three-body abrasive wear where free hard particles rotate or slide between the sliding surfaces. (b) Two-body abrasive wear where abrasive grit is embedded in the top surface and abrades the lower surface. (c) Two-body abrasive wear where asperities of a rough, hard surface abrade a softer surface.*

2. *Two-body abrasion*—Wear is caused by the hard protrusions on one surface scrapping and gouging the other surface. Cutting tools and sand paper are examples of two-body abrasive wear.

Since in three-body abrasion the loose particles are relatively free to roll or slide between the sliding interfaces, typically they are only in a suitable position to abrade surfaces about 10% of the time. Consequently, wear rates for three-body abrasion tend to be an order of magnitude lower than for two-body abrasion.

Models for two-body abrasion have been more extensively developed as it is much easier to model the fixed geometry of an abrading asperity. A simple way to model such a two-body abrasion process is to consider a cone of a hard material gouging a groove into a softer material, as shown in Figure 12.7. (This model is closely related to the model for plowing friction discussed in Section 4.2.2.) If the depth of the groove is h and the slope of the cone is $\tan \theta$, the volume V of material displaced when this asperity slides a distance s is given by

$$V = sh^2 \tan \varphi = sh^2 \cot \theta. \tag{12.12}$$

If we assume that the surface being gouged is completely ductile so that the contact pressure underneath the cone equals the hardness H, and, since the loading force L is supported by just the front half the cone asperity (area $= \pi a^2 /2$), we have

$$L = \frac{1}{2} H \pi a^2 = \frac{1}{2} H \pi h^2 \cot^2 \theta. \tag{12.13}$$

Combining eqs. (12.12) and (12.13) to eliminate h, we obtain the following expression for the wear rate:

$$W = \frac{V}{s} = \frac{2 \tan \theta}{\pi} \frac{L}{H}. \tag{12.14}$$

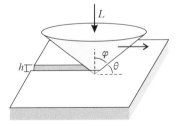

Figure 12.7 *A cone-shaped asperity plowing through a softer surface to form an abrasive wear groove.*

Though derived from somewhat different assumptions, eq. (12.14) has the same form as the Archard wear equation (eq. (12.7)), with a wear coefficient for abrasive wear of $K_a = 2 \langle \tan \theta \rangle / \pi$, where $\langle \tan \theta \rangle$ is the average slope of the asperities on the hard surface. From eq. (12.14), we see that the wear coefficient is proportional to the average slope of the asperities doing the abrading.

One problem with eq. (12.14) is that it predicts wear rates much higher than actually observed. An obvious source for this overestimation is the assumption that all the material displaced is lost from the surface. In reality, only a fraction of the displaced material detaches as wear debris with each sliding pass, while the remainder piles up along the edges of the groove (this micro-plowing is illustrated in Figure 12.8). Equation (12.14) also predicts that the wear rate should be inversely proportional to hardness, but the hardness often fails to predict the abrasive wear resistance of materials as it neglects the degree to which fracture is important to the formation of wear debris during abrasion.

Figure 12.8 illustrates some of the underlying mechanisms of abrasive wear (Zum Gahr 1998):

- *Plowing*—During a single pass, a hard asperity or particle plastically displaces surface material in front of the asperity and to the sides of the wear grove. For ideally ductile material, the volume of displaced material equals the groove volume and no material is removed.

- *Fatigue*—Subsequent passes by other abrading bodies repeatedly replow the material deformed during prior passes. Low cycle fatigue eventually causes material to detach as wear debris.

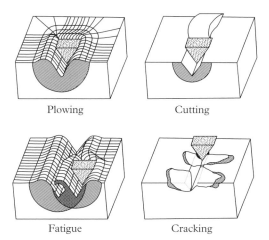

Plowing   Cutting

Fatigue   Cracking

Figure 12.8 *Schematic representation of the possible ways that a single hard abrasive asperity or particle can abrade a surface. Reproduced with permission from Zum Gahr (1998). Copyright 1998, Elsevier.*

- *Cutting*—For ductile materials, increasing the attack angle (θ in Figure 12.7) of the asperity transitions the abrasion mechanism from plowing to cutting, where material is pushed up the front face of the abrasive body. At higher friction coefficients an intermediate *wedging* stage occurs between plowing and cutting (Figure 12.9). Eventually, a chip breaks off that becomes the wear particle.

- *Cracking*—For brittle materials, the concentration of stresses during abrasion leads to cracks forming in the near-surface region. When the cracks become sufficiently large and numerous, delamination of wear particles occurs.

Figure 12.9 shows a simple map for how some these mechanisms depend on the interfacial shear strength and the depth of penetration of the abrading species for a hard spherical indenter sliding against unlubricated metals. Since this map is for a single pass, fatigue is omitted since it involves repeated passing. Cracking is also omitted, as embrittlement did not occur for the materials under consideration. Figure 12.9 shows the ranges of penetration and interfacial shear strength over which the different abrasive mechanisms of plowing, cutting, and wedge formation occur. Below a critical penetration depth (i.e., below a critical attack angle) the metallic material is mainly displaced into ridges up along the groove edges. The wear coefficient from plowing is relatively low ($<10^{-3}$) in this region as wear debris formation occurs through fatigue during subsequent passes. When the shear strength is relatively large ($f > 0.5$), the frictional traction in the contact zone pushes the softer material out in front of the indenter in a wedge shaped formation, and material is eventually removed by the propagation of a

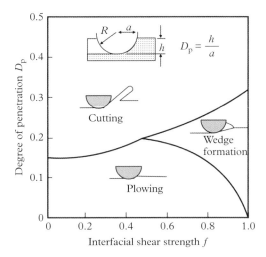

Figure 12.9 *The wear mechanism map for a spherical indenter abrading an unlubricated metal. The degree of penetration* D_p *is defined as the penetration depth* h *divided by the radius of contact* a. *The dimensionless shear strength* f *is the ratio of the interfacial shear strength and the shear yield stress of the material being abraded. Reproduced with permission from Hokkirigawa and Kato (1988). Copyright 1988, Elsevier.*

crack. For degrees of penetration > 0.2, the cutting mechanism dominates where the displaced material forms micro-chips out in front of the indenter that are easily detached as wear particles. In this cutting regime, wear coefficients over 0.1 can be achieved.

12.3.4 Oxidative wear

12.3.4.1 Metals

Many materials, such as metals, react with oxygen in air to form oxide layers on their surfaces. Due to its universality, the role of oxidation in the wear of metals has been studied extensively.

A fairly typical case is the oxidative wear of steel. The conditions under which this happens are illustrated in the wear-mechanism map for steel shown in Figure 12.10. For unlubricated steel surfaces sliding at speeds below about 1 m/s, wear occurs predominately through plastic deformation of the near surface region, to generate mainly metallic debris.

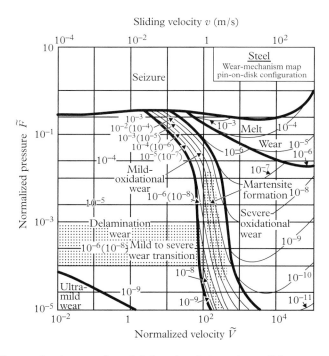

Figure 12.10 *Wear-mechanism map for unlubricated steels in a pin-on-disk geometry. The axes are the normalized contact pressure (defined as the loading force divided by the nominal contact area times the hardness of the softer material) and a normalized velocity (defined as the sliding velocity divided by the velocity of heat flow). The thick lines delineate different wear regimes, while the thin lines are contours of the normalized wear rate, defined as the wear rate divided by the nominal (apparent) contact area. Reprinted with permission from Lim (1998). Copyright 1998, Elsevier.*

A sliding speed of 1 m/s works out to be just sufficient to generate the flash temperatures of about 700°C needed to cause oxidation at contacting asperities on the steel surfaces. Rapid oxidation of these contacts is promoted by the mechanical action of sliding, which generates numerous defects (voids, vacancies, dislocations, etc.) in the near surface region that provide channels for easy diffusion of oxygen atoms into the metal. Oxide films reduce the wear rate of metals by decreasing the shear strength of the sliding interface, which reduces the large subsurface shear strains that are necessary for the plastic deformation wear mechanisms.

At sliding speeds just above the oxidation threshold, the oxide is thin, patchy, and brittle; mild oxidational wear occurs as these patches are scraped off and the exposed metal is reoxidized.

With increased frictional heating at higher sliding speeds, oxidation becomes more widespread, resulting in a thicker, more continuous oxide film. The higher interface temperature at these higher speeds also promotes plastic flow of the oxide film, leading to more severe oxidational wear.

12.3.4.2 *Carbon overcoats*

In the past few decades, amorphous carbon films (also called *diamond-like carbon* or DLC) have come into common use as protective films that are deposited on many types of surfaces (disks in disk drives, razor blades, etc.) to reduce corrosion and wear (Erdemir and Donnet 2006). In unlubricated sliding, these carbon overcoats suffer from oxidative wear when the high temperatures generated by frictional contacts cause the carbon to burn off as carbon dioxide (Marchon et al. 1990, Yen 1996).

Similarly, oxidative wear is the main form of wear for the carbon–carbon composite material used in high performance brakes such as those in the wheels of jet airplanes.

12.3.4.3 *Ceramics*

Non-oxide ceramics (such as silicon nitride, silicon carbide, titanium nitride, and titanium carbide) form oxide layers when exposed to air or water vapor, and these layers can have a strong influence on the wear of the surfaces. The friction and wear of these materials are therefore sensitive to the presence of water and the relative humidity in the sliding environment.

For example, Fischer and Tomizawa (1985) showed that silicon nitride forms an oxide layer in the presence of water via the reaction

$$Si_3N_4 + 6H_2O \rightarrow 3SiO_2 + 4NH_3.$$

Subsequently, the surface of this oxide film is hydrated via the reaction

$$SiO_2 + 2H_2O \rightarrow Si(OH_4).$$

The presence of water on ceramic surfaces is also known to promote crack propagation and plasticity of ceramics (the Rehbinder effect, Andrade et al. 1950). For silicon nitride, the presence of a hydrated surface layer reduces its shear strength, resulting in lower wear rates at high humidity. Increasing the sliding speed increases the wear rate as the higher temperatures generated by frictional heating decrease the hydration layer and its protective influence. For some ceramics such as silicon carbide, alumina, and zirconia, the presence of water increases wear rates, rather than decreasing them, due to the increased plasticity of the surface from the Rehbinder effect.

12.4   Wear maps

A useful way for better understanding the interactions of different wear mechanisms and for summarizing wear data is to construct a *map* of wear behavior based on experimental observations and physical modeling (Lim et al. 1987, Hsu and Shen 1996, Adachi et al. 1997, Lim 1998, Lim and Lim 2004, Hsu and Shen 2005). Once established for a particular set of contacting materials, these maps can provide valuable guidance for how the wear rates and mechanisms change with sliding conditions.

Several wear-mechanism maps have already been presented in this book:

- The map in Figure 2.1 delineates the regimes of surface roughness where severe wear and mild-to-insignificant wear occur for a stainless steel ball sliding over a lubricated stainless steel plate. Figure 2.1 highlights one the main ways that wear maps are valuable: displaying the ranges of parameters over which it is safe to operate with low wear.

- Figure 12.9 illustrates the regimes of depth penetration and shear strength where the different abrasion wear mechanisms of plowing, cutting, and wedge formation dominate for a spherical indenter abrading unlubricated metals. Figure 12.9 highlights another way that wear maps are valuable: displaying the regions of operating conditions where different wear mechanisms are prevalent.

- Figure 12.10 shows the different wear regimes for unlubricated steel. Section 12.3.4.1 discussed the oxidative wear regions of this wear map, while Section 12.4.1 will cover the other wear mechanisms for unlubricated steel and other metals that are represented on this wear map.

A wear-mechanism map indicates the wear regime's dependence on two parameters that make up the coordinate axes, while all other parameters are kept constant (though sometimes a third parameter, like the loading force in Figure 2.1, is also varied and the result displayed as contours within the diagram.) This is obviously a simplification, as typically more than two parameters influence wear. However, if the two parameters are the most dominant ones, then the resulting map can provide for a good guide when wear transitions from one mechanism to another.

As discussed in the review papers by Hsu and Shen (1996), Adachi et al. (1997), Lim (1998), Williams (2005), and Hsu and Shen (2005), wear-mechanism maps have been developed for metals, ceramics, metal–matrix composites, polymers, coatings, fretting, and erosion. In the following sections, we discuss a few notable wear-mechanism maps to further illustrate the utility of these maps.

12.4.1   Unlubricated steel wear-mechanism map

Figure 12.10 shows one of the best known examples of a wear-mechanism map. This map for unlubricated steel sliding in air was developed by Lim et al. (1987), and is based on a large number of published test results using the rotating pin-on-disk test configuration, as well on theoretical analysis calibrated to experiment. For this map, the two parameters for the coordinate axes are load and sliding speed; normalized wear rates are also plotted as curves of constant wear rates.

Knowing how wear rates trend with the plotted parameters (as shown with contours in Figure 12.10) is particularly useful when developing an accelerated wear test. The purpose of an accelerated wear test is to validate that the total wear of a particular tribological design will not exceed some wear specification over the expected operational life, but with a test period that is orders of magnitude shorter than the operational life. Consequently, for the accelerated test, the operating parameters are shifted towards that area of the map where the wear rates are sufficiently higher for the accelerated test be completed in the desired test time, but not so close to a transition zone on the map that other wear mechanisms would start to dominate.

Figure 12.10 illustrates several features of how wear occurs on unlubricated steels that are also generally applicable to the wear unlubricated other metals:

- At low sliding speeds, the wear of unlubricated steel is dominated by mechanical processes that are determined more by mechanical stresses than by sliding speed.

- Increasing the loading force generates higher contact pressures and shear stresses; this causes more plastic deformation in the metal, which leads to more mechanical damage.

- The temperature rise at a contacting junction depends on the rate of frictional heating and on how fast the heat is conducted away from the contact point. At slow sliding speeds, the sliding is considered isothermal as the frictionally generated heat is quickly conducted away (rate at which heat is generated = sliding speed × friction force). At high sliding speeds, heat is generated much faster than it is conducted away, resulting in high interface temperatures; in the limit of high sliding speeds, the sliding is considered adiabatic. Increasing the loading force increases the frictional force resulting in more frictional heating and higher interface temperatures.

- For metals, these high temperatures can lead to mechanical softening of the asperities and, in the extreme, melting of the near surface region. For ice, it is well known that much of its slipperiness comes from frictional heating melting the ice near the surface to form a thin film of water that lubricates the sliding surfaces (Rosenberg 2005).

- High interface temperatures from frictional heating promote chemical reactions at the sliding surfaces, such as oxidation of metal surfaces in air or degradation of lubricant films.

- For steel sliding in air, oxidational wear dominates those regions of the wear map where frictional heating is sufficient to promote surface oxidation. In this regime, the somewhat counterintuitive wear behavior occurs whereby wear rates decrease with increasing velocity as the formation of an oxide film provides a protective layer against further wear.

12.4.2 Ceramic wear map

Another example of a highly cited wear map is the one shown in Figure 12.11 for ceramics. This wear map was developed by Adachi et al. (1997) and further modified by Kato and Adachi (2002) for advanced ceramics, such as silicon nitride used in lightweight bearings and silicon carbide used in highly durable mechanical seals. Such ceramics are usually polycrystalline with a specific characteristic grain size.

For the experiments on which the Figure 12.11 wear map is based, the wear is characterized as either "mild" or "severe."

- *Mild regime for ceramics*—The wear of these ceramics is considered mild if the wear generates debris that is relatively small in size. In the experiments of Kato and Adachi (2002), this corresponded to a wear particle size < 10% of the size of grains in the ceramics. This mild wear also results in a surface roughness in the same size range; so, if surfaces are originally rough, this mild wear acts to smooth the

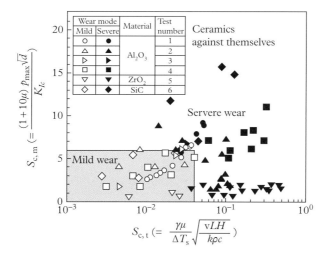

Figure 12.11 *Wear map of various ceramics sliding unlubricated against themselves in air. Individual data points are from experiments that showed either mild wear (open symbols) or severe wear (closed symbols). The data are plotted against the dimensionless parameters of the mechanical severity of contact $S_{c,m}$ and the thermal severity of contact $S_{c,t}$. Reprinted with permission from Kato and Adachi (2002). Copyright 2002, Elsevier.*

surface. In this mild regime, the wear rate is less than 10^{-6} mm³/(N m). Wear in this regime occurs by the mechanical removal or by the dissolution of a soft, thin surface layer that is formed by tribochemical reactions. The specific tribochemical reaction underpinning the wear process depends on the ceramic, but these reactions tend to belong to the family of stress-assisted corrosive reactions whereby exposure to water and applied stress leads to room-temperature oxidation and hydrolyzation of surfaces. For example, the reaction pathway for silicon nitride is shown in Section 12.3.4.3 and produces a hydrolyzed silica surface that is rather weak and easily removed.

- *Severe wear regime*—In the severe regime, the stresses are large enough for inter-granular fracture and/or the delamination of the tribofilms to occur, producing wear debris with sizes bigger or equal to those of the grains in the ceramics. Thus, with severe wear, surfaces will roughen to the scale of the grain size, and the wear rates are greater than 10^{-6} mm³/(N m).

Since fracture is necessary for severe wear of these ceramics to take place, Adachi et al. (1997) and Kato and Adachi (2002) define the two parameters that are used to construct the wear map in Figure 12.11 as corresponding to two different ways that cracks can form:

- *Mechanical severity of contact* $(S_{c,m})$—This parameter describes the mechanical conditions required to propagate pre-existing vertical surface cracks to drive fracture and removal of the grains. It is defined as:

$$S_{c,m} = \frac{(1 + 10\mu)\,p_{max}\sqrt{d}}{K_{Ic}} \tag{12.15}$$

where μ is the friction coefficient, p_{max} the maximum Hertzian contact pressure, d the length of pre-existing vertical surface cracks, and K_{Ic} is the opening mode fracture toughness of the ceramic (which describes how fracture-resistant the material is). High contact pressure, high friction, long cracks, and low fracture toughness all increase the mechanical severity parameter. The fracture toughness parameter is discussed further in Section 12.6.3.

- *Thermal severity of contact parameter* $(S_{c,t})$—This parameter describes the tendency for vertically-oriented surface cracks to be propagated by thermal stresses generated by frictional heating, that is, how likely it is that the thermal shock from the flash temperature will fracture the ceramic. It is defined as:

$$S_{c,t} = \frac{\gamma\mu}{\Delta T_s}\sqrt{\frac{vWH}{k\rho c}} \tag{12.16}$$

where ΔT_s is the ceramic's thermal shock resistance (the maximum temperature jump a brittle material can sustain without cracking), γ the heat partition ratio (the fraction of generated heat entering the body whose wear is being considered), v the sliding velocity, L the normal load, H the ceramic's hardness, k the ceramic's thermal conductivity, ρ the ceramic's density, and c the ceramic's specific heat.

As seen in Figure 12.11, experiments under different sliding conditions in air reveal a well-defined boundary between the mild and severe wear regimes. In particular, for self-mated ceramics to slide in the mild wear regime (the shaded box in Figure 12.11), it is required that

$$S_{c,m} \leq 6 \text{ and } S_{c,t} \leq 0.04. \tag{12.17}$$

The result is striking given that very little correlation if the experimental data is plotted in other ways, for example, the wear rate vs. the friction coefficient. Moreover, the data of Adachi et al. (1997) indicate that the wear rate can be predicted if $S_{c,m}$, and $S_{c,t}$ are known.

12.5 Atomic attrition and the transition state theory of wear

So far we have talked about wear mechanisms that generate wear debris consisting of many atoms. It is also possible for wear to occur where atoms are removed individually from a surface or in very small clusters consisting of just a few atoms. This type of wear occurs by an entirely different mechanism from those that generate microscopic wear particles. For this wear by atomic attrition, the mechanism involves the contact forces promoting the reaction of atoms and their subsequent attachment to the opposing surface.

One general idea of how atom-by-atom wear occurs is illustrated in Figure 12.12. In this illustration, as an asperity apex slides over a flat surface thermally activated bonds forms between the apex atoms and the flat surface, as illustrated in Figure 12.12(c). At this point there are three possible options: (1) either this bond with the surface breaks when the asperity moves on across the surface and the atom remains bonded to the asperity (no wear occurs); (2) this single new bond with the substrate is sufficiently strong so as to pull the atom off the asperity (unlikely); or (3) additional thermally activated bonds form between the atom and surface (Figure 12.12(d)) and this combination of bonds is strong enough to pull the atom off the asperity (Figure 12.12(e)).

Atomic attrition or atom-by-atom wear can be quantitatively modeled using *transition state theory*, which describes the kinetics of chemical bond formation and breaking. Originally, transition state theory (which is also known as *reaction rate theory*) was formulated by Eyring to model chemical kinetics. The classic formulation involves describing a set of reactants that undergo a process (the reaction) where they transform from their initial state (which is a locally stable equilibrium point) to their final state (also a locally stable equilibrium point) where they have formed a reaction product. Figure 12.13(a) illustrates how the energy of the system varies along the reaction

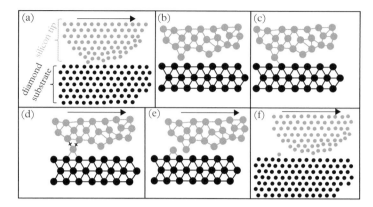

Figure 12.12 *Reaction pathway for the wear of an individual atom on an asperity in the form of a silicon AFM tip contacting a flat diamond surface (Jacobs and Carpick 2013). (b) shows a magnified view of (a) where a single tip atom protrudes close enough for the initial bond to form to the flat surface (c). (d) shows the case where more bonds form, which causes the bonds with the tip to break. In this case, the atom is transferred from the tip to the flat surface (e), and the tip moves on with another atom positioned closest to the flat surface (f). Reproduced with permission from Jacobs (2013).*

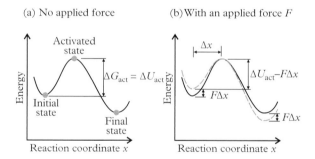

Figure 12.13 *Schematic diagram showing the energy as a function of a reaction coordinate for a system as it undergoes a thermally activated process. (a) The reaction pathway when no external force is applied to the reaction system. The dashed line in (b) shows how the energy potential is modified when a mechanical stress is applied to the system causing a force F to be applied along the direction of the reaction coordinate x.*

pathway. The *reaction coordinate* represents the degree of progress of the reactants along the reaction pathway and is usually a geometric parameter that changes most during the reaction. The height of the energy barrier between the initial and final states is the *Gibbs free energy of activation* ΔG_{act}.

The Boltzmann distribution describes what fraction of reactants have sufficient energy to overcome the activation barrier ΔG_{act} due to thermal fluctuations. Using the Boltzmann distribution, one can show that the rate of a forward reaction k_f is given by

$$k_f = r_0 \exp\left(-\frac{\Delta G_{act}}{k_B T}\right),\tag{12.18}$$

where r_0 is an effective attempt frequency, k_B is Boltzmann's constant, and T the absolute temperature. (Equation (12.18) is the Arrhenius rate law.) In addition to describing chemical reactions between atoms and molecules, transition state theory has also been successfully extended to many thermally activated processes in solids such as atomic diffusion (Vineyard 1957), dislocation motion (Gibbs 1965, Hirth and Nix 1969), adsorption/desorption, electronic carrier concentration in semiconductors and insulators (Hänggi et al. 1990), and mass transport (Hänggi et al. 1990).

Within solids and liquids, thermally activated processes are impacted when a mechanical stress is applied. In this book we have already discussed the impact of a shear stress on the thermally activated process of the flow of liquids (the Eyring viscosity in Section 9.2.3) and on friction (Section 11.2.4.2). Some of the other thermally activated processes in solids where mechanical stress modifies the energy landscape are plastic deformation (Prandtl 1928), the strength in solids (Zhurkov 1984), rubber friction (Schallamach 1953), and biomolecule interactions (Bell 1978).

The application of an external force does mechanical work on the atoms that provides part of the energy needed to overcome the activation energy barrier. Figure 12.13(b) illustrates how a mechanical force influences a thermally activated process by modifying the activation barrier. In this case the reaction coordinate corresponds to a lateral displacement x of the group of atoms undergoing the wear process. The distance along the reaction coordinate from the initial state to the activated state is the *activation length* Δx.[1] When a force F is applied to the system, the intrinsic energy barrier ΔU_{act} is lowered by the amount of work $F\Delta x$ done by the force moving the system over the barrier; this results in a new activation barrier height: $\Delta G_{act} = \Delta U_{act} - F\Delta x$. Since we typically have better knowledge of the average stress σ being applied to interface rather than the force being applied to the particular group of atoms involved in the wear process, it is more convenient to rewrite $-F\Delta x$ as $-\sigma \Delta V_{act}$ where ΔV_{act} is called the *activation volume*. So, with the application of mechanical stress to eq. (12.18), the rate of contacting atoms removed per second becomes

$$k_f = r_0 \exp\left(-\frac{\Delta G_{act}}{k_B T}\right) = f \exp\left(-\frac{\Delta U_{act}}{k_B T}\right) \exp\left(-\frac{\sigma \Delta V_{act}}{k_B T}\right).\tag{12.19}$$

From this equation, we see that the reaction rate for atomic wear k_f increases with the stress σ.

---

[1] This Δx is slightly different than the Δx used in eq. (9.13) for the Eyring viscosity model: there Δx corresponded to the distance between the initial and final state, rather than the distance from the initial state to the top of the energy barrier as used here.

Park et al. (1996) were the first report experimental results showing a rate of atomic-scale wear that depended exponentially on the applied force. In that experiment, a silicon nitride atomic force microscope (AFM) tip was used to cause atomic-scale wear at single atomic step edges of a calcite crystal in an aqueous environment. In addition to applying the force to cause wear, the AFM was also used to obtain high-resolution topographic images, from which Park et al. showed that the growth rate of the wear track depended exponentially on the applied loading force, consistent with transition state theory. Since the work of Park et al. (1996), numerous other wear experiments using the AFM have also reported evidence of atom-scale wear (Sheehan 2005, Gotsmann and Lantz 2008, Bhaskaran et al. 2010, Jacobs and Carpick 2013).

An example of these atom-scale wear results is shown in Figures 12.14 and 12.15. In these experiments by Jacobs and Carpick (2013), a silicon tip slides against a diamond surface, and the tip is periodically imaged with atomic resolution in a transmission electron microscope (TEM). In the TEM images (Figure 12.14), the tip is observed

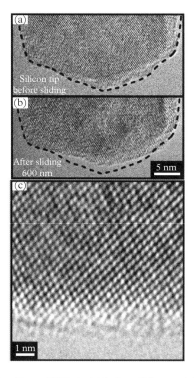

Figure 12.14 *TEM image of a silicon AFM tip (a) before sliding and (b) after sliding 600 nm over a diamond surface. During sliding, ~ 1 nm of material is removed from the end of the tip, but, as shown in the magnified TEM in (c), there is no evidence of dislocations and defects in the silicon lattice at the surface that would be present if the wear process involved plastic deformation or fracture. Reproduced with permission from Jacobs (2013).*

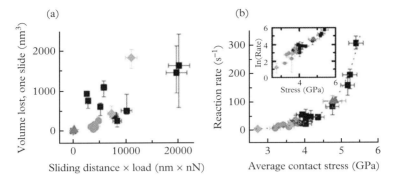

Figure 12.15 *In the experiments by Jacobs and Carpick (2013), a silicon AFM tip slides over a diamond surface, as illustrated in Figure 12.12. No external load was applied, but adhesion forces (typically a few tens of nanonewtons) produce contact stresses acting perpendicular to the sliding interface. The wear volume is periodically measured from TEM images, as shown in Figure 12.14. The different symbols corresponded to experiments with different AFM tips. The left-hand plot shows the wear volume plotted against sliding distance; as the data does not fall on a straight line when plotted in this manner, this indicates that the wear process does not follow the Archard wear equation. The right-hand plot shows that the wear rate depends exponentially on the contact stress and that the data is well fitted by the transition state theory of eq. (12.19). Reproduced with permission from Jacobs and Carpick (2013).*

to shorten after each sliding interval, often by increments of 1 nm or less, and with no observable plastic deformation or fracture of the tip. These observations are consistent with a wear process that removes atoms individually.

Figure 12.15(a) shows that the wear process is inconsistent with the Archard wear model as expressed by eq. (12.11). In this plot, the tip wear volume is plotted against the product of sliding distance and load; but the data points are scattered rather being on the straight line predicted by the Archard wear model. However, when the data is plotted as the wear rate versus the average applied contact stress as in Figure 12.15(b), the data points clearly follow an exponential dependence that is well fitted by transition-state theory as expressed by eq. (12.19). Assuming an attempt frequency of $10^{13\pm1}$ s^{-1}, the fitting yields $\Delta V_{act} = 6.7 \pm 0.3$ Å^3 and $\Delta U_{act} = 0.85 \pm 0.06$ eV, both of which are physically reasonable values, as ΔV_{act} is on the order of a size of an atom and ΔU_{act} is on the order of the energy of a covalent bond.

Ideally in eq. (12.19) we should use the component of stress acting in the same direction as the reaction coordinate x. Generally, however, it is difficult to ascertain the particular direction along which the rate-limiting atomic process takes place; that is, it may be along the direction normal or parallel to the sliding interface, or at some angle to it. If the shear stress is proportional to the normal stress (i.e., Amontons' law is locally obeyed), then, when fitting the wear rate data to the transition state theory equation to the wear rate data, we can use either the stress perpendicular (as done for Figure 12.15(b)) or tangential to the sliding interface in eq. (12.19), and the fitted activation volume ΔV_{act} will only differ by a scaling factor.

12.6   Hardness, plasticity, and fracture at the nanoscale

12.6.1   Hardness and nanoindentation

Since plastic deformation underlies many wear mechanisms, hardness appears in many models of wear, such as the Archard wear equation (12.7) and abrasion wear equation (12.14). While wear is also influenced by many other material parameters besides hardness—such as, thermal conductivity, toughness, chemical stability, oxidation resistance, and friction coefficient—given the prominent role of hardness in many wear models, we will discuss it more detail.

There are three ways of measuring hardness, each of which probe different aspects of hardness:

1. *Indentation hardness*—In an indentation test, a sphere or a sharp indenter is compressively loaded without sliding onto a sample with a sufficient load to cause plastic deformation and/or fracture. When the indenter is withdrawn, a hardness value is calculated from the geometry of the resulting permanent indentation. Examples of macrohardness indentation tests include Rockwell, Brinell, Vickers, Shore, and Knoop (Broitman 2017).

2. *Scratch hardness*—A scratch test measures the resistance to fracture and plastic deformation during sliding of a sharp object across the sample due to the combination of friction and loading forces. Since a scratch test also applies a tangential stress in addition to a normal load, it is more representative of what happens in sliding wear situations. For wear resistant coatings, the critical load of a scratch test is the force where the sliding indenter pierces through the coating. The Mohs scale used by mineralogists is another example of a scratch hardness.

3. *Rebound or dynamic hardness*—In a rebound hardness test, an impact body with an indenter is spring loaded and then shot towards the sample so that the indenter impacts the sample with a well-defined impact energy. The Leeb rebound hardness value is derived from the energy loss after the impact body rebounds and is defined as $1000v_r/v_i$, where v_i is the incident velocity of the impact body and v_r is the rebound velocity. The surface damage from a rebound test is more representative than an indentation test for what happens during high speed impacts. The rebound test is typically used for measuring the hardness of metals, where most of the impact energy lost is through plastic deformation.

The most commonly used hardness values are those from indentation tests. As illustrated in Figure 12.16, in an indentation experiment, a sharp indenter is pushed into a sample, which responds by undergoing both elastic and plastic deformation. As discussed in Section 3.3.4, the indentation hardness H is typically defined as $H = L/A_p$, where L is the loading force used to make a permanent indentation and A_p is the area of the permanent indentation.

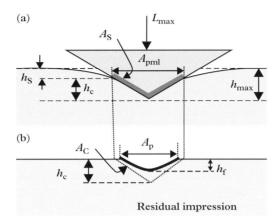

Figure 12.16 *(a) A surface undergoes elasto-plastic deformation as a sharp indenter is pushed against it until a maximum load of L_{max} is applied. (b) After unloading the indenter, a permanent indentation is remains due to plastic deformation. Reprinted from Broitman (2017); copyright 2017, Springer Nature; licensed under a Creative Commons Attribution 4.0 International License.*

While measurements of *macroscale hardness* (where the indentation area is many square microns to square millimeters) have existed since the work of William Wade (1856), techniques for measuring *nanoscale hardness* (indentation area = a few square microns to as small as few square nanometers, and with a penetration depth $h < 200$ nm) were only developed beginning in the mid-1970s. Since then, great strides have been made in nanoindentation techniques for measuring the elasticity and the yield stresses in very localized regions. Since the nanoscale hardness indentation is much more localized to the near surface region than for a macrohardness measurement, the material properties extracted from nanoindentation are much more relevant to understanding the surface damage caused by wear.

Since the indentation area in a nanoindentation experiment is so small, the area cannot be as conveniently measured using optical microscopy as for macroscale indents. This is commonly overcome by using an indenter geometry that is known to a high precision. Then, by recording the depth of penetration of the indenter during the indent process, the area can be calculated from the depth data and the known geometry of the indenter. A Berkovich diamond indenter is often used in nanoindentation instruments, as it terminates in a sharp point with a well-defined geometry, as shown in Figure 12.17(a). An example of an indent made in a ductile material by a Berkovich indenter is shown in Figure 12.17(b).

Figure 12.18 shows a generic plot of the loading force L on the indenter and displacement h of the indenter relative to the initial undeformed surface. From this load–displacement curve, not only can the indentation area can be calculated using the known indenter geometry, but also the hardness and elastic modulus can be calculated using the following relationships (Oliver and Pharr 1992, 2004):

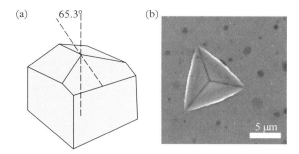

Figure 12.17 *(a) Schematic diagram of the tip of an Berkovich indenter. The sides of the Berkovich indenter are inclined 65.3° relative to the normal. (b) The permanent indent made by a Berkovich indenter in a ductile material (aluminum) where no pile-up occurs; reprinted with permission from Pharr (1998); copyright (1998), Elsevier.*

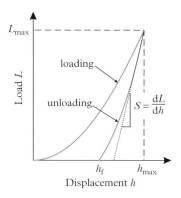

Figure 12.18 *A typical plot of the data collected during an indentation experiment on a ductile material. During loading, both elastic and plastic deformation take place, while during unloading only the elastic portion of the deformation is recovered. From the parameters extracted from this load versus displacement curve it is possible to calculate the hardness and elastic modulus.*

$$h_c = h_{max} - \varepsilon \frac{L_{max}}{S} = \sqrt{\frac{A_{pml}}{24.5}}, \tag{12.20}$$

$$E_c = \frac{1}{\beta} \frac{\sqrt{\pi}}{2} \frac{S}{\sqrt{A_{pml}}}, \tag{12.21}$$

$$H = \frac{L}{A_{pml}} = L_{max} \left(\frac{24.5}{h_c}\right)^2, \tag{12.22}$$

where for a Berkovich indenter $\varepsilon = 0.75$ and $\beta = 1.034$. In these equations, the projected area A_{pml} is used rather than permanent area of indentation to determine the hardness.

Also, the slope of the unloading curve $S = dL/dh$ is the contact stiffness during the initial unloading; as indicated by eq. (12.21), it is proportional to the composite elastic modulus E_c of the indenter and sample (eq. (3.8)).

One shortcoming of eqs. (12.20)–(12.22) is the assumption of no pile-up of material occurring around the indent, which can happen in many elastic–plastic situations. When pile-up does occur, the contact area is larger than predicted by eq. (12.20), and the elastic modulus and hardness may be up to one third less than that predicted by eqs. (12.21) and (12.22) (Oliver and Pharr 2004). So far there are no satisfactory methods for correcting for pile-up.

12.6.1.1 *Example: load–displacement curves on single crystal Cr_3Si*

Figure 12.19 shows two load–displacement curves obtained in a nanoindentation experiment on a single crystal of Cr_3Si by Bei et al. (2005).

The left-hand plot shows the case where the maximum load L_{max} is below the threshold for nucleating or propagating dislocations. This particular crystal has a very low density of defects, so few to no dislocations exist in the very small contact zone, and the deformation is entirely elastic; consequently, the unloading curve retraces the loading curve. The line shows the fit to an elastic indentation model where the exponent 1.69 lies between the values for a spherical (1.5) and a conical (2.0) indenter (even for the sharpest Berkovich indenters, the apex is always slightly rounded, though this can be as small as 10 nm in radius). So, for the few millinewtons of loading force used in this case, the contact area is still mostly supported by the rounded portion of the tip apex rather than the pyramid faces, making the contact behavior intermediate between spherical and conical.

The right-hand plot in Figure 12.19 shows the case where the loading force is increased above the threshold for nucleating dislocations. At this threshold of $L_c = 3.9$ mN, the material underneath the indenter becomes elastically unstable and abruptly undergoes plastic flow causing a "pop-in." As further evidence that plastic deformation

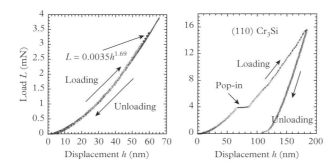

Figure 12.19 *Load versus displacement curves for a diamond Berkovich indenter pushing against a (110) oriented Cr_3Si single crystal. In the left-hand plot, only elastic deformation occurs as the maximum load is below the threshold for nucleating dislocations. In the right-hand plot, the plastic deformation is initiated by the formation of dislocations during the pop-in event. Reprinted with permission from Bei et al. (2005). Copyright 2005, American Physical Society.*

has occurred, the unloading curve for the right-hand plot does not retrace the loading curve, but instead unloads to a finite h_f, which is the depth of the residual indentation hole. From the geometry of the indenter and the value of L_c, Bei et al. estimate that the maximum shear stress underneath the indenter at the onset of the pop-in is close to the theoretical maximum in eq. (12.24), which, as discussed below, is when the maximum shear stress τ_{max} reaches the value needed to slide atoms over the crystal planes with the lowest shear strength. (Bei et al. (2004) determined a value of $\tau_{max} = G/2\pi = 20.5$ GPa for Cr_3Si.) Once this slippage starts, it creates dislocations that enable atomic motion along the other crystal planes, and the process cascades with rapid plastic flow during a pop-in event until the stress is reduced to the yield stress determined by the motion of the dislocations.

12.6.1.2   Example: fracture and plasticity in a hard carbon film

Amorphous carbon films are frequently used as wear resistance coatings in tribological applications. Since films with a high degree of sp^3 bonding between the carbon atoms exhibit high hardness values, these films are often referred to as DLC films.

As the amount of sp^3 bonding is nearly 100% for tetrahedral amorphous carbon (ta-C), these films are among the hardest types of carbon films. The high hardness of ta-C films, however, also makes them susceptible to fracture. One way of characterizing the fracture properties of films and other materials is by indentation. During an indentation experiment, fracture cracks initiate and grow from the corners of the indenter tip as shown in Figure 12.20(b) for an indentation on a 110 nm thick ta-C film (Jungk et al. 2006).

In the load–displacement curve in Figure 12.20(a), the maximum load is high enough for plastic deformation to take place in the film, so the unloading curve occurs at higher

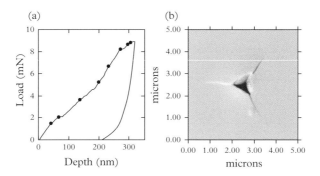

Figure 12.20 *(a) Load versus penetration depth during the nanoindentation into a tetrahedral amorphous carbon (ta-C) film. In this experiment, an indenter with the geometry of a corner of a cube was used rather than a Berkovich indenter to lower the load threshold for fracture (Pharr 1998). The dots correspond to the detection of acoustic emissions during a crack extension event. (b) AFM image of the resulting indent in the film after the indentation experiment. In addition to the triangle shaped indent from plastic deformation, fracture cracks are also visible growing from the corners of indentation. Reprinted with permission from Jungk et al. (2006). Copyright (2006), Elsevier.*

displacement depths than the loading curve. In the AFM image in Figure 12.20(b) taken after unloading, not only is the characteristic triangle shaped indent from the plastic deformation visible, but also visible is a fracture crack at each of the three corners of the indent. From the length of the cracks, Jungk et al. (2006) were able to determine that, for ta-C films prepared with different levels of compressive residual stress, the fracture toughness ranged from 3.3 to 3.7 MPa m$^{1/2}$.

12.6.2 Plasticity

Within ductile materials like metals, plasticity occurs via crystallographic planes sliding over each other, usually over the close packed atomic planes where the resistance is weakest. The maximum shear stress needed to initiate slippage occurs when all the atoms in one plane slide over an adjacent plane in unison, as illustrated in Figure 12.21. If we assume that the tangential force on the individual atoms varies sinusoidally with their lateral displacement x, then the relationship between the shear stress τ and the displacement is

$$\tau = \left(\frac{Gb}{2\pi a} \right) \sin \left(\frac{2\pi x}{b} \right), \qquad (12.23)$$

where G is the shear modulus and a and b are, respectively, the interplanar and interatomic distances. This relationship reduces to the expected relationship of $\tau = Gx/a$ for small values of x/b, and predicts a maximum shear stress $\tau_{\max}$ at $x = b$ where slippage occurs with all the atoms in the slip plane moving in unison:

$$\begin{aligned} \tau_{\max} &= \frac{Gb}{2\pi a} \\ &\approx \frac{G}{2\pi} \text{ for } a \sim b. \end{aligned} \qquad (12.24)$$

As was mentioned for the example in Section 12.6.1.1, nanoindentation experiments on low defect density crystals often find that the critical shear stress for the first pop-in event is close to the $G/2\pi$ value predicted by eq. (12.24) (Bei et al. 2004). By low defect density, we mean that the dislocation density is low enough that the size of the region undergoing plastic flow is much smaller than the typical distance between dislocations.

Often, though, other factors act to lower the ideal maximum shear strength for low defect density materials below the $G/2\pi$ value. In particular, Mackenzie (1949) has shown that $G/30$ is a reasonable value for the ideal shear strength for face-centered

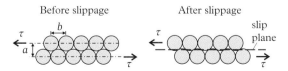

Figure 12.21 *Slippage of one plane of atoms over another due to an applied shear stress τ.*

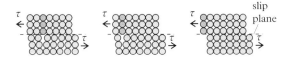

Figure 12.22 *The motion of an edge dislocation (dark gray atoms) from right to left under applied shear stress τ. This moves the atoms in bottom half of the crystal one row of atoms at a time from left to right along the slip plane.*

cubic (fcc) crystals, when one considers the actual form of the intermolecular forces (rather than using the sinusoidal approximation in eq. (12.23)) and considers the other configurations of mechanical stability available to the crystal lattice. As discussed by MacMillan (1972), the value of $\tau_{max} = G/30$ has been observed experimentally for fcc metals when shearing the {111} crystal plane in the $<11\bar{2}>$ direction.

Only in the case where the density of defects (such as dislocations) is insufficient to assist plane slippage does the actual shear strength approach the ideal maximum value (which ranges of $G/30$ to $G/2\pi$ depending on the situation.) For most materials, plastic flow is mainly due to the formation and motion of defects in the crystal structure, and this causes the experimental values for τ_{max} to be one to three orders of magnitude less than the ideal maximum shear strength.

12.6.2.1 *Dislocations*

The most important defects for promoting plasticity in solids are dislocations. Figure 12.22 illustrates the motion of one type of dislocation, call an edge dislocation, that is generated by inserting an extra half plane of atoms into the solid. Moving atoms on the slip plane at the end of an edge dislocation is relatively easier than moving them all past the opposing atoms all in unison, as the atoms on the lower side of the slip plane only have to move past the end atom one at a time. This explains why mobile dislocations can make a crystal very plastic, as the shear stress needed to move a dislocation through the solid can be quite low, certainly much lower than moving all the atoms in the slip plane in unison.

Pure crystals of metals and of many other materials tend to have moderate densities of dislocations making them very plastic or ductile. Immobilizing the dislocations reduces this plasticity (and increases yield strength), which can be done intentionally to increase the hardness of the bulk material. Similarly, the action of sliding contact and wear can often harden a surface by immobilizing the dislocations (Zhu and Li 2010). Common ways that dislocations are immobilized are as follows:

- *Mechanically blocking dislocation motion*—This is often done by introducing small particles of a hard material into the crystal. An example of this is the use of iron carbide in iron to harden steel.
- *Pinning dislocations on impurity atoms*—Since each dislocation has regions where the atoms are expanded or contracted relative to the crystal lattice, small impurity atoms will preferentially reside in the contracted areas and larger impurity atoms in

the expanded areas. Since the bonding of these impurities around the dislocation lowers the free energy, a higher stress needs to be applied to the dislocation to move the dislocation away from the impurity.

- *Increasing the dislocation density*—Work hardening or strain hardening is where the shear strength increases with the amount of plastic deformation due to the increasing density of dislocations. This effect occurs when the dislocation density is high enough for the strain fields of adjacent dislocations to overlap and become entangled.

- *Grain boundary strengthening (or Hall–Petch strengthening)*—Most materials are polycrystalline, that is, they are composed of many small, individual crystal grains with different orientations. During plastic deformation, dislocations move through these individual grains until they encounter a grain boundary, at which point they tend to become pinned to the grain boundary rather than migrating into a neighboring crystal. Consequently, the smaller the average grain size, the harder the material will be; this is the Hall–Petch effect. Eventually, the grain size gets so small that sliding at the grain boundaries dominates over dislocation motion for plastic deformation and the strength decreases with decreasing grain size; this is referred to as the inverse Hall–Petch effect.

Given the key role that dislocations play in plastic flow, it is not surprising that their generation and motion also play a key role at the atomic level for many wear mechanisms. Some examples of the roles that dislocations play in wear:

- As mentioned in Section 12.2.2, the delamination wear process is thought to start by the strains induced by the sliding contact generating large numbers of dislocations that then coalesce under the action of the shear stresses to form voids and microscopic cracks.

- Plasticity and hardness depend to a large degree on the motion of dislocations. So, for wear modes involving plastic deformation like adhesive wear or the plowing mode of abrasive wear, the plastic flow of the material originates at nanoscale from dislocation motion.

- During fatigue wear, the strains induced with each sliding contact generate more dislocations each cycle. This initially causes strain hardening, then eventually embrittlement, and finally fatigue fracture. This particularly true at high friction coefficients where the high tangential shear stress induces a large amount of strain deformation in the near surface.

12.6.3 Fracture

Both ductile and brittle materials fail when excessive loading causes fracture, which occurs when the stress exceeds the level needed to separate a solid into two or more pieces. A material is characterized as *ductile* if the fracture involves a large amount of plastic deformation. A material is considered to be *brittle* if the fracture is accompanied

(a) Opening mode   (b) Sliding mode   (c) Tearing mode

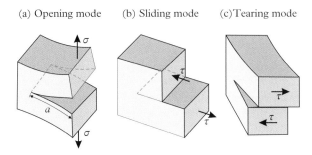

Figure 12.23 *The three principal modes of fracture: (a) opening mode due to a tensile stress, (b) sliding mode due to an in-plane shear stress, and (c) tearing mode due to an out-of-plane shear stress.*

by very little plastic deformation; brittle fracture is typically characterized by the cracks spreading very rapidly, resulting in sudden separation.

Fracture can be caused by both tensile and shear stresses. As illustrated in Figure 12.23, tensile, in-plane, or out-of-plane shear stress results respectively in the opening, sliding, or tearing modes of fracture. Fracture involves first the creation of a crack, followed by the propagation of the crack, and finally separation of the material into separate pieces.

During World War I, British aeronautical engineer A. A. Griffith realized that in most instances fracture failure initiates on preexisting flaws or cracks in the material. The way that he demonstrated this was true was by conducting a series of experiments fracturing glass specimens where he had first created an artificial crack on the surface with an extended length a into the specimen. He found that the stress σ_f needed to propagate this crack obeyed the relationship

$$\sigma_f \sqrt{a} = \text{constant}. \tag{12.25}$$

This relationship indicates that the larger the initial crack or flaw, the smaller the stress needed to fracture the material.

Griffith determined the value of the constant in eq. (12.25) by computing the change in a free energy that consisted of two components: (1) the increase in surface energy when the crack increases in size and (2) the reduction in stored elastic energy when crack propagation relaxes the stresses in the material. Spontaneous fracture occurs when the free energy reaches its maximum value; at this point the crack length has reached a critical value, beyond which the reduction in elastic energy is greater than the increase in surface energy during crack formation. From this analysis, Griffith determined what is now called the *Griffith energy criterion*:

$$\sigma_f \sqrt{a} = \sqrt{\frac{2E\gamma}{\pi}}, \tag{12.26}$$

where E is the elastic modulus and γ is the surface energy of the exposed material in the fracture.

While the Griffith energy criterion works well for brittle materials like glass, for more ductile materials the value of surface energy γ predicted by eq. (12.26) can be several orders of magnitude too high. This occurs because, for ductile materials, a plastic zone exists around the tip of the crack; as the stress increases, some of the strain energy goes into increasing the plastic deformation around the crack tip rather than being stored as elastic energy or released to create new surface areas. G. R. Irwin of the U.S. Naval Research Laboratory modified Griffith's analysis by including a term in the dissipated energy for the energy G_p dissipated via plastic deformation and other sources.[2] With this modification, eq. (12.26) becomes

$$\sigma_f \sqrt{a} = \sqrt{\frac{EG}{\pi}}, \tag{12.27}$$

where $G = 2\gamma + G_p$ is the energy dissipated during fracture. The surface energy terms dominate for a brittle material such as glass; in this case $G_{glass} \sim 2\gamma_{glass} = 2 \text{ J/m}^2$. For ductile materials, the plastic dissipation term dominates; for the example of steel, $G_{steel} \sim G_{p\text{-}steel} = 1000 \text{ J/m}^2$.

Another way of writing eq. (12.27) is

$$\sigma_f = \frac{\beta K_{Ic}}{\sqrt{a}}, \tag{12.28}$$

where K_{Ic} is the fracture toughness for an opening mode fracture and β is a dimensionless constant that depends on crack geometry. (The fracture toughness for the sliding and tearing modes are labeled K_{IIc} and K_{IIIc}, respectively.) Equation (12.28) tells us how much normal stress σ can be applied to a crack of length a before catastrophic facture. The fracture toughness corresponds to the upper limit of the stress intensity factor K, which is a proportional to the stresses in the material around the tip of the crack. So, a material with a high fracture toughness can tolerate high amount of internal stress before fracture occurs.

· ·

12.7 PROBLEMS

1. Consider a sphere of radius R in contact with a plane experiencing adhesion. The composite contact modulus of the sphere and plane is E_c, as defined in eq. (3.8). Show that the JKR equation relating contact area and load can be derived by applying linear elastic fracture mechanics to this contact. In other words, show that in the JKR model, which was discussed in Section 6.2.2.1, the contacting interface is equivalent to the solid–solid interface between two materials with an interfacial work of adhesion W_{AB} and is surrounded by what is called an "external circular crack," as illustrated in

---

[2] The dissipated energy is labeled G to honor the earlier work by Griffith. Note that the symbol G is also used in the chapter for the shear modulus.

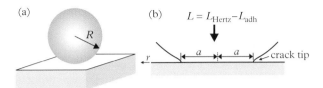

Figure 12.24 *For a sphere-on-flat contact geometry shown in (a), adhesion at the solid–solid interface can cause significant elastic displacement away from the Hertzian model, as discussed in Section 6.2.2.1 and illustrated in (b) and Figure 6.4.*

Figure 12.24. In this case, the variable a represents the radius of the unopened part of the crack rather than the length of the crack. As seen in Figure 12.24, this geometry is essentially that of a contact of width $2a$, as illustrated in Figure 6.4.

To do this, follow these steps:

- Assume the total load L acting over the contacting interface has two contributions: a compressive force L_{Hertz} from elastic deformation of the bodies as given by the Hertz model in eq. (3.9), and a tensile force L_{adh} from the adhesion between the surfaces. Thus $L = L_{Hertz} - L_{adh}$.
- For a JKR contact, since as the adhesive force acts over a distance that is much shorter than the elastic displacements caused by the adhesion, this adhesion is considered infinitely short-range. This implies that the adhesive force is exerted uniformly everywhere within the contact, so it acts like a rigid flat punch, but one that exerts a uniform tensile deformation rather than a compressive deformation. This leads to the stress profile given in eq. (3.20) for a flat punch contact, with $p_{min} = L_{adh}/2\pi a^2$.
- We then apply the Griffith criterion, eq. (12.26), where we set $\sigma_f = p_{min}$, $E = E_c$, and $\gamma = W_{AB}$. As stated above, a is not the crack length, but instead is the radius of the unopened part of the crack.
- You will now have an equation relating L and a in terms of E_c, R, and W_{AB}. Rearranging terms and using the quadratic formula will lead to the JKR equation.

2. Consider the paper by Gotsmann and Lantz (2008), where a conical shaped tip develops the truncated shape (Figure 12.25) with a contact radius a after sliding a distance d against the surface at a load L. In eq. (1) of their paper, the Gotsmann and Lantz propose $a \propto d^m L^n$ as a modified form of Archard's Law for the wear of a conical tip.

    (a) Show that the exact form of this equation is $a(d) = k(\tan\theta Ld)^{1/3}$ where θ is the cone angle of the tip (Figure 12.25) and k is a dimensional wear coefficient. Note that the variable k in this equation is *not* the same k that shows up in Archard's law. Also, note that the tip is assumed to start from a sharp point. In the figure,

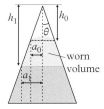

Figure 12.25 *The apex of a conical AFM tip undergoing different levels of wear.*

the flattened end has a radius a_0 after a height h_0 has been removed, and then later has a radius of a_1 after a height h_1 has been removed. So, the "worn volume" shown refers to the gray area *and* the white area above it.

(b) Gotsmann and Lantz go on to derive a different equation to describe atomic-scale wear, given by eq. (4) in their paper; this is the equation for atomic attrition. There are several differences in the starting assumptions between their modified Archard model for atomic attrition, and the original Archard wear model described in Section 12.2.1. For example, one such difference is that the original Archard model assumes the two surfaces are nominally flat blocks, while the atomic attrition model assumes one surface has a conical shape. List three other differences. For one of these three differences, consider the fact that the original Archard model assumes the physical process of material removal is from adhesive junctions as shown in Figure 12.2, which leads to the wear volume being proportional to the distance slid; describe the contrasting assumption used in the atomic attrition model.

3. A set of wear experiments is conducted where an AFM tip made of single crystal silicon nitride is slid against a flat diamond surface at a range of applied normal stresses and temperatures. The following wear rates of the SiN are obtained:

Stress (MPa)	Temperature (K)	Rate of atom loss per second of sliding (s^{-1})
500	250	0.660
750	250	2.11
1000	250	6.80
500	300	103
750	300	275
1000	300	726
500	350	8850
750	350	2040

Make an appropriate plot with fits to show that the data are consistent with atom-by-atom wear; from these fits, determine the activation energy in eV/atom and the activation volume in nm^3. Assume that the effective attempt frequency for wear events is 10^{13} s^{-1}.

· ·

12.8   REFERENCES

Adachi, K., K. Kato and N. Chen (1997). "Wear map of ceramics." *Wear* **203**: 291–301.

Aghababaei, R., D. H. Warner and J.-F. Molinari (2016). "Critical length scale controls adhesive wear mechanisms." *Nature Communications* **7**: 11816.

Andrade, E., R. Randall and M. Makin (1950). "The Rehbinder effect." *Proceedings of the Physical Society, Section B* **63**(12): 990.

Archard, J. F. (1953). "Contact and rubbing of flat surfaces." *Journal of Applied Physics* **24**(8): 981–8.

Argibay, N. and W. G. Sawyer (2012). "Low wear metal sliding electrical contacts at high current density." *Wear* **274–275**: 229–37.

Arrazola, P. J., T. Özel, D. Umbrello, M. Davies and I. S. Jawahir (2013). "Recent advances in modelling of metal machining processes." *CIRP Annals* **62**(2): 695–718.

Bei, H., E. P. George, J. L. Hay and G. M. Pharr (2005). "Influence of indenter tip geometry on elastic deformation during nanoindentation." *Physical Review Letters* **95**(4): 045501.

Bei, H., E. P. George and G. M. Pharr (2004). "Elastic constants of single crystal Cr_3Si and $Cr–Cr_3Si$ lamellar eutectic composites: a comparison of ultrasonic and nanoindentation measurements." *Scripta Materialia* **51**(9): 875–9.

Bell, G. (1978). "Models for the specific adhesion of cells to cells." *Science* **200**(4342): 618–27.

Bhaskaran, H., B. Gotsmann, A. Sebastian, U. Drechsler, M. A. Lantz, M. Despont, P. Jaroenapibal, R. W. Carpick, Y. Chen and K. Sridharan (2010). "Ultralow nanoscale wear through atom-by-atom attrition in silicon-containing diamond-like carbon." *Nature Nanotechnology* **5**(3): 181–5.

Blanchet, T. A. and F. E. Kennedy (1992). "Sliding wear mechanism of polytetrafluoroethylene (PTFE) and PTFE composites." *Wear* **153**(1): 229–43.

Broitman, E. (2017). "Indentation hardness measurements at macro-, micro-, and nanoscale: a critical overview." *Tribology Letters* **65**(1): 23.

Deuis, R. L., C. Subramanian and J. M. Yellup (1997). "Dry sliding wear of aluminium composites—a review." *Composites Science and Technology* **57**(4): 415–35.

Erdemir, A. and C. Donnet (2006). "Tribology of diamond-like carbon films: recent progress and future prospects." *Journal of Physics D: Applied Physics* **39**(18): 311–27.

Fischer, T. E. and H. Tomizawa (1985). "Interaction of tribochemistry and microfracture in the friction and wear of silicon nitride." *Wear* **105**(1): 29–45.

Gibbs, G. B. (1965). "Thermodynamics of thermally-activated dislocation glide." *Physica Status Solidi* **10**(2): 507–12.

Gotsmann, B. and M. A. Lantz (2008). "Atomistic wear in a single asperity sliding contact." *Physical Review Letters* **101**(12): 125501.

Hänggi, P., P. Talkner and M. Borkovec (1990). "Reaction-rate theory: fifty years after Kramers." *Reviews of Modern Physics* **62**(2): 251–341.

Hirth, J. P. and W. D. Nix (1969). "An analysis of thermodynamics of dislocation glide." *Physica Status Solidi* **35**(1): 177–88.

Hokkirigawa, K. and K. Kato (1988). "An experimental and theoretical investigation of plowing, cutting and wedge formation during abrasive wear." *Tribology International* **21**(1): 51–7.

Holmberg, K. (2001). "Reliability aspects of tribology." *Tribology International* **34**(12): 801–8.

Hsu, S. and M. Shen (2005). "Wear mapping of materials." In: *Wear: materials, mechanisms and practice*, G. W. Stachowiak, Ed. Chichester, UK: John Wiley & Sons, pp.369–423.

Hsu, S. M. and M. C. Shen (1996). "Ceramic wear maps." *Wear* **200**(1–2): 154–75.

Independent Investigation Board (2006). *The train derailment at Hatfield: a final report by the independent investigation board*. Report for the Office of Rail Regulation.

Ivester, R. W., M. Kennedy, M. Davies, R. Stevenson, J. Thiele, R. Furness and S. Athavale (2000). "Assessment of machining models: progress report." *Machining Science and Technology* **4**(3): 511–38.

Jacobs, T. D. B. (2013). *Imaging and understanding atomic-scale adhesion and wear: quantitative investigations using in situ TEM*. PhD Thesis, University of Pennsylvania, USA.

Jacobs, T. D. B. and R. W. Carpick (2013). "Nanoscale wear as a stress-assisted chemical reaction." *Nature Nanotechnology* **8**(2): 108–12.

Johnson, K. L. (1985). *Contact mechanics*. Cambridge: Cambridge University Press.

Jost, H. P. (2005). "Tribology micro & macro economics: a road to economic savings." *Tribology & Lubrication Technology* **61**(10): 18.

Jungk, J. M., B. L. Boyce, T. E. Buchheit, T. A. Friedmann, D. Yang and W. W. Gerberich (2006). "Indentation fracture toughness and acoustic energy release in tetrahedral amorphous carbon diamond-like thin films." *Acta Materialia* **54**(15): 4043–52.

Karthikeyan, S., H. J. Kim and D. A. Rigney (2005). "Velocity and strain-rate profiles in materials subjected to unlubricated sliding." *Physical Review Letters* **95**(10): 106001.

Kato, K. and K. Adachi (2002). "Wear of advanced ceramics." *Wear* **253**(11): 1097–104.

Lim, S. C. (1998). "Recent developments in wear-mechanism maps." *Tribology International* **31**(1–3): 87–97.

Lim, S. C. and C. Y. H. Lim (2004). "Wear mapping and wear characterization methodology." In: *Mechanical tribology*, G. E. Totten and H. Liang, Eds. Boca Raton, FL: CRC Press, pp. 247–68.

Lim, S. C., M. F. Ashby and J. H. Brunton (1987). "Wear-rate transitions and their relationship to wear mechanisms." *Acta Metallurgica* **35**(6): 1343–8.

Mackenzie, J. K. (1949). *A theory of sintering and the theoretical yield strength of solids*. PhD Thesis, University of Bristol, UK.

MacMillan, N. H. (1972). "The theoretical strength of solids." *Journal of Materials Science* **7**(2): 239–54.

Marchon, B., N. Heiman and M. R. Khan (1990). "Evidence for tribochemical wear on amorphous-carbon thin films." *IEEE Transactions on Magnetics* **26**(1): 168–70.

Meng, H. C. and K. C. Ludema (1995). "Wear models and predictive equations: their form and content." *Wear* **181–183**: 443–57.

Oliver, W. C. and G. M. Pharr (1992). "An improved technique for determining hardness and elastic modulus using load and displacement sensing indentation experiments." *Journal of Materials Research* **7**(6): 1564–83.

Oliver, W. C. and G. M. Pharr (2004). "Measurement of hardness and elastic modulus by instrumented indentation: advances in understanding and refinements to methodology." *Journal of Materials Research* **19**(1): 3–20.

Park, N. S., M. W. Kim, S. C. Langford and J. T. Dickinson (1996). "Atomic layer wear of single-crystal calcite in aqueous solution using scanning force microscopy." *Journal of Applied Physics* **80**(5): 2680–6.

Pharr, G. M. (1998). "Measurement of mechanical properties by ultra-low load indentation." *Materials Science and Engineering A* **253**(1–2): 151–9.

Prandtl, L. (1928). "Ein Gedankenmodell zur kinetischen theorie der festen Körper." *Zeitschrift für Angewandte Mathematik und Mechanik* **8**(2): 85–106.

Rigney, D. A. (2000). "Transfer, mixing and associated chemical and mechanical processes during the sliding of ductile materials." *Wear* **245**(1–2): 1–9.

Rigney, D. A. and S. Karthikeyan (2010). "The evolution of tribomaterial during sliding: a brief introduction." *Tribology Letters* **39**(1): 3–7.

Rosenberg, R. (2005). "Why is ice slippery?" *Physics Today* **58**(12): 50–5.

Schallamach, A. (1953). "The velocity and temperature dependence of rubber friction." *Proceedings of the Physical Society, Section B* **66**(5): 386.

Sheehan, P. E. (2005). "The wear kinetics of NaCl under dry nitrogen and at low humidities." *Chemical Physics Letters* **410**(1–3): 151–5.

Spikes, H. (2004). "The history and mechanisms of ZDDP." *Tribology Letters* **17**(3): 469–89.

Suh, N. P. (1973). "The delamination theory of wear." *Wear* **25**(1): 111–24.

Suh, N. P. (1977). "An overview of the delamination theory of wear." *Wear* **44**(1): 1–16.

Totten, G. E., Ed. (2017). *ASM handbook. Volume 18. Friction, lubrication, and wear technology.* Materials Park, OH: ASM International.

Vineyard, G. H. (1957). "Frequency factors and isotope effects in solid state rate processes." *Journal of Physics and Chemistry of Solids* **3**(1–2): 121–7.

Wade, W. (1856). Hardness of metals. In: *Reports on experiments on the strength and other properties of metals for cannon with a description of the machines for testing metals, and of the classification of cannon in service.* Philadelphia, PA: Henry Carey Baird, pp. 259–75 and 313–4.

Williams, J. A. (2005). "Wear and wear particles—some fundamentals." *Tribology International* **38**(10): 863–70.

Yen, B. K. (1996). "Influence of water vapor and oxygen on the tribology of carbon materials with sp^2 valence configuration." *Wear* **192**(1–2): 208–15.

Zhang, J. and A. T. Alpas (1993). "Delamination wear in ductile materials containing second phase particles." *Materials Science and Engineering A* **160**(1): 25–35.

Zhu, T. and J. Li (2010). "Ultra-strength materials." *Progress in Materials Science* **55**(7): 710–57.

Zhurkov, S. N. (1984). "Kinetic concept of the strength of solids." *International Journal of Fracture* **26**(4): 295–307.

Zum Gahr, K. H. (1998). "Wear by hard particles." *Tribology International* **31**(10): 587–96.

Index